GAYLORD			PRINTED IN U.S.A.

Published 1994

ISBN: 0-89181-701-8

Association Editor: Kevin T. Biddle
Science Director: Gary D. Howell
Publications Manager: Cathleen P. Williams
Special Projects Editor: Anne H. Thomas

Related Titles from AAPG:

- Computer Modeling of Geologic Surfaces and Volumes (AAPG Computer Applications in Geology, No. 1), edited by D. E. Hamilton and T. A. Jones
- Basic Well Log Analysis for Geologists (AAPG Methods in Exploration, No. 3), by G. B. Asquith with Charles Gibson
- Log Evaluation of Shaly Sandstones: A Practical Guide (AAPG Continuing Education Course Note Series, #31), by G. B. Asquith

These publications, and all other AAPG titles, are available from:

The AAPG Bookstore
P.O. Box 979
Tulsa, OK 74101-0979
Telephone (918) 584-2555; (800) 364-AAPG (USA— book orders only)
FAX: (918) 584-0469; (800) 898-2274 (USA—book orders only)

To my wife, Mary

ACKNOWLEDGMENTS

Many individuals helped me during this book project and share much of the credit for the good points of this volume. As usual, as author I take sole responsibility for any shortcomings. James Schmoker, John Cubitt, and Steve Cannon were the technical reviewers and provided substantial and effective critiques that laid the groundwork for an improved revision. John Davis, Ricardo Olea, and Geoff Bohling also made many useful suggestions in their reviews of Chapters 4, 5, 6, and 7. I benefited from discussions with Bill Guy, Al Macfarlane, and Lynn Watney. Dave Collins produced the colored gray-level image of regional geology in Chapter 2. Jan Harff introduced me to regionalization theory and ran the Dakota cluster analysis in Chapter 4. Steve Prensky provided electronic copies of his monumental logging bibliography as well as references that were difficult to track down. Cora Cowan typed much of the original manuscript when it was still under its working title of "Drink, Puppy, Drink." Jo Anne DeGraffenreid undertook the major task of copy-editing the book, and readers should be grateful to her for unraveling some of the more murky and convoluted passages. Comments by Bill Brownfield led to important improvements on some of the figures. I thank Cathleen Williams and Anne Thomas at AAPG Headquarters for guiding the book through its production phase.

Acknowledgment is made for illustrations used or adapted from the following sources: Society of Professional Well Log Analysts (Figures 2.20–2.23, 3.8, 3.14–3.15, 4.5, 4.25–4.26, 5.15); American Association of Petroleum Geologists (Figures 2.29–2.31, 4.9, 4.22, 4.30, 5.18, 6.1, 6.3, 6.13, 6.14–6.16, 7.6–7.7); Society of Petroleum Engineers (Figures 2.25–2.26, 3.1, 3.13, 4.1, 4.3–4.4, 4.6–4.7, 4.13, 6.18–6.20, 7.3, 7.9–7.10); Society of Exploration Geophysicists (Figures 5.24, 6.9, 6.11); Geological Society of London (Figures 6.12, 6.23); International Association for Mathematical Geology (Figures 6.4–6.5); Society for Sedimentary Geology (Figure 5.11); Canadian Well Logging Society (Figures 3.9–3.11); *Bulletin* of the Canadian Petroleum Geology (Figures 5.26–5.30); *The Mountain Geologist* (Figure 3.2); Academic Press (Figures 6.21–6.22); Wiley-Interscience (Figures 3.7, 5.25); Pergamon Press (Figure 7.2); Enslow Publishers (Figure 5.13); Marathon Oil Company (Figure 6.7).

John H. Doveton

TABLE OF CONTENTS

ABOUT THE AUTHOR

John H. Doveton received his M.A. degree from Oxford and his Ph.D. from Edinburgh University, both in geology. Following work as an exploration geologist for Mobil Oil Canada, he joined the Kansas Geological Survey and is now senior scientist, with numerous publications in the area of computer applications to geology and petrophysics. Doveton has taught log-analysis courses since 1975 for universities and industry in the United States, Canada, Europe, West Indies, and the Middle East. He is an author of the widely-used KOALA/TERRALOG interactive computer log-analysis system. In 1981 he received the Best Paper Award from *The Log Analyst.* He has been a Distinguished Speaker for the SPWLA. Doveton is author of the textbook, *Log Analysis of Subsurface Geology* (Wiley-Interscience, 1986).

PROLOG

The earliest paper on computer applications to log analysis was probably that published in 1961 by Tuman and Bollman, who described programs to discriminate sandstones and shales, estimate formation water resistivity, find porosity from an acoustic velocity log, and compute the formation factor from porosity. They speculated that computers would ultimately replace slide rules and charts. If digitizing costs could be brought down, entire sections could be analyzed quickly and efficiently. This would be a great advance over manual procedures that tended to focus, by necessity, on specific zones of interest. This constraint reflected the great length of most logs and the limited time generally available at the well site to make test decisions. The potential value of the computer was seen as an extension of its success in other areas. It performed admirably as a tireless, giant calculating machine that could process large amounts of data at blinding speeds. Tuman and Bollman (1961) estimated that a comprehensive log analysis of 10,000 ft of section should take somewhere between 30 minutes and an hour of computer time.

Computers are now used routinely for log processing, both at the well site and in the office. Their principal economic justification is still the tremendous savings in labor over that which would be involved if the same calculations were done by hand. This consideration is more true than ever as the numbers and types of logs continue to expand. Geologists today, however, are slowly becoming aware of the potential for much broader applications of computers to petrophysical logs. There are several reasons for this. Early electric logs were just that: measures of the electrical properties of rocks. Quantitative log analysis seemed to be a step removed from the concerns of traditional geology, because the electrical properties were tied almost exclusively to the salinity, volume, and shape of fluids within the pore space. With the exception of enigmatic conductivity effects of clay minerals, most rock frameworks appeared to be electrical insulators that were indistinguishable from one another. That apparent limitation changed dramatically when a wide range of nuclear tools were deployed to measure both natural and induced radioactivities of subsurface formations. Many of these properties are keyed directly to rock framework composition, with implications that range across depositional origin, diagenetic history, and geochemical aspects. The recording of acoustic wave forms, borehole electrical imaging, and other logging techniques have also added a significant dimension to geological analysis.

In society as a whole, computers have come to be recognized not as mere calculating engines but as devices for graphic design, simulation, database management, and a whole host of operations that were the science fiction of a previous generation. Most geologists in both industry and academia now have ready access to modern computers. Their dealings with computers are usually no longer the batch procedures with remote mainframe machines that was formerly the norm. Instead, geologists typically work with computers in an interactive style, using either a microcomputer, workstation, or terminal served by a larger machine. So, the computing environment is now generally one of dialog in which a geologist can explore and analyze logging data from extensive stratigraphic sections.

It is the purpose of this book to review many of the current methods and paradigms that have been developed for the computer analysis of petrophysical logs for geological applications. In Chapter 1, we review the fundamental equations of classical log analysis and the role of statistical inference methods for calibration, prediction, and error analysis of reservoir rock properties. The focus of Chapter 2 is on graphical pattern recognition techniques. The presentation of traditional crossplots on the computer terminal screen is described and leads to discussion of graphical methods of multidimensional projection and color transformations that would be unthinkable without the use of a computer.

In Chapter 3 we examine the progress that has been made in the calculation of compositional profiles of stratigraphic sequences from various log combinations. The spirit of these methods is a "top-down," model-driven approach that attempts to give acceptable quantitative estimates in spite of uncertainties in the analytical models and tool measurements. The methods contrast with those of Chapter 4, which are "bottom-up" and data-driven. Multivariate statistical classification techniques are used to analyze distinctive associations of log responses that represent facies. Supervised techniques base their discrimination on logs from sequences where facies types are already known from core data or other information. Results are then used to classify zones of logged sections of unknown character. Unsupervised methods attempt to uncover inherent natural clusterings of log data that should signify different types of lithofacies or petrofacies.

In Chapter 5, logging data are considered from the viewpoint that they form sequences ordered with time. The applications of time series methods are described, ranging from analysis of simple trends to unraveling complex repetitive patterns. The possible implications of the results are discussed in the light of various geological models. Chapter 6 discusses the use of logs for stratigraphic correlation and reviews the different automated methods that have been devised for log correlation by computer. Most commonly, the results of correlation form a framework of linked stratigraphic tops that sketch out the geometry of formation boundaries. Chapter 6 also discusses two approaches to the interpolation of sequences between wells. In the first, moments of logs are interpolated as broad spatial trends. In the second, the finer scaled features of laterally discontinuous units are simulated by models of stochastic lithologies.

The final chapter, Chapter 7, discusses the two major schools of thought in artificial intelligence and their applications to log analysis. The symbolic approach is known more commonly as the expert system and has enjoyed some success when the logging problem is fairly clear-cut. Connectionist theories are used to design neural networks that have the capacity to learn complex patterns and associations. This characteristic suggests a potential for innovative and powerful log-processing techniques, although current working examples are rudimentary at best.

Almost all of the methods described in this book can be run on a microcomputer using moderately priced software. This means that the programs are widely accessible to geologists for use in calibration, pattern recognition, classification, prediction, and other purposes. An axiom of this book is that the active participation of the geologist is crucial to the success of these computer methods. Hopefully, narrow attitudes about computers are increasingly a thing of the past. The complex problems posed by heterogeneous reservoirs, exotic lithologies, and subtle traps require all the native wit and experience of the geologist to direct the computer analysis. Those with a more mature appreciation of the modern computer see it as an extraordinary power tool for the mind. A geological analysis of logging data then becomes a cooperative venture between human and machine. Both the methods and the applications are relatively young, so they hold great potential for significant contributions to the larger field of geology.

REFERENCE CITED

Tuman, V. S., and D. Bollman, 1961, Application of computers to the interpretation of well logs: Journal of Petroleum Technology, v. 23, no. 4, p. 311-318.

Chapter 1

Statistical Methods For Log Analysis Of Reservoir Properties

The critical reservoir properties that define a productive unit are pore volume, hydrocarbon saturation, and permeability. Their evaluation is a primary task of log analysis in all phases of petroleum exploration and production. Even today, the equations established by Archie (1942) are the touchstones for quantitative calculations of hydrocarbon saturation, in spite of the fact that the controlling parameters are poorly understood and often difficult to estimate. In addition, they describe models which are highly simplified representations of the complex structure of real petrophysical relationships. A common form of the composite Archie equation is written as:

$$S_w^n = \frac{aR_w}{\Phi^m R_t}$$

where R_t is the resistivity of the rock, R_w is the resistivity of the connate water, S_w is the water saturation, Φ is the porosity, while a, m and n are constants that are controlled by the pore geometry and fluid properties (as will be described later). Although resistivity is actually measured by a logging tool, porosity itself must be deduced from the acoustic velocity, electron density, or neutron absorption of the zone of interest.

This situation becomes more complex when one considers the heterogeneity of most reservoir rocks and that the systematic signal in the logging data is confounded with errors associated with the tool measurement process. But in spite of all these problems, the equations for computing both porosity and water saturation have stood the test of time as reasonable functional relationships, provided that analysts are judicious in their choice of values drawn from published data, experience, and their own logging pattern-recognition skills. Attempts at permeability prediction from logs have generally been judged to be far less successful, although some progress continues to be made as controls of permeability become better understood.

The techniques of classical statistical analysis can be valuable aids to basic log analysis procedures. There are some similarities and differences in practice and goals that should be recognized from the outset. Ideally, statistical inference should be applied to experimental data recorded from treatments specifically designed to test hypotheses under carefully controlled conditions. The experiments use predetermined sample sizes and numbers of replicates. By contrast, logging data are mostly indirect measurements of the formation properties of interest, recorded in less than ideal conditions. The rock formation is itself the product of complex, unobservable processes that occurred over periods of remote geologic time.

However, many of the aims of log analysis and statistical inference are similar. These include the need to make systematic predictions of key variables under conditions of uncertainty. Oil exploration is sometimes described as gambling; in reality it is a high-risk business. The risk-averse character of inferential statistics is in general agreement with a fiscally responsible strategy of oil exploration and production. Even so, there are times when the intrinsic conservatism of statistics should be considered carefully. Inferential statistics are best at characterizing the typical by estimates of parameters that represent samples of observations. By contrast, atypical observations can often be discriminated in statistical analysis as outliers, but their comparative rarity impedes conclusions as to whether they represent systematic effects or freak errors. However, there are occasions when these same anomalies may flag the occurrence of commercial pay zones. An isolated 10-m pay zone in a 10,000-m section may make a well commercial. The philosophy of basic statistical inference and its application to petrophysical data are discussed in the narrative and case studies of the following sections.

FIRST THINGS FIRST: THE STATISTICS OF DISTRIBUTIONS

Almost all of the statistical methods described in this and later chapters are parametric—that is, they are based on properties that summarize the distribution characteristics of the measured variables, taken both singly and jointly. The most basic property of a distribution is some measure of location. This is given by one of the statistics of central tendency, which are

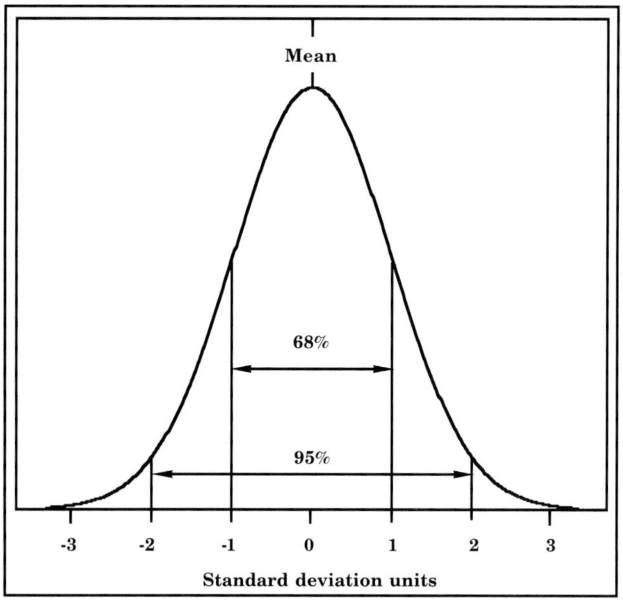

Figure 1. The standard normal distribution showing the percentage of observations that would be expected to occur within ranges of one and two standard deviations from the mean.

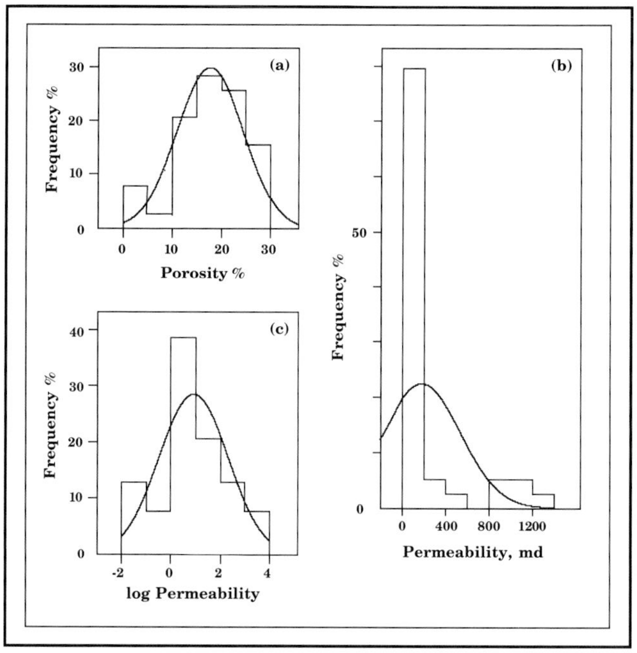

Figure 2. Histograms of (a) porosity, (b) permeability, and (c) logarithmically scaled permeability of Miocene sandstones in the Picaroon field (Texas), together with fitted normal distribution curves.

the mode (most frequent value), median (the value of the 50th percentile), and the mean (the arithmetic average). The mean ($\overline{X}$) of a variable X is the arithmetic average:

$$\overline{X} = \frac{\sum X_i}{n}$$

based on a sample of n observations. The mean also functions as the expected value of the distribution, which is symbolized by E(X). In the absence of any other information, the mean would be the best guess of the value of an observation taken at random from the distribution. This is because the average error would be the minimum possible, based on squaring the differences between the observations and the mean value.

The second descriptor of a distribution is its relative degree of spread or dispersion, which is given by the variance. Only a restricted sample is generally available from the total hypothetical population, so this statistic, called the variance (s^2), is estimated from the sample by:

$$s^2 = \frac{\sum \left(X_i - \overline{X}\right)^2}{n-1}$$

where X_i is the ith observation of x. The standard deviation, s, is the square root of the variance and represents the expectation of the average square-root difference of any observation from the mean value.

The two statistics of the mean and variance are therefore fundamental descriptors of a distribution, giving its location and dispersion. These same parameters are the only statistics necessary to define a "normal" distribution, which graphs as the familiar bell-shaped curve (Figure 1). Although the normal distribution was originally a theoretical abstraction, Karl Gauss (1777–1855) showed that it described the scatter of random measurement errors about their true value. This was to be expected if errors were modeled as the summation of minute arithmetic displacements from the true value. The errors then conform to a binomial distribution that, when taken to the continuous limit, becomes the normal distribution. Although the "normal law of error" could be developed from theoretical reasoning, the normal curve was suggested to fit the distribution of many kinds of observational data with no particular justification other than appearance and a vague appeal to some higher "universal law." As Jules Henri Poincaré (1854–1912) observed about the normal distribution in 1898: "Everyone believes in it because the mathematicians think it is a fact of observation and the observers assume it is a mathematical law."

In the parametric statistical analysis of logging data, an approximately normal distribution is generally assumed for all variables, or failing this, non-normal variables are transformed appropriately in order to meet this assumption. The reason for this concern is very simple. Although all of the data points are considered in manual log analysis, either taken singly or collectively, classical statistical inference procedures base their calculations on the sample estimates of the descriptive parameters. Consequently, in place

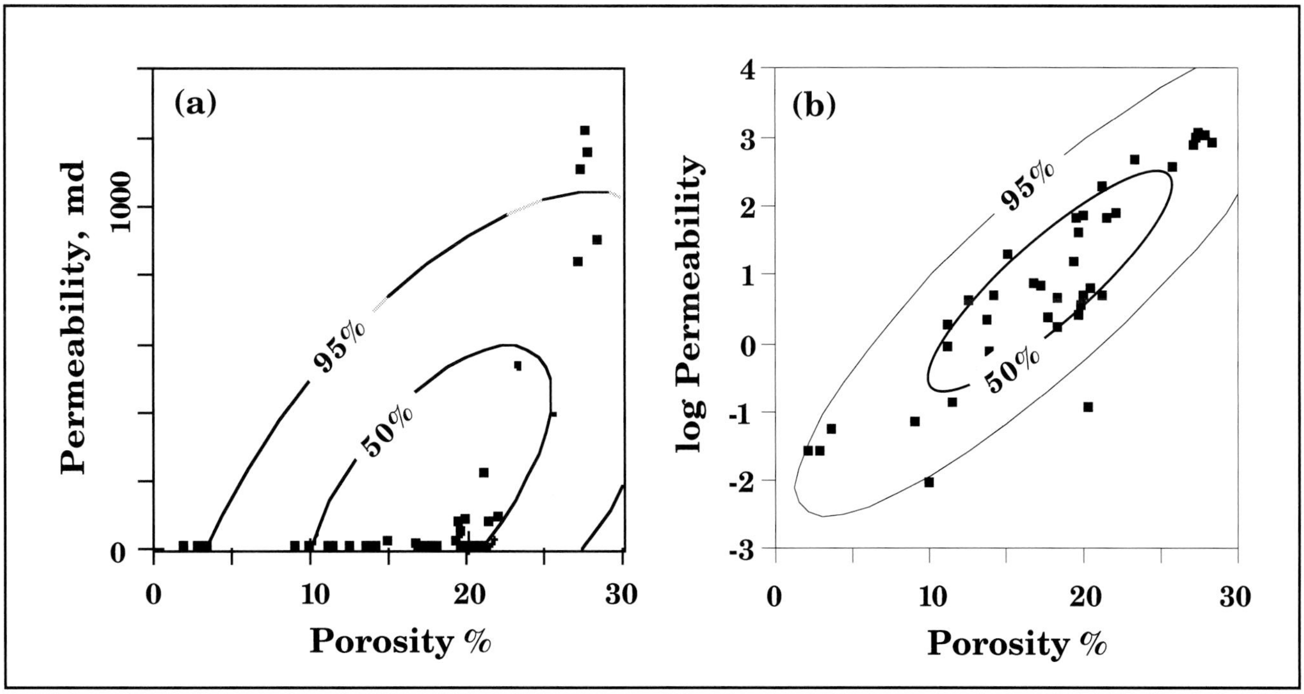

Figure 3. Crossplots of (a) porosity versus permeability and (b) porosity versus logarithmically scaled permeability of Miocene sandstones in the Picaroon field (Texas), fitted with bivariate normal contours for population densities of 50 and 95%.

of the total raw variability, the statistical analyses substitute a form that is specified by the mean and variance, and this is the normal distribution. If the distribution of the variable is poorly matched by a normal curve, then analysis based on the mean and variance may be misleading. Exact conformation to normality is unnecessary. If the data are approximately symmetrical about the mean, are unimodal, with about two-thirds of the data spread within one standard deviation of the mean, then the data are effectively normal (see Figure 1). In simple terms, if it looks like a duck and quacks like a duck, considered it to be a duck!

As a practical illustration of these points, histograms of the porosity and permeability distributions of Miocene sandstones in the Picaroon field (Texas) are shown in Figure 2 based on data from Taylor (1990). Normal curves fitted to these variables'arithmetic means and variances show a good match for porosity, but poor representation for permeability. However, the fit of the normal distribution is improved markedly when applied to logarithmically transformed permeabilities. This implies that the raw permeabilities conform approximately to a lognormal distribution, which is generated by processes that are multiplicative in character. The lognormal distribution commonly has been used as a simple model to describe particle sizes generated by splitting processes of crushing and breakage.

The porosity, permeability, and logarithmic permeability data are crossplotted in Figure 3. In each case the central location is matched with the coordinates of the two variable means. The dispersion of the data is captured by the variance of each variable and the covariance between each pair of variables. Each variance gives a measure of spread parallel to the variable axis. The covariance absorbs the covariation of the data across the plane of each variable pair. The covariance is estimated from the sample by the formula:

$$\mathrm{cov}_{XY} = \frac{\sum (X_i - \overline{X})(Y_i - \overline{Y})}{n-1}$$

The bivariate normal distribution is defined by the means, variances, and covariance of the two variables and is expressed by elliptical isocontours of density centered on the bivariate mean (see Figure 3). The bivariate mean is simply the coordinate pair that is given by the means of the two variables. The benefit of the logarithmic transformation of the permeability is shown on the porosity–log permeability plot, where the data trend is linearized effectively and the bivariate normal contours are a good representation of the data cloud. The shape of the ellipse is controlled by the correlation coefficient:

$$r_{XY} = \frac{\mathrm{cov}_{XY}}{s_X s_Y}$$

For the Picaroon field Miocene sandstones, the correlation was estimated to be 0.88. In the event of zero correlation, the bivariate normal ellipse becomes a circle; for a perfect correlation, the ellipse becomes extended to a straight line with an orientation whose slope is given by the sign of the coefficient.

The correlation coefficient is a measure of the strength of the linear trend between two variables. When it is believed that this line matches an underlying functional relationship, then the scatter of data about the line constitutes error. As mentioned earlier, Gauss was able to demonstrate that random error should be normally distributed. In practice, the deviations of logging data from idealized trends are compounded by errors from a variety of sources, and are both systematic and random. This potentially irksome situation is helped considerably by the Central Limit Theorem. The theorem states that even when individual observations are not normally distributed, their means will tend towards the normal distribution. In the present context, if the deviations from the true log values are the sum of errors from different sources, then the collective effect will often tend to normalize them.

In the remainder of the chapter we will examine statistical techniques that extract systematic linear trends from logging data for the purposes of both prediction and estimation of constants used in functional petrophysical equations. The target variables of these analyses are porosity, water saturation, and permeability, which are primary goals of traditional log analysis.

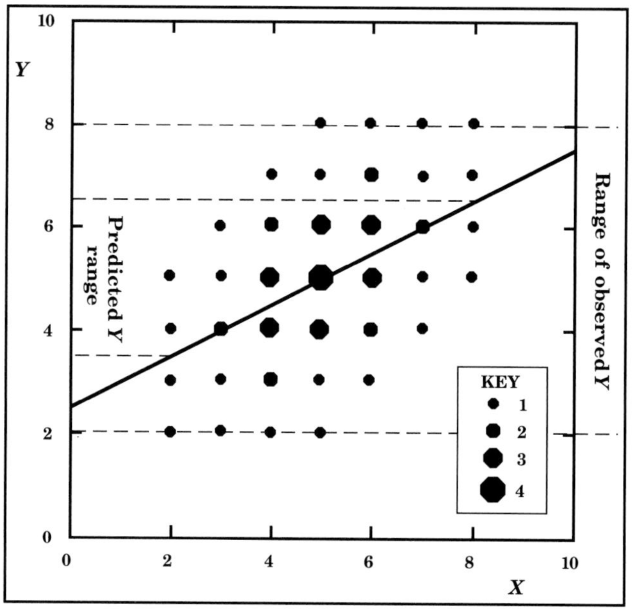

Figure 4. Regression of *Y* on *X* for artificial data, which shows the effects associated with the best prediction on average for any given *X* value, including the contraction of *Y* prediction range compared with observed values. The size of the symbols represents the number of points at each location as defined in the key. Data from Campbell and Stanley (1966).

STATISTICAL ESTIMATION OF LINEAR TRENDS

The statistical fitting of lines to petrophysical data has been discussed by a number of authors (e.g., Collins and Pilles, 1979; Etnyre, 1984a, b; Collins, 1984; Whitman, 1986), who are all in substantial agreement, but who have much to say on this subtle and perplexing topic. The most widely used method for prediction purposes fits a line to the data by the method of least squares. The variable to be predicted is the dependent variable (conventionally denoted as *Y*), based on values of an independent variable (*X*). The regression of *Y* on *X* is calculated such that the sum of the squared deviations of *Y* are minimized about the line. The predicted value of *Y* is its average expectation, given any value of *X*, i.e. E(*Y*/*X*).

The equation of the line can be written as:

$$\hat{Y} = a + bX$$

where a is the intercept and b is the slope. The "hat" on the dependent variable, *Y*, signifies that it is an estimate or prediction based on an equation, rather than a measured value. We will use this convention at other points in the book. The slope (b) is computed from:

$$b = \frac{r \cdot s_y}{s_x}$$

where r is the correlation coefficient and s_y and s_x are the standard deviations of the *X* and *Y* variables. In common with other best-fit lines, the regression of *Y* on *X* passes through the bivariate mean, $\overline{X}$ and $\overline{Y}$. The intercept is then given by:

$$a = \overline{Y} - b\overline{X}$$

The implications of this model can be seen by reference to Figure 4, where a regression line of *Y* on *X* is fitted to an artificial data set. At first glance, the orientation of the line would appear to be counter-intuitive. The symmetry of the distribution suggests that the regression line is skewed to a more shallowly sloping angle than the true trend. Closer examination shows that the line passes exactly through the mean value of *Y* for all possible values of *X*. The prediction of *Y* is therefore true on average for any given value of *X*. The averaging effect is also shown by the fact that the range of possible predictions of *Y* is smaller than the actual range of observed *Y* values. This results in a built-in tendency to underestimate extremely high values of *Y* and overestimate extremely low ones. Nevertheless, the regression provides the best estimate, because on average, the squared deviation error associated with any prediction will be the smallest possible.

Notice that this same line is not appropriate for predictions of *X* given values of *Y*. Instead, the regression of *X* on *Y* takes a much steeper trend that

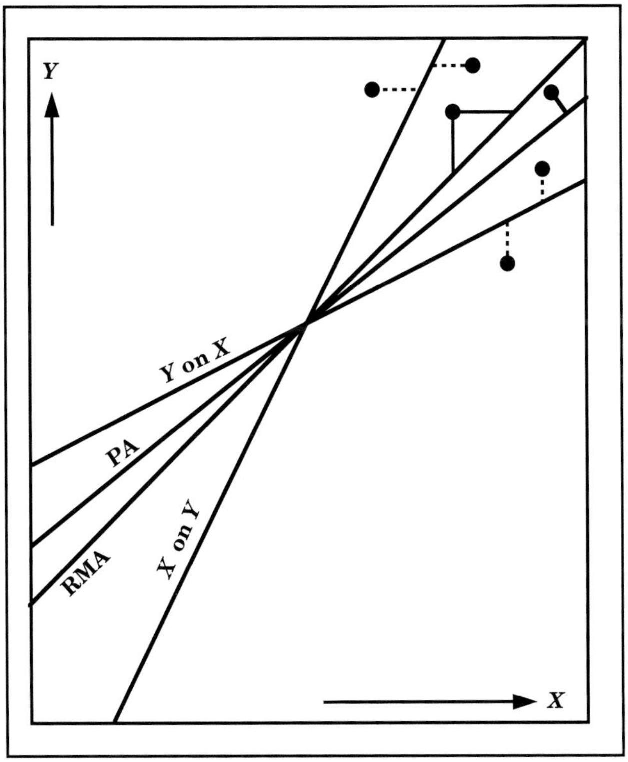

Figure 5. Simplified diagram of the relative position of alternative line-fits to data and their fitting criteria: *Y* on *X* regression, *X* on *Y* regression, the reduced major axis (RMA), and the principal axis (PA).

passes through the means of *X* that are associated with each value of *Y* (Figure 5). The slope is related to the *Y* on *X* regression line slope by the equation:

$$b' = \frac{b}{r^2}$$

The apparent paradox of two regression lines can be understood when one realizes that for each alternative, the regression estimate is predicated on the notion that the independent variable is given without error.

There is a broad consensus on the choice of appropriate line-fitting procedures for geological and petrophysical data. This consensus is dictated both by the nature of the data and the purpose intended for the results. If prediction of either true or observed values is the goal, then a regression should be made using the variable to be predicted as the dependent variable. According to Lindley (1947) and many others, this applies whether or not the independent variable is in error. Other authors qualify this rule, and specify an *X* on *Y* regression, when the dependent variable (Y) is held to have no error, and all the error is associated with the independent variable (X).

The difference between these two choices can be seen on Figure 5. The divergence between the two regression lines results from the fact that each one minimizes the average squared deviation parallel to either the *X* or *Y* axis. When the variables have a perfect correlation, the two alternatives coincide. At the other extreme, that of zero correlation, the lines are at right angles when the *Y* on *X* regression predicts a mean value of *Y*, regardless of the value of *X*, and the *X* on *Y* regression locks onto the mean value of X. In all situations, the angle between the two regression lines is controlled by the correlation coefficient, as can be seen in the last equation.

Because the correlation between logging variables can often be moderate to low, evaluation of the coefficients of the alternative regression equations can be problematical if the equations have a functional petrophysical meaning. When the relationship between the two variables is subject to a proven (or at least, accepted) physical model or known natural constraints, the goal becomes *functional analysis*, rather than simple prediction. The two regression lines are then seen to be the two limiting extremes, where all the error is attributed to either one or other of the two variables. The real functional line should lie somewhere in between, with its slope controlled by the relative amount of error assigned to each variable. The error in question is due to random measurement effects that result from both the tool characteristics and the fluctuations in the borehole environment. The issue in question is the precision of the measurement rather than its accuracy.

The tight focus on regression theory in most introductory statistics texts appears to have created a widespread impression that functional analysis is not accepted by professional statisticians, who would restrict the choice to one or other of the two possible regression lines. The potential misunderstanding is clarified by Kendall and Stuart (1979, p. 402), who state:

> One consequence of the distinctions we have been making has frequently puzzled scientists. The investigator who is looking for a unique linear relationship between variables cannot accept two different lines, but he was liable in the early days of the subject (and perhaps even today) to be presented with a pair of regression lines. Our discussion should have made it clear that a regression line does not purport to represent a functional relation between mathematical variables or a structural relation between random variables: it either exhibits a property of a bivariate distribution or, when the regressor variable is not subject to error, gives the relation between the mean of the dependent variable and the value of the regressor variable. The methods of this chapter, which our references will show to have been developed largely since 1940, permit the mathematical model to be more precisely fitted to the needs of the scientific situation.

The error variance of a variable *X* gives measurement precision and can be determined by repeating the measurement for *n* replicates of the same observation, when:

$$e_X^2 = \frac{\sum\left(X_i - \overline{X}\right)^2}{n}$$

If the error is independent, then the error variance can be determined for each variable separately. The ratio between error variances:

$$\lambda = \frac{e_Y^2}{e_X^2}$$

can be used to estimate the true functional line that takes into account the relative amount of measurement error associated with both variables. When λ=0, then the Y values are known without error and the appropriate solution is an X on Y regression. At the other extreme, when λ is infinite, all the error is linked with the Y variable and the choice must be a regression of Y on X. In all intermediate cases, the line will be located between and its slope can be calculated from the slope of the Y on X regression by:

$$b_f = \frac{\left(\frac{b^2}{r^2} - \lambda\right) + \sqrt{\left(\frac{b^2}{r^2}\right) + 4\lambda b^2}}{2b}$$

Heseldin (1968) recommended the use of the error ratio in least squares fitting of data from log analysis and demonstrated the improvement in performance when compared with standard regression or other line-fit procedures.

How can one determine the error variances or even estimate λ in practice? Collins and Pilles (1979) pointed out that random error of logging data is apparent when contrasting repeat runs of properly calibrated instruments with main runs, provided that there is no bias in the measurement. If there is a distinctive bias, then this is the concern of standard quality-control procedures, where differences in the main and repeat sections reveal systematic effects that can be attributed to tool problems, depth discrepancies, or poor hole conditions. Good examples of the recognition of logs with these systematic errors were described by Farnan and McHattie (1984), based on their extensive experience in the digital comparisons of repeat and main runs. Logs of acceptable quality have errors with a relatively small, unbiased scatter that is a function of the physics of the tool, its response characteristics, and the borehole environment. In particular, the nuclear tools are subject to statistical counting error, because they record stochastic atomic processes of radioactive decay and particle generation. By contrast, electrical measurements are deterministic, but are still subject to error, determined by the precision of the instrument under borehole conditions.

The overlay of a typical density log main and repeat sections is shown in Figure 6, where the close match of the curves demonstrates good quality in the

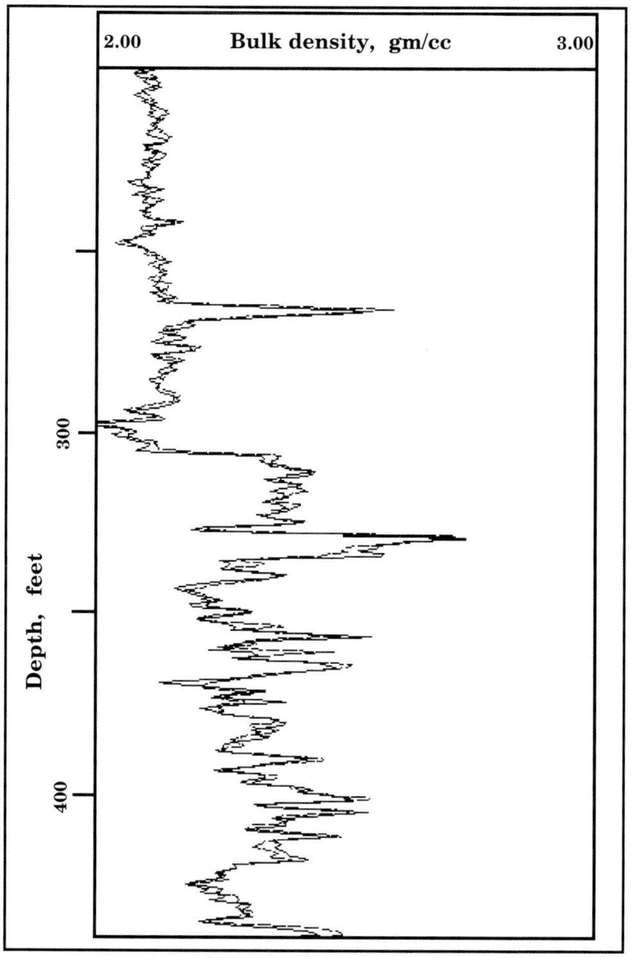

Figure 6. Example main (solid) and repeat (dashed) runs of bulk density log.

logging operation. The effective symmetry of the crossplotted runs (Figure 7) suggests that the deviations can be attributed to random measurement error.

If the deviations are indeed random, then they would be expected to be normally distributed with a mean value of zero. A histogram of halved deviations between the main and repeat runs (Figure 8) shows a broadly symmetrical form with a mean of 0.002 gm/cc. A normal distribution fitted to the sample mean and variance suggests that the deviations are a little more peaked than would be expected. However, some of this effect may be caused by the change in error magnitude with density. Etnyre (1984a, b) showed from the physics of the density measurement that errors in density estimation increase exponentially with bulk density. This effect was evaluated and confirmed by a weak, positive correlation of 0.24 between absolute error and bulk density. However, the difference in average error over the range of the logs was fairly minor, ranging between the extremes of 0.007 and 0.026 gm/cc.

These diagnostic steps in error evaluation are necessary in order to discriminate systematic bias (the target of quality-control procedures) from normal random measurement error, which represent, respec-

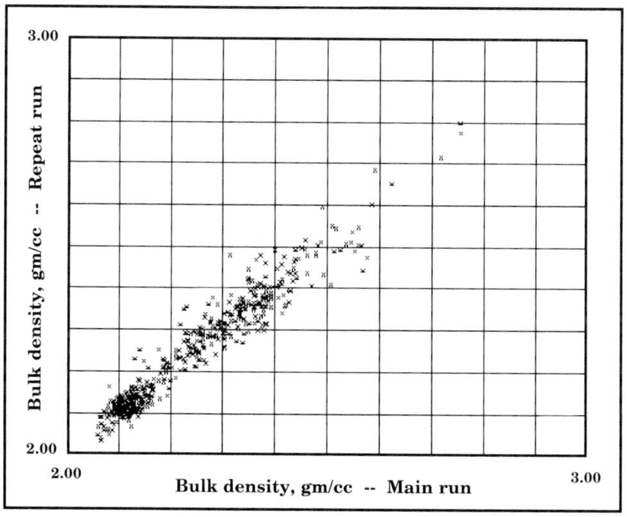

Figure 7. Crossplot of main and repeat runs for the density log example.

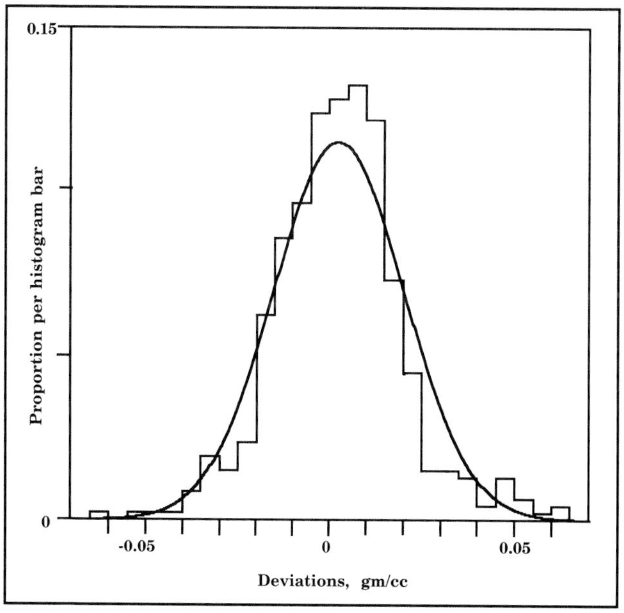

Figure 8. Deviations from mean of main and repeat runs of an example density log fitted with a normal distribution.

tively, the two separate but important concerns of relative accuracy and precision. For logging data that have been checked to be unbiased, the estimate of random measurement error (and so precision) is given by the standard deviation of the errors between main and repeat runs. In the density log repeat example, the error standard deviation is 0.018 gm/cc, which is equivalent to an average expected error of 1.09 percent equivalent sandstone porosity. The square of this standard deviation is the error variance. The error variance estimate can then be used as contributory information to the selection of an appropriate fitted line for functional relationships between the density log and other petrophysical measurements.

In many instances, data will not be readily available to compute the error variances directly. However, they can be estimated approximately if some concrete notion of precision can be associated with each variable. So, for example, if a known resolution, *U*, of a measurement device can be considered as equivalent to a 95% statistical confidence limit for the observed value, then the error variance is:

$$e^2 = \left(\frac{U}{1.96}\right)^2 = 0.26U^2$$

(Mark and Church, 1977) because 95% of the normal distribution is contained within 1.96 standard deviations of the mean. When the resolution of a variable measurement is a matter of opinion based on experience, then the numbers in this formula are themselves overly precise! However, the form of the equation gives a useful rule-of-thumb guide to the effect that the error variance is about a quarter of the squared resolution.

When no data analysis or prior knowledge can be brought to bear on the problem of error variance, then the error-variance ratio, λ, is often estimated following one of two assumptions. The first assumption considers that the best estimate of the error-variance ratio of two variables is given by the ratio of their total variances, i.e.:

$$\lambda = \frac{s_X^2}{s_Y^2}$$

This choice was advocated by Dent (1935) as the maximum likelihood estimate of λ in the absence of any other information. The method minimizes the areas between the points and the best-fit line, which is known as the reduced major axis (RMA). In common with the other best-fit lines, the reduced major axis passes through the bivariate mean and its slope is simply the ratio of the standard deviations of the two variables, with the appropriate sign given by that of the correlation coefficient (see Figure 5).

The alternative second assumption states that the error-variance ratio is unity, i.e., λ=1. This stipulation implies that the two variables have equal errors if the measurements are made in the same units. The best-fit line that is generated by this assumption is the principal axis, which minimizes the squared deviations, measured perpendicular to the line (see Figure 5). It corresponds to the principal eigenvector of the variance-covariance matrix that is computed by principal components analysis (see Chapter 4). Unlike with the other best-fit lines, the solution is sensitive to the units of measurement. Consequently, the principal axis is usually calculated for standardized data and then transformed to the original units.

In summary, the choice of best-fit line is first determined by the purpose of the procedure. If the intent is only to make predictions of one variable on the basis of measurements of another, then regression is the preferred choice. The variable to be predicted is the dependent variable; the predictor is the independent variable. Alternatively, when a functional analysis is the goal and where the controlling parameters have both meaning and utility, the best-fit line should incorporate estimates of the random errors associated with each variable. Whenever possible the error variances should be computed from replicate samples, which in the case of wireline logs are provided by the consideration of both main and repeat runs. At the other extreme, the error variance ratio can be assumed to be linked with the total variance in the computation of either the reduced major axis or the principal axis.

Yet another option is available in functional analysis, when one realizes that the selection of the most appropriate fitted line should provide the most reasonable error-variance ratio and, simultaneously, the equation intercepts and slopes matching the rock properties and physical constraints of the functional relationship. These considerations and the general problems of prediction and characterization are explored in the following examples. In these analyses of real logging data, several initial remedial steps should be taken in order to maximize the validity and value of the analysis results:

(i) the data should be environmentally corrected for systematic borehole effects;
(ii) the logs should be shifted wherever necessary to ensure common depth registration;
(iii) the logs should have a common vertical resolution, a compatibility requirement that may involve smoothing of finer resolution measurements (see discussion later in this chapter);
(iv) preferably, the logs should be zoned, with data sampled from peak and trough extremes, in order to reduce extraneous error introduced by transitional curve features (zonation methods are described in Chapter 5, under Markov Chain Analysis of Electrofacies Sequences).

ESTIMATION OF POROSITY

In this example we examine the problem of transforming transit times from a sonic log to a porosity equivalent using core measurements of porosity. The consequences of which of the alternative line fits one chooses are by no means of purely academic interest. It is now common practice for estimations of porosity to be tied to core–log calibrations in unitized fields. By this means, porosities can be calculated in uncored wells and used for estimating volumetrics on a field-wide basis. Even minor differences in line slope can cause significant changes in the allocation of reserves between the participating operators. This was widely appreciated at the equity hearings of the '80s, when there was considerable debate as to the relative merits of alternative statistical line-fitting strategies.

The data consist of 44 measurements of sonic log transit time (Δt) of a sandstone reservoir, matched with core porosities (Φ) at equivalent depths (Table 1).

Table 1. Core Porosities (%) and Sonic Log Transit Times (microseconds per ft) from a Sandstone Reservoir.

Φ	Δt	Φ	Δt
6.8	63.8	13.4	75.1
9.3	6.6	13.4	74.4
8.5	68.1	13.4	72.1
10.9	68.3	13.5	69.9
10.3	69.9	14.3	72.8
10.4	70.5	15.1	72.9
10.2	71.6	15.1	74.5
10.1	72.2	15.1	75.1
10.1	72.7	15.3	77.9
8.6	72.2	15.2	78.5
11.0	72.7	15.2	80.0
11.0	72.3	15.2	81.0
10.9	71.6	16.9	83.8
11.0	71.1	17.6	81.1
11.1	70.0	17.6	79.5
11.7	70.9	16.9	76.0
11.8	71.6	17.8	77.3
11.8	73.3	18.6	76.7
11.8	73.9	18.5	79.1
11.8	74.8	20.1	82.3
12.6	75.6	20.3	83.5
12.7	73.8	19.6	84.5

The core porosities were previously smoothed by a moving average filter resulting from 1-ft sample increments and 2-ft measurement spans of the sonic log, ensuring an approximately common vertical resolution between the two measurement types. Failure to do this initial remedial step would result in data incompatibility, causing both distinctive error and bias, as will be demonstrated later.

Initially the problem can be seen to be one of simple prediction: given a transit time from a sonic log, how would the calculated porosity of the zone compare with a core analysis? A linear relationship between porosity and transit time is commonly assumed to be a usable approximation (Wyllie et al., 1956). The prediction equation is then:

$$\hat{\Phi} = a + b\Delta t$$

which is a regression of porosity on transit time. The result corresponds to the most shallowly sloping line on the crossplot of porosities and transit times on Figure 9. The regression equation is:

$$\hat{\Phi} = -33.3 + 0.63\,\Delta t$$

and has a coefficient of determination of 0.76, meaning that the linear prediction accounts for 76% of the total variability, with the remaining 24% left in the

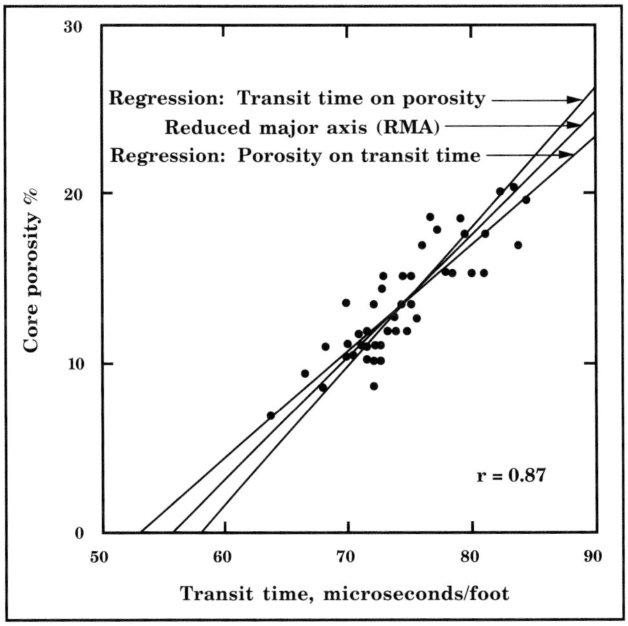

Figure 9. Best-fit lines to core porosities and transit times for the example sandstone reservoir.

residual squared deviations about the line. The coefficient of determination is equal to the square of the correlation between core porosity and transit time, which is 0.87.

This regression model of porosity on transit time ascribes all the error to the core porosity and none to the transit time. The consequences can be seen on the crossplots of Figure 10, where the errors in predicted porosity are random when graphed against transit time, but show a tendency for underprediction at the high end and overprediction at the low end when plotted against measured porosity. This effect was discussed earlier in connection with the artificial data example, and simply shows that any prediction of porosity is the best on average for any given value of transit time. However, as pointed out by Collins (1984), the choice of this line would be resisted in unit operating negotiations by an owner whose property had porosities that tended to be higher than the average.

The alternative regression of transit time on porosity results in the steepest fitted line of Figure 9 and allocates all the error to the transit time with none to core porosity. The descriptive equation is:

$$\hat{\Delta t} = 58.1 + 1.2\,\Phi$$

This line-fit solution would be welcomed by the owner of a property with higher than average porosity, for the opposite reasons attached to the other regression line: now the result would appear to enhance higher porosities, while further downgrading lower porosities. Traditional regression texts would reject this alternative out of hand, since they view the situation as one of predicting the best estimate of porosity on average, based on a given value of transit time. However, others would argue that this is a calibration problem in which the core porosities must be honored as the calibration standard, and so effectively considered as free of error.

Finally, the prediction line that is often used as a compromise between the two regression extremes is the reduced major axis (RMA). The line passes through the bivariate mean and its slope is determined by the ratio of the standard deviations of the core porosity and transit time. The equation of the RMA line is then:

$$\hat{\Phi} = -40.0 + 0.72\,\Delta t$$

The selection of the RMA is sometimes based on intuitive appeal, since it typically has the visual appearance of best fit. The reason for this is that a best-fit line drawn by eye will usually minimize scatter about the line in a direction normal to the line. This is the criterion for the principal axis, but is also closely approximated by the reduced major axis. The comparative simplicity of the equation parameters and its failure to include cross product terms between the two variables puts a strain on its credibility. However, if the measurement error-variance ratio is closely approximated by the total variance ratio, then the RMA will be the optimal solution.

In reality, errors are inevitably associated with both measurements and the problem can be recast as one of functional analysis if the descriptive equation can be considered as a functional relationship. The intercept and slope are then no longer abstract parameters, but quantities that have a physical meaning and so should honor the constraints and expectations of the real world. When the prediction equation is rewritten as a functional relation, it becomes the time-average equation of Wyllie et al. (1956), usually in the form of:

$$\Phi = \frac{\left(\Delta t - \Delta t_{ma}\right)}{\left(\Delta t_f - \Delta t_{ma}\right)}$$

where Δt_{ma} and Δt_f are the transit times of the matrix mineral and the fluid. The equation is functional to the extent that empirical observations have shown that it is a reasonable linear approximation to real variation. If this is the case, then lines that are fitted to porosity–transit time data should result in transit times for matrix and fluid that are considered to be reasonable values. In this example, the lithology of the reservoir is a sandstone, so that the matrix transit time is that of quartz (conventionally 55.6 microseconds per foot) and the fluid transit time corresponds to mud filtrate in the borehole wall (conventionally 189 microseconds per foot).

As discussed earlier, a key step in functional analysis is the estimation of the error variance associated with each variable or the ratio of their variances.

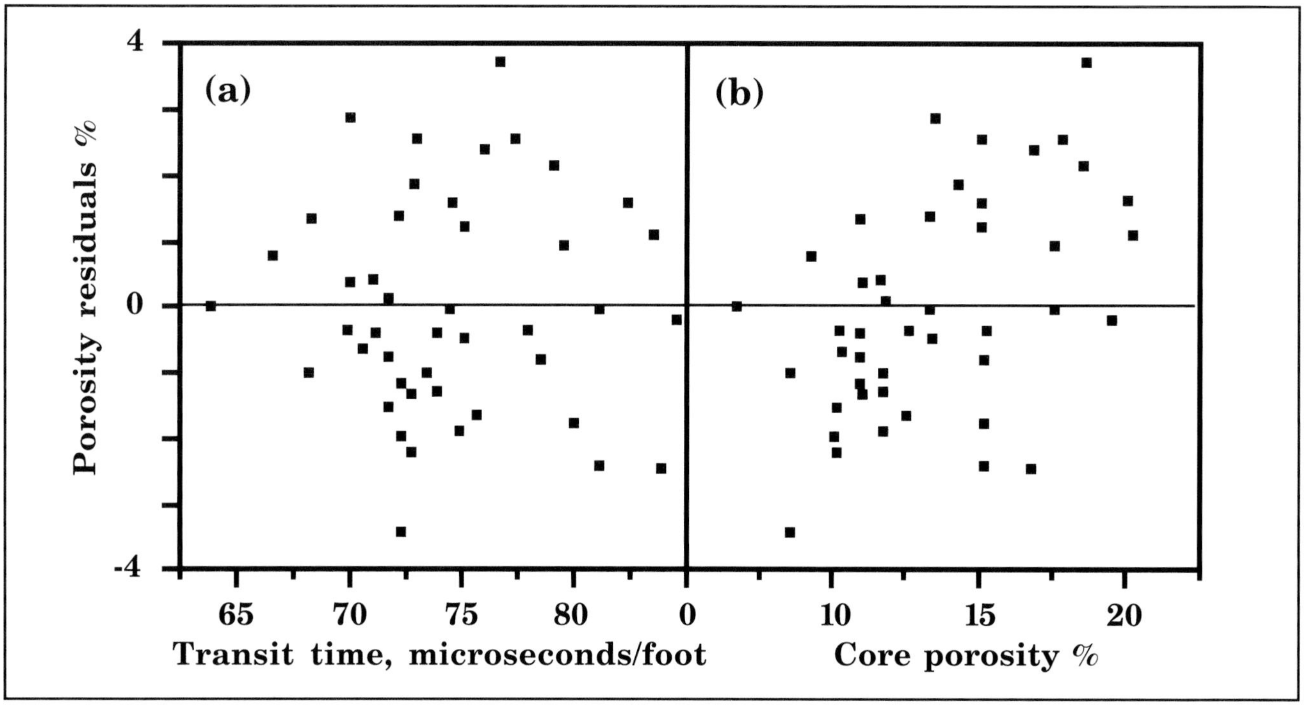

Figure 10. Residual differences between core porosities and predictions based on the regression of porosity on transit time versus (a) transit time and (b) core porosity.

Although this information was not available for this example, the error variances could have been estimated as a contributory part of the line-fit analysis. The error variance of the transit times would be estimated by analysis of the deviations of the main sonic log from its repeat section in a procedure similar to that applied to the density log example described earlier. The error variance of core samples comprises two sources of variability. The first is controlled by the resolution of the laboratory method of porosity measurement, which can be deduced from repeated analysis of the same core samples. This procedure is a fundamental quality check and is widely practiced by laboratories on standard core samples to gain information on relative precision and to check for bias when comparing with alternative methods or different laboratories. The results of this type of work are now reported more widely, as in the statistical summary of the data quality assurance tests at Amoco described by Thomas and Pugh (1989). However, the integration of such data as a part of standard log analysis is still a rare event. The second source of core variability is caused by the fact that measurements are most commonly made on small plugs sampled at intervals of 1 foot. These are only estimates of the porosities represented in whole-core measurements. The smaller volume causes plug measurements to have higher variances than those of the larger whole-core samples. This possibly surprising assertion will be supported a little later in a discussion of work by Baker (1957), whose results are shown in Figure 13.

In the absence of specific information on the error variances of the two variables, functional analysis proceeds by evaluating the consequences of alternatives in search of an optimum line-fit. The criteria to be met are that the joint estimates of the functional parameters and the error-variance ratio, λ should be judged the most reasonable combination. The range of possibilities are bound by the regression line at each extreme, where the total error is attributed to one or other of the variables. However, it should be remembered that these extremes are only estimates of the true regression lines, because they are based on a sample size of 44 observations. All the possible best-fit lines pass through the bivariate mean, so that the trace of possible parameter solutions is a straight line as shown on the crossplot of matrix and fluid transit time of Figure 11. The reduced major axis (RMA) is close to the idealized transit times of quartz and fresh water. If this line is the best choice, then the error-variance ratio would be 0.52. Converting transit times to porosity equivalences, the number would suggest that the sonic log and core data estimate the porosity to about the same accuracy. This conclusion is credible when it is remembered that the core measurement is based on a small plug and is only an estimate of the whole rock sample. This would probably cause the standard deviation of both estimates to be approximately the same at about one porosity unit.

When evaluating the results of functional analysis, clear distinctions must be made between useful descriptive models and functional relations that are actual mathematical descriptions of processes. In this present example, the Wyllie equation is a descriptive functional relationship that should only be honored

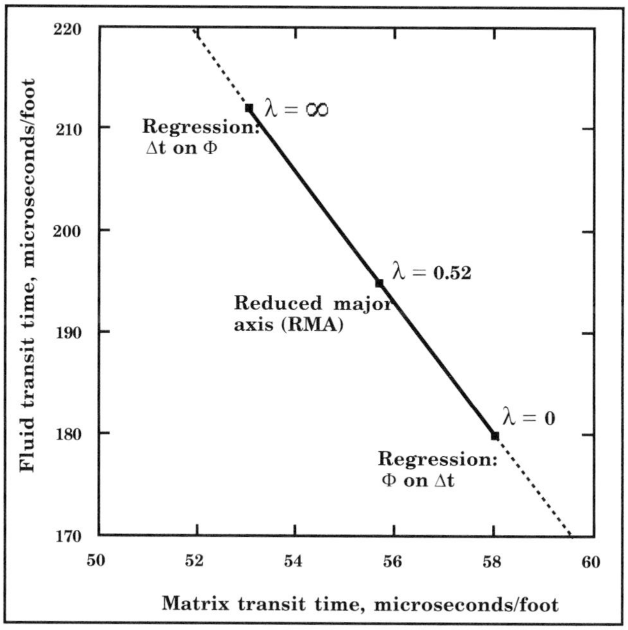

Figure 11. Crossplot of the matrix and fluid transit times for the range of all possible best-fit lines from the sandstone reservoir transit time–porosity example. λ is the error-variance ratio.

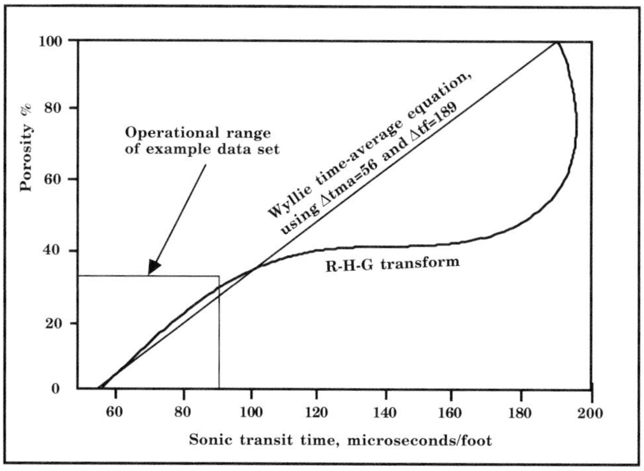

Figure 12. The Raymer-Hunt-Gardner (R-H-G) sonic transit time-to-porosity transform contrasted with the Wyllie time-average equation.

to the extent that it models reality. The shortcomings of the time–average relation were understood at the outset by Wyllie et al. (1956), and modifications have been proposed over the years, of which the most widely adopted is that of the Raymer-Hunt-Gardner transform. From a study of many sandstones, Raymer et al. (1980) established a generalized transit time–porosity relationship, which is shown in Figure 12. Since the curve is a closer representation of functional reality than the equation used up to now, how does this affect our conclusions? Examination of the figure shows that for the data range of the example, the curve is closely approximated by the time–aver-

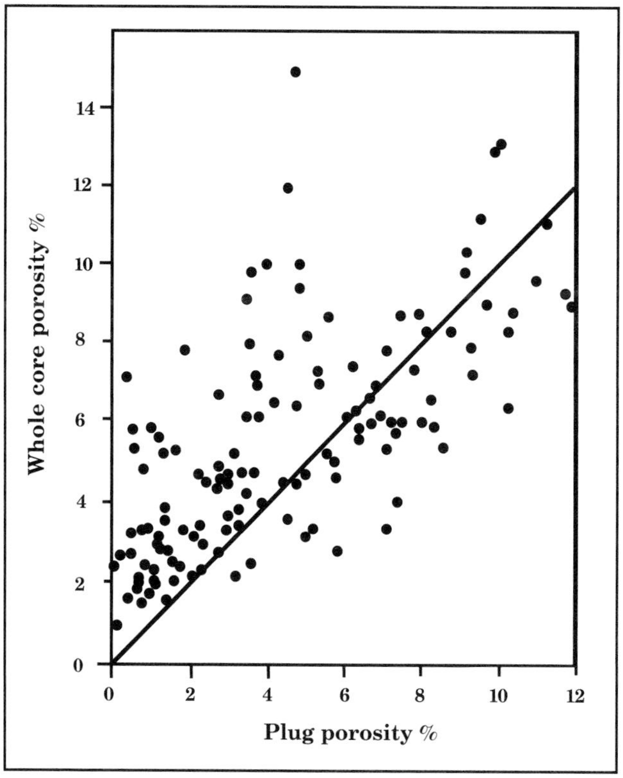

Figure 13. Crossplot of porosities measured from core plugs with porosities of whole-core samples. Adapted from Baker (1957). Copyright SPE.

age equation, with expectations of matrix and fluid transit times that would be close to their physical values. However, if the samples had been drawn from a higher porosity range, then a linear trend would have been tangential to the curve, with an expected apparent matrix transit time artificially lower than its real value. Consequently, the expectations of credible parameters would need to be modified appropriately. These considerations do not invalidate the approach of functional analysis, but instead remind the analyst that reality takes precedence over models when they reach their limitations. It should also be noted that this same warning applies to the Archie equation and several other fundamental log analysis relationships.

In the earlier discussion of this example, it was stated that comparisons of petrophysical variables must be made in terms of a common vertical resolution. The necessity for this rule has been discussed widely in the log analysis literature so that, for example, Runge and Powell (1967) stated: "Different logging devices and sampling techniques have different spans and when a comparison is desired, the differences in span lead to an incompatibility between these modes of measurement." Ideally, a deconvolution of the coarser resolving measurement to a finer scale would be desirable, but generally this is not practical. Problems such as the nonlinear response of the induction tool and the stochastic nature of nuclear measurements make effective deconvolution very difficult (Looyestijn, 1982). In practice, measurements are smoothed to an equivalent common scale with the variable with the coarsest vertical resolution. In the current example, the core measurements of porosity were smoothed by a running-average filter to give an approximate common resolution with the 2-ft span of the sonic log.

There are consequences that follow from the failure to correct the incompatibility of measurement scale by appropriate smoothing. These can be better understood by consideration of the relationship between porosities of plugs and whole-core samples. In experiments with the early density tool, Baker (1957) contrasted porosities measured from 1-in.-diameter plugs with porosities measured from their whole-core samples. A crossplot of the results is shown on Figure 13, which demonstrates that the error in predicting the porosity of any 1-ft interval on the basis of a plug measurement has a highly distinctive bias. At higher porosities the plug measurement will tend to overestimate the average porosity, while at lower values the plug measurement tends to be an underestimate. The relationship is inevitable, because the porosity of the whole core represents an average of all the potential plugs it contains. Consequently, although a set of plugs and whole core should have the same mean value, the variability of the whole core will be less than that of the smaller plugs. The decreased variance on the whole-core porosity axis, compared with the variance on the plug porosity axis, gives the appearance of rotating the data cloud from a simple diagonal trend. The mechanism for this effect is simply the aggregation process of measurements from samples of larger volume, in which the extremes in the smaller samples are averaged out. Although these arguments have been developed from the perspective of core measurements, they apply equally to logging data.

ESTIMATION OF WATER SATURATION

All current methods for estimating water saturation in reservoir zones are variants based on the Archie equation:

$$S_w^n = \frac{aR_w}{\Phi^m R_t}$$

Using the known porosity of a zone and the resistivity, R_t, read from a log, the water saturation can be solved, provided that values for the saturation exponent, n, the cementation exponent, m, and the formation water resistivity, R_w, are all available.

Methods for the deduction of the cementation exponent and water resistivity from log data have had a long history—particularly with regard to graphical pattern-recognition techniques, which will be reviewed in the next chapter. However, the availability of digital log data, increasing access to computer software, and the nature of the problem have encouraged the development of direct numerical solutions based on statistical analysis. As we shall see, a variety of difficulties needs to be addressed in a convincing manner.

For a sequence of completely water-saturated zones, the Archie equation is:

$$\frac{R_o}{R_w} = \frac{a}{\Phi^m}$$

where R_o is the bulk resistivity. The equation can be linearized to:

$$\log R_o = \log(aR_w) - m \log \Phi$$

which describes a line with a negative slope of the cementation factor and an intercept of the logarithm of the water resistivity (when the constant a is considered to be unity, as in the original Archie equation). In working with this relationship, the goal is neither to predict zone resistivity given porosity nor vice-versa. Instead, the initial purpose is to establish a "water line" whose slope and intercept are consistent with reasonable values of the cementation exponent and formation-water resistivity. The line estimates of these parameters will then be used in the Archie equation for calculation of water saturation in zones that are potentially productive. This formulation of the problem is clearly an application for functional analysis.

The fact that most line-fitting methods have used a regression procedure would appear to conflict with the preceding remarks, since the goal is not prediction. However, in a detailed discussion of the theoretical sources of error associated with the common logging tools, Etnyre (1984a, b) considered the error component to be greater for porosity measurements than resistivity. In the case of porosities inferred from density and neutron tools, the counts of nuclear particles are confounded with statistical error. This stochastic character contrasts with the deterministic nature of resistivity, although some degree of measurement error is inevitable, particularly at the higher resistivity range. This conclusion is further supported by Farnan and McHattie (1984) from their observations contrasting main and repeat runs of porosity and resistivity logs. They reported better repeatability of resistivity logs, although they noted discrepancies at higher resistivity values. The implication follows that the error-variance ratio is dominated by errors attributed to porosity. Therefore, the line favored for functional analysis is weighted close to the regression of the logarithm of porosity on the logarithm of resistivity.

If this regression model is appropriate, then the descriptive equation is rewritten as:

$$\log \Phi = \left(\frac{1}{m}\right) \log R_w - \left(\frac{1}{m}\right) \log R_o$$

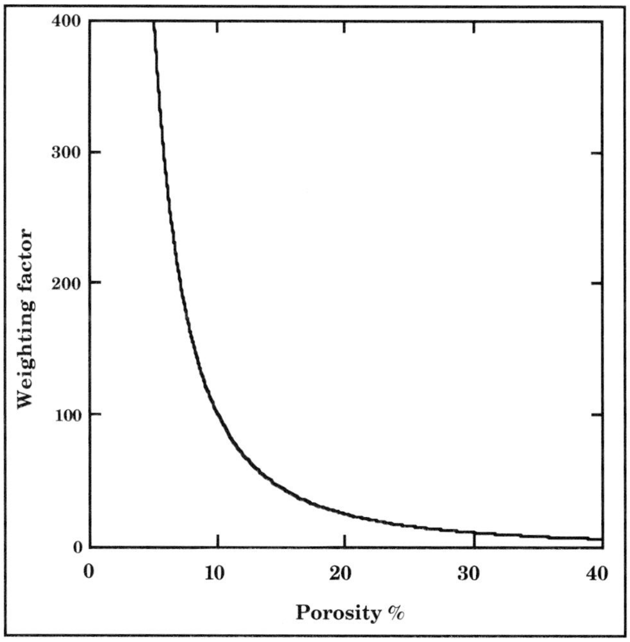

Figure 14. Weighting of resistivity porosity observations due to logarithmic transformation as a function of porosity.

where all errors are assigned to porosity. These errors would be expected to be normally distributed about the porosity estimates. A slight drift in error ranges as a function of porosity in neutron and density measurements is caused by the physics of the particle-counting process (see Etnyre, 1984a, b), but the effect is fairly minor. Of much greater concern is the result of logarithmic transformation of the measurement variables, which results in preferential weighting in the low range of porosity and exaggerated influence of errors at these levels. The potential distortions are a function of the porosity, so that the relative weight assigned to any given zone is given by:

$$w = \left(\frac{1}{\Phi^2}\right)$$

A graph of the weighting factor that is assigned implicitly to each point in the logarithmic transformation is shown in Figure 14. The most marked distortions are caused by readings with porosities less than 10%. Etnyre (1984a, b) presented equations to compensate for this unequal weighting, but also concluded that failure to make compensations was less important than the appropriate choice of the dependent variable in the regression equation. In fact, zones with porosities of less than 10% should be treated cautiously for petrophysical reasons. Focke and Munn (1987) found that core measures of carbonate cementation exponents show a systematic decline in value in this range. At higher porosities, cementation exponents appear to be asymptotic to a value of two in carbonates with intercrystalline and/or intergranular porosities (see Figure 15).

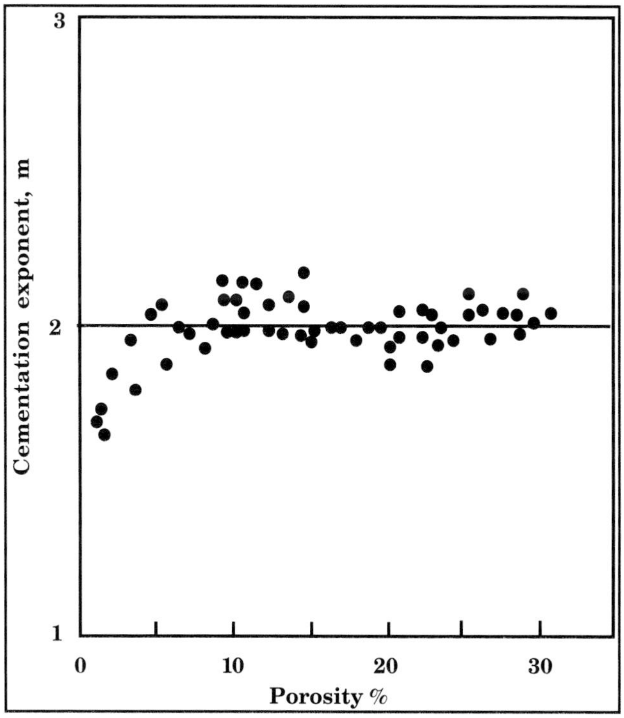

Figure 15. Relationship between core measurements of cementation exponent and porosity in carbonates with intercrystalline and/or intergranular porosities. Data from Focke and Munn (1987). Copyright SPE.

Table 2. Log porosities (%) and Induction Resistivities (ohm-m) from a Carbonate Gas Well in Oklahoma (data collected by G.R. Pickett and tabulated by Whitman, 1986.

Φ	R_t	Φ	R_t
1.5	75.0	11.5	5.0
4.0	16.0	11.0	7.5
6.5	8.0	12.5	3.0
1.5	90.0	15.5	2.4
2.0	80.0	15.5	3.5
1.5	110.0	16.5	3.0
3.0	60.0	14.5	3.5
1.0	100.0	19.0	2.8
4.5	23.0	21.0	1.7
1.0	160.0	18.5	2.6
5.0	60.0	12.5	6.0
22.0	70.0	6.0	15.0
20.0	22.0	8.5	21.0
13.0	18.0	2.0	80.0
10.5	7.0		

The practical business of fitting a line to resistivity–porosity data is now discussed in an application to logging measurements from a gas well in Oklahoma.

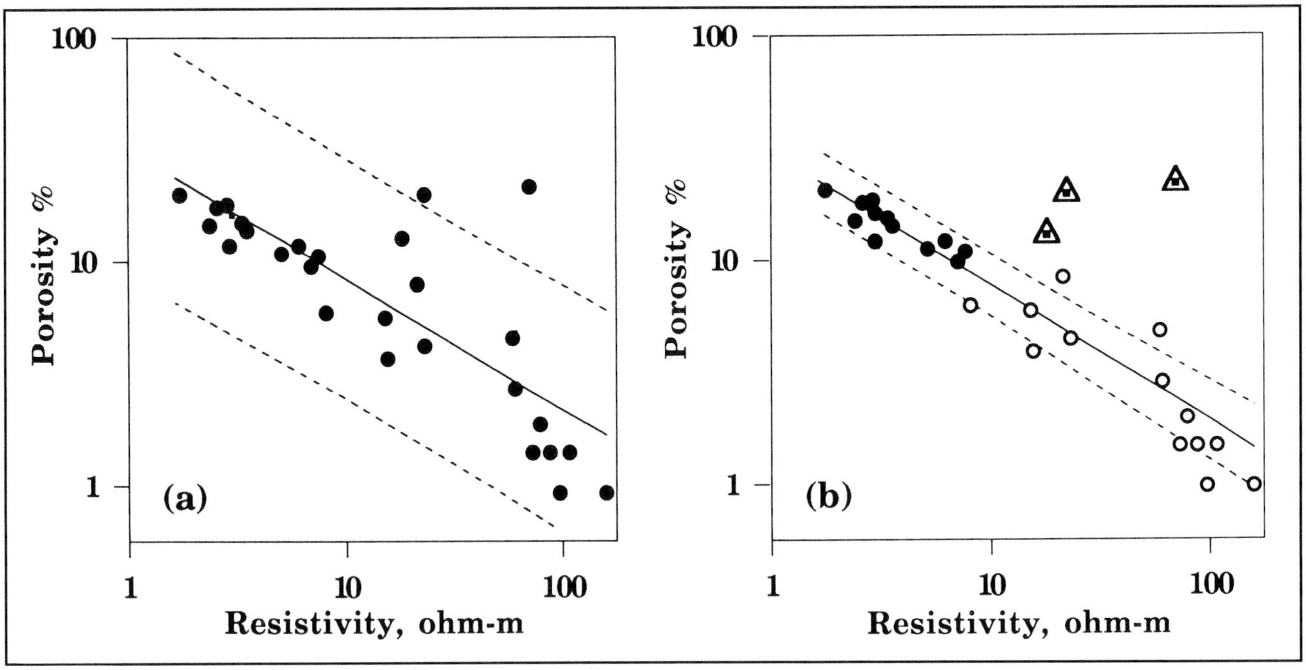

Figure 16. Iterative line fitting of regression of logarithmic porosity on logarithmic resistivity with associated confidence intervals for (a) initial fit using all data and (b) final fit on fourth iteration generated through the selective exclusion of zones. Black circles are final water zones, open circles are excluded low-porosity zones, and triangles are outliers indicated to be hydrocarbon zones.

The data were used by George Pickett for various teaching demonstrations and are reproduced in Table 2 from Whitman (1986). The data set contains both water-saturated and productive zones. Up to this point, oil and gas zones have been excluded from the restricted discussion of appropriate line-fits for water-saturated rocks. However, these same prospective zones will be the target of most routine resistivity–porosity log analyses. Hydrocarbon saturation causes systematic increases in resistivity away from the hypothetical water line that should be more anomalous than the normal statistical scatter of water-saturated zones about the ideal water line. In the terms of conventional regression analysis, the anomalous zones constitute distinctive "outliers." These considerations led Whitman (1986) to apply a modified regression-analysis procedure that would simultaneously estimate a valid water line and identify potential hydrocarbon zones. A simplified version of his approach is described in the following section.

An initial, provisional regression line was fitted to the resistivity–porosity data from all the zones and is shown in Figure 16a. The line has a coefficient of determination of 0.67, from which it follows that 33% of the total variation remains in the deviations of points about the line. Anomalous points to the right of the line signify zones where the resistivity appears to be markedly higher than the average trend, and so are candidates for productive intervals. Which of these can be considered to be distinctive outliers caused by systematic effects of hydrocarbon saturation and which are indistinguishable from the statistical scatter about the estimate of the water line? If the regression model is fitted to a sample consisting entirely of water zones, then the deviations of the zones would be expected to be approximately normally distributed about the line. If there are productive zones, then these should be marked by deviations that pull them outside the range of typical error scatter.

Confidence intervals are shown as dashed-line limits which parallel the regression line (Figure 16a) and outline the theoretical range of occurrence of 95% of the observations, based on a normal error distribution. It should be realized that the inclusion of potential outliers will inflate these bounds to some degree. However, at least one point is clearly outside the confidence interval and is judged to be a productive zone. As an improvement on the initial estimate of the water line, both the outlier and zones with porosity less than 10% were removed from the data set, and the regression analysis repeated on the remainder. The exclusion of the low-porosity zones in this revised analysis follows from both Etnyre's (1984a, b) observation of the excessive weighting applied to them by the logarithmic scaling, and the low cementation exponents commonly observed in this range which would tend to distort the line.

The process of regressing fitted lines was iterated several times. At each iteration, the elimination of a positive outlier resulted in the shrinkage of the confidence limits about the line on the successive iteration. On the fourth iteration, the regression model was considered to be the stable, final solution shown on

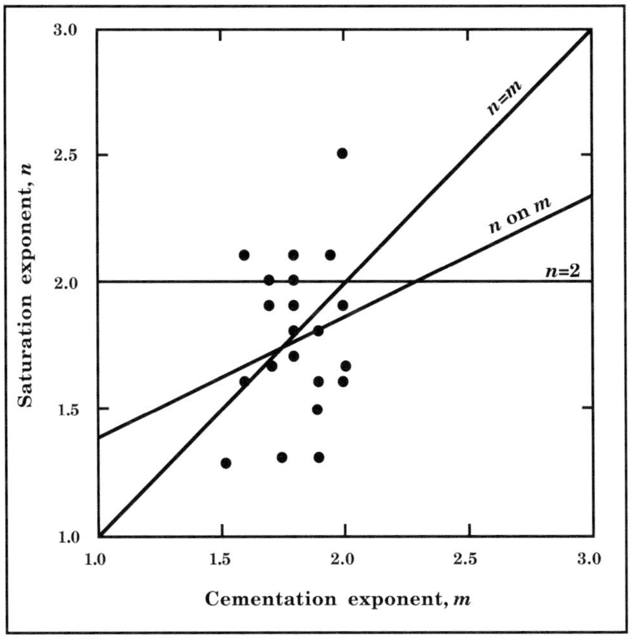

Figure 17. Crossplot of cementation and saturation exponents for sandstone data reported by Coates and Dumanoir (1974), together with regression line of *n* on *m*, line of equivalent *n*-*m*, and line of constant *n* of value 2.

Figure 16b. The criteria used were pragmatic, but reasonable. All the outliers that could be attributed to hydrocarbons appear to have been identified, while the remaining zones are equably distributed about the line and within the confidence limits. At porosities below 10% the distribution is more ragged, but this is an inevitable consequence of the exaggerated scaling of the logarithmic transform.

The fitted regression line can now be evaluated in terms of functional analysis. The coefficient of determination of the regression has improved substantially to 0.95. This means that there is a correlation of 0.98 between the logarithms of porosity and resistivity in the higher porosity water zones. Consequently, the issue of the measurement error variance of the two logs turns out to be of little significance in this example, because the difference between the alternative fitted lines will be very small. Attention should now be focused on the match of the model and its parameters with the expectations of the real world. Notice that the deviations at the lower porosity observations are all below the water line. This is a natural consequence if values of cementation exponent are lower in this range. However, zones with porosities between 5 and 10% appear to be symmetrically distributed about the line although the scatter is stretched by the logarithmic scaling. It therefore appears that data in this intermediate range are usable, but must be treated cautiously in view of the high weighting factors associated with these zones. The slope and intercept of the regression water line result in estimates of cementation exponent of 1.97 and formation-water resistivity of 0.07 ohm-m. The cementation exponent is a highly acceptable estimate of a value that would typify carbonates with intercrystalline and/or intergranular porosity types.

In the regression analyses of these same data by Whitman (1986), cementation exponents were calculated typically at about 1.6, because of the inclusion of the low porosity zones. He attributed the cause of the low values to a failure to correct the sidewall neutron porosity log for lithology effect. While the current analysis suggests that this interpretation is unnecessary, it demonstrates a correct line of functional analysis reasoning. The most likely natural explanation of a carbonate cementation exponent of 1.6 would be the presence of significant fracturing. If this situation is unlikely, then the most reasonable alternative would be some systematic bias in the porosity or resistivity logs, indicating that a miscalibration should be corrected.

The computation of water saturation from the Archie equation requires only one other unknown to be resolved, and this is the saturation exponent, *n*. In his original work, Archie (1942) concluded that the saturation exponent value was typically about 2 for sandstones; this is still the most widely used value today, even though the saturation exponent is known to show wide variability. Numerous laboratory measurements have been made over the years and their sometimes puzzling results and complex implications are summarized well by Edmundson (1988). Typically, saturation exponents range between a lower limit of 1 and a more diffuse higher limit of about 4. Fundamental controls include the pore structure of the rock and its wettability with respect to oil or water (Keller, 1953). The value is also influenced markedly by core sample preparation techniques (Worthington et al., 1988). Von Gonten and Osoba (1969) concluded that predictions of the saturation exponent were best estimated by a trend of increasing values linked with decreasing water resistivity (increasing salinity) and increasing permeability.

Some authors have suggested that the saturation exponent could be assumed to be equal to the cementation exponent (e.g. Schulze et al., 1985). There appears to be no evidence for this relationship other than the intuitive appeal that the exponents are both affected by pore geometry to unknown degrees. However, if true, the equivalence of the two exponents would simplify the Archie equation considerably. The idea was first suggested by Coates and Dumanoir (1974) who collected a sample of laboratory measurements of *m* and *n* and stated: "Although the table shows distinct cases where m and n are different, there are many cases where they have approximately the same values. Assuming this to be acceptable, we set *m* and *n* equal and call them both *w* (*m*=*n*=*w*)." This conclusion was disputed by Ransom (1974) in the ensuing discussion of the paper, based on experience with a broader sample.

A crossplot of the saturation exponents and cementation exponents is shown in Figure 17 for the sandstone data reported by Coates and Dumanoir (1974). A regression line prediction of *n* based on *m*

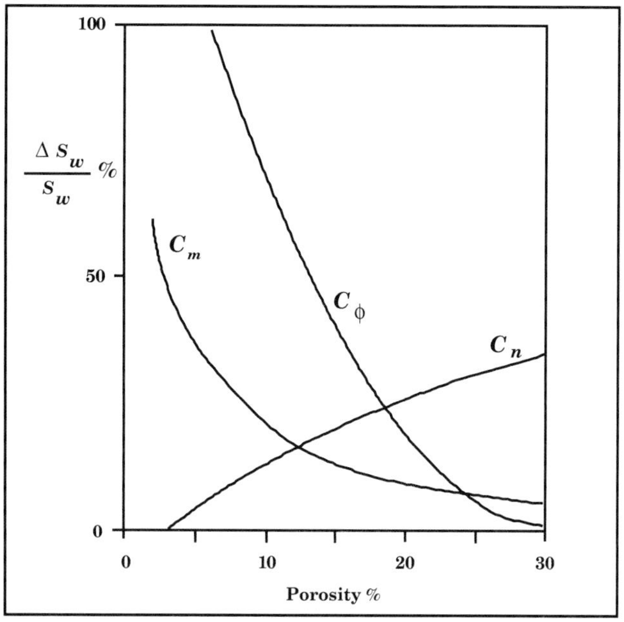

Figure 18. Example of the results of a sensitivity analysis calculation of fractional error contributions of porosity (C_Φ, cementation exponent (C_m), and saturation exponent (C_n) to water saturation estimations related to porosity. In this unequal uncertainty mode example, uncertainties of 10% were assigned to *m* and *n*; the uncertainty associated with porosity ranged from 2% at a porosity of 30 units to 50% at a porosity of 2 units. From Chen and Fang (1986).

shows a generalized positive trend. However, the possible trend implied by the scatter of the raw data appears to be less than convincing. Is there a statistically significant relationship between the two exponents? The question can be examined by proposing a null hypothesis that there is no correlation. The correlation coefficient of the cementation and saturation exponents for this sample is 0.23. This is clearly not zero, but could it be generated in a random sample drawn from a hypothetically infinite population of laboratory measurements? The potential significance of the correlation coefficient may be examined through the use of Student's *t* test (Davis, 1986, p. 66-67). The test value of *t* is given by:

$$t = \frac{r\sqrt{n-2}}{\sqrt{1-r^2}}$$

where *r* is the correlation coefficient and *n* is the number of observations. This quantity does not exceed 2.07 for the data of Figure 17, which is the critical value of *t* at (*n*–2) degrees of freedom and a conventional significance level of 5%. Therefore, the null hypothesis is not rejected and no significance is attached to any linear trend that can be drawn through this sample of data.

It should be noted that the *t*-test did not prove that there was no relationship between the two exponents. It determined simply that there was certainly not sufficient evidence to conclude that there was a relationship with a chance of less than one in twenty of being wrong. It is possible that a significant correlation might emerge from a larger sample of observations. However, based on the scatter in this sample, the prognosis is not good for the usable predictive relationship of *n* based on *m*. If these data are typical, then a saturation exponent of 2 is a reasonable choice for the default value, unless more explicit information is available.

Obviously, this fallback solution is only partially satisfactory because of the potentially large errors in hydrocarbon saturation that can result when the saturation exponent is substantially different from a value of 2 (see Dorfman, 1984). The clear remedy is to make extensive laboratory measurements of saturation exponents from core, but this may be difficult to justify in many instances for either economic or technical reasons. In the next chapter we will examine some crossplot conventions that may be useful as pattern-recognition aids for finding appropriate and useful values of *n*. The most reasonable alternative is to make sensitivity tests of the errors that will be associated with the best estimates of water saturation. The information can then be used after the fact in the computation of best-case and worst-case scenarios linked with probability levels selected by the user. Alternatively, the error analysis can be used to guide the analyst before the fact in the selection of a critical value for exploration or production decisions.

In actuality, statistical errors are associated with all the parameters of the Archie equation. It is therefore important to assess the relative degree to which they influence water saturation estimates, and whether the separate effects change over different porosity ranges. Sensitivity analysis is the ideal approach to evaluate the results of small perturbations in parameter values within the Archie equation. In studies of Archie-equation errors, the critical parameter was identified variously as cementation exponent (Walstrom et al., 1967), porosity (Rosepiler, 1981), and saturation exponent (Dorfman, 1984). Chen and Fang (1986) were able to reconcile these apparent conflicts of opinion in a comprehensive sensitivity analysis in which they considered the simultaneous interplay of errors in all the parameters. Because the errors are independent random variables, the total variance in the estimate of the water saturation can be partitioned as the sum of the variances associated with the six parameters in the Archie equation, namely: *a*, *m*, *n*, *f*, R_w, and R_t. The variance relationship allows the effects of perturbations to be evaluated methodically under a variety of conditions of both equal and unequal uncertainties.

The results from one of the experiments of Chen and Fang (1986) are illustrated in Figure 18, which shows that the relative importance of the parameters varies with porosity magnitude. Small errors in porosity constitute a relatively higher proportion of lower porosities than of higher porosities. The effect of changes in the cementation exponent is greater at lower porosities than higher ones. The reverse situation is true in the case of the saturation exponent, which becomes especially important at higher porosities. The utility of these types of studies is that they allow the investigator to focus on the parameters which play the most important role in water-saturation estimations for specific reservoir types. They also allow margins of error to be computed around reservoir properties, based either on core measurements or hypothetical error variances.

ESTIMATION OF PERMEABILITY

The only wireline logging tool that is capable of gauging permeability in a direct fashion is the repeat formation tester. The measurements of other logging devices are used to infer permeabilities based on links deduced between formation physical properties and fluid flow. Although the basic physical theory of permeability has been understood for some time, reliable estimations are difficult to make in all but the simplest rock types. Multiple processes of sedimentation and diagenesis can create complex and variable pore systems. Consequently, successful methods are generally those that are based on a mixed strategy that uses basic theoretical concepts, insights drawn from the geological history of the formation, and an empiricism that is guided by statistical analysis.

The most simple quantitative methods to predict permeability from logs have been keyed to empirical equations of the type:

$$K = A\,\Phi^B$$

where A and B are constants determined from core measurements, and applied to log measurements of porosity (Φ) to generate predictions of permeability (K). When applied to special cases of homogeneous sandstones, the results may be adequate, but prediction errors are often large in typical sandstones and the errors in predicted permeability commonly range across orders of magnitude when applied to carbonates. The reason for this is that permeability is not exclusively determined by pore volume, but is also controlled by internal surface area, pore network tortuosity, pore throat geometry and other variables.

Timur (1968) developed an equation which linked permeability with both porosity and irreducible water saturation (S_{wi}) in sandstones, based on laboratory measurements of core. The results showed a considerable improvement in permeability estimation over those based on porosity values alone. The use of irreducible water saturation as an input variable restricts the method to virgin hydrocarbon reservoir zones. Further, the log calculations of irreducible water saturation are themselves estimates, whose error magnitude is controlled by choices of values for Archie's cementation factor and saturation exponent as discussed earlier. In spite of these limitations, the success of the Timur equation (whose generic form is):

$$K = \frac{A\Phi^B}{S_{wi}^C}$$

can be understood when it is compared with the classic Kozeny-Carman equation (Carman, 1937):

$$K = \frac{A\,\Phi^3}{(1-\Phi)^2 S^2}$$

which incorporates the specific surface area, S. The specific surface area is the ratio of surface area to volume of framework solid and is difficult to measure directly by conventional methods. However, the specific surface area is inextricably linked with pore size, which in turn controls irreducible water saturation. The irreducible water saturation term in the Timur equation therefore functions as a powerful surrogate variable for specific surface area, and this accounts for the improvement in permeability estimates when incorporated with porosity.

The morphology of the rock framework is the exact complement of the internal geometry of the pore network. It therefore follows that measures which are sensitive to characteristics such as grain size or crystal size, will ultimately be useful indicators of specific surface, and thus estimators for permeability. The control of permeability by grain size and sorting in sandstones has been documented for many years (e.g. Krumbein and Monk, 1942). Analogous relationships which link porosity and crystal or grain size in carbonates have also been widely reported (e.g. Choquette and Traut, 1963; Bebout et al., 1987). However, modern logging measurements are sensitive to framework compositions rather than direct responses to grain size. Herron (1987) pointed out that rather than being a limitation, this property has definite advantages. So, within clastics, permeability is ultimately controlled by a complex association of disparate factors. These are texture (grain size, shape, and sorting), and the quantity and types of both cementing minerals and clay minerals. Textural maturity, cements, and clays are all reflected in mineral compositions and these can be deduced from several types of logs.

Herron (1987) has applied this concept successfully in an adaptation of the Kozeny-Carman equation that substitutes mineral assemblage estimates from geochemical logs for the specific surface term:

$$K = \frac{A\,\Phi^3}{(1-\Phi)^2 \exp(\Sigma B_i M_i)}$$

where A now represents a textural maturity term that is dependent on feldspar, M is the abundance of an ith mineral and B_i is a constant for the ith mineral.

Table 3. Logarithmic Core Permeabilities, Log Porosities (%), Uranium (ppm), and Potassium (%) Contents from a Well in the Lower Permian Chase Group, Southwestern Kansas.

log *K*	Φ	Ur	K	log *K*	Φ	Ur	K
–0.01	9.8	1.1	1.2	0.03	11.3	2.3	0.3
–1.56	1.9	2.7	0.5	0.90	16.4	3.4	0.5
–1.94	5.6	2.5	0.7	0.17	13.2	2.6	0.2
–1.59	6.5	2.7	0.6	0.52	15.1	3.4	0.3
1.33	15.1	0.5	0.2	0.06	13.4	4.3	0.4
0.71	11.1	0.9	0.2	0.57	16.1	3.7	0.5
1.03	14.7	0.5	0.3	0.07	14.8	2.8	0.2
0.60	11.5	0.7	0.4	0.53	17.2	2.8	0.8
0.93	13.5	0.7	0.4	0.09	16.7	3.5	0.8
0.58	8.8	0.7	0.3	–1.30	15.0	2.3	0.7
1.20	12.5	1.0	0.3	–1.03	15.8	2.4	0.7
1.35	15.0	0.7	0.2	–1.57	16.5	2.4	0.8
0.76	8.8	0.9	0.4	–1.23	16.1	2.2	1.0
1.35	10.8	0.8	0.4	–2.00	8.9	3.2	0.6
1.31	13.5	2.3	0.4	–0.13	13.2	4.3	0.9
–0.46	0.7	3.7	0.6	–1.58	13.5	4.8	1.4
–0.94	1.6	4.9	0.2	–1.89	12.1	4.8	1.0
–0.07	5.5	0.4	0.7	–1.53	11.2	5.4	1.0
–0.71	9.6	0.9	1.0	–1.43	2.9	1.3	0.1
0.53	11.9	0.7	0.7	–0.65	5.2	1.4	0.1
0.28	9.3	0.8	0.3	–0.83	5.4	1.7	0.1
0.93	9.1	0.9	0.2	–1.03	6.0	1.8	0.2
0.66	4.8	0.7	0.2	–1.69	7.2	4.0	0.5
1.36	9.7	1.4	0.5	–0.94	11.0	4.3	0.5
1.40	14.6	1.7	0.3	0.16	15.3	4.7	0.4
–0.66	3.8	3.0	0.1	0.19	14.3	2.7	0.2
0.47	17.9	4.4	0.5				

The relationship can then be linearized to:

$$\log K = \log A + 3 \log \Phi - 2 \log(1 - \Phi) + \sum B_i M_i$$

In experiments with geochemical logs of clastic successions from Venezuela, Oklahoma, and California, Herron (1987) was able to develop useful predictive relationships for permeability based on mineral-abundance estimates. The coefficients that were resolved for each hypothetical mineral component matched general petrophysical expectations. Clay minerals all had negative loadings, ranging from a high for smectite to a low for kaolinite. Calcite had a smaller, but still negative coefficient as a consequence of its role as a cementing agent.

These ideas can be extended to carbonates in models which incorporate concepts drawn both from depositional facies and diagenetic processes. Several log measures should be useful, particularly since diagenesis is often fabric-selective and frequently linked with changes in mineral composition. The following example uses a data set of core permeabilities and logging measurements from the Lower Permian Chase Group of the giant Hugoton gas field in southwestern Kansas (Table 3). The raw measurements of permeability were smoothed with a 5-point (2.5 ft) binomial filter to give approximately equivalent vertical resolution with the wireline logging measurements. The data set consists of zoned readings of permeability, porosity computed from a density–neutron log combination, and uranium and potassium measures from a spectral gamma-ray log.

Predictions of permeability from logs are most commonly keyed to porosity measurements, either from core or logs, using the relationship:

$$\log K = A + B \cdot \Phi$$

The difference in permeability and porosity scaling is an implicit recognition of the observation that porosities appear to be approximately normally distributed and permeabilities tend to be lognormally distributed. The distribution form has a direct bearing on line-fit procedures, because the slope is controlled by

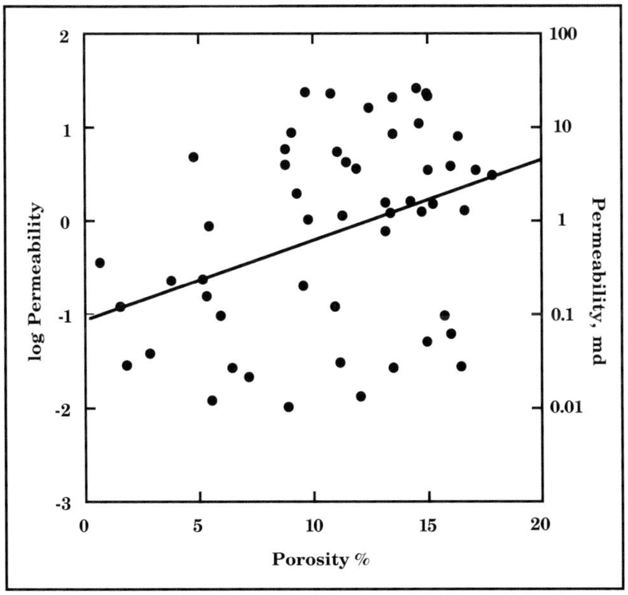

Figure 19. Regression of the logarithm of permeability on porosity for zones in the Chase Group example.

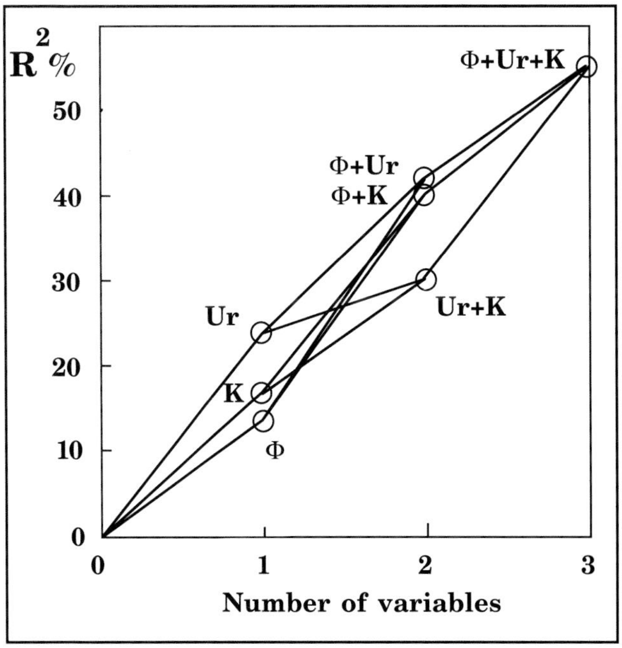

Figure 20. Coefficients of determination (R^2) associated with alternative regressions of logarithmic permeability on combinations of porosity (Φ), potassium (K), and uranium (Ur) in the Chase Group example.

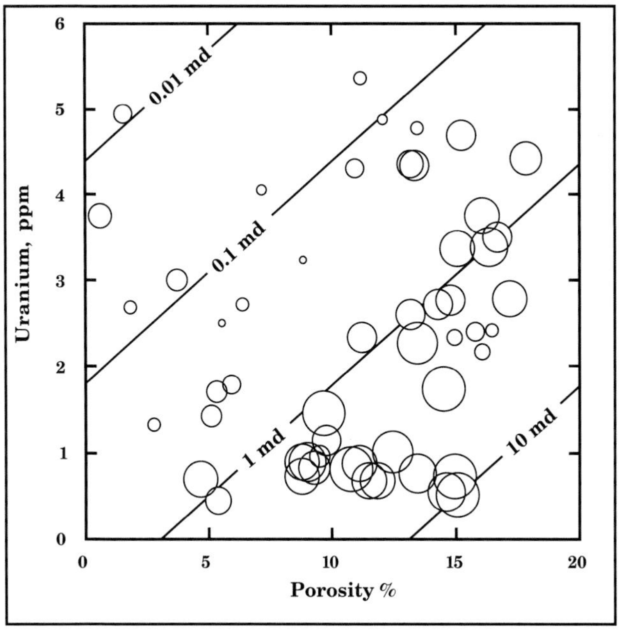

Figure 21. Bubble plot indexed with contour lines of the regression of permeability on porosity and uranium in the Chase Group example. The diameter of the bubbles is proportional to the measured permeabilities.

the variance–covariance estimates, as discussed earlier. Notice that the use of the functional relationships that are based on the Kozeny-Carman equation would lead to a linear relationship between permeabilities and porosities that are both logarithmically scaled.

The goal is to predict permeability from log measurements, which suggests that a regression analysis is the appropriate procedure. The alternative of functional analysis is hamstrung by the difficulties in giving petrophysical meaning and expected numerical values for the parameters of the equation. In addition, the measurement error variance is likely to be higher for permeability measurements than for porosities. This would be an additional argument for a choice of a regression of permeability on porosity, in which all the error is assigned to permeability.

The regression line of permeability on porosity for the Chase Group data is shown in Figure 19. The line picks up the broad trend of increasing permeability with porosity, but accounts for only 14% of the total variability. The low fit causes the slope to be markedly shallow, with an accentuation of the innate tendency to underpredict high permeabilities and overestimate low permeabilities. The logarithmic scaling of the permeabilities also results in a selective weighting in the low-permeability range and so influences the fitted line at the expense of permeabilities in the higher ranges. Wendt et al. (1986) have discussed these problems at length in their attempts to optimize permeability predictions from logs in the Sadlerochit sandstone at Prudhoe Bay. They concluded that the cumulative effect could be offset to some degree by weighting the high and low extremes. They also noted that the alternative regression using porosity as the dependent variable gave substantially similar results, because both methods caused a pivoting of the line about the mean value.

Whether or not these additional steps should be taken depends on the purpose of the permeability prediction. If the predicted permeabilities are to be averaged for reservoir description or mapping purposes, then the basic regression procedure is preferable because it is the best estimate on average, with a minimum squared error. If the predictions are to be used to locate high-permeability streaks, then a modified regression may be applied to extend the prediction range of permeability. However, this represents a less risk-averse approach, in which false indications of high permeabilities will be generated in addition to correct predictions. Thus the appropriate test criterion should be changed from a minimum squared error to a probability threshold of misclassification that matches the requirements of the reservoir-engineering project.

Multiple regression is an extension of the regression analysis, as described so far, that incorporates additional independent variables in the predictive equation. By this means, permeability predictions may be improved through the inclusion of log measurements which are indirectly related with pore geometry—principally the internal surface area. The form of the expanded regression model is:

$$\log K = A + B \cdot \Phi + C \cdot L_1 + D \cdot L_2 + \ldots$$

where L_1, L_2 etc. are additional log measurements. The choice of useful log variables is helped through the procedure of stepwise regression where different combinations of variables are used in an iterative process to determine the set that provides the best estimate and where the contribution of each variable is judged to be statistically significant.

Figure 20 shows the coefficients of determination for the alternative regressions of core permeability on all possible combinations of log measurements of porosity, uranium, and potassium for the Chase Group example. There is a systematic improvement in prediction power with the inclusion of additional variables. The regression equation that links permeability with porosity and uranium represents a plane of predictions mapped on to the two dimensions of the independent variables (Figure 21). When potassium is included as a third independent variable, the equation describes a hyperplane of predicted permeabilities in the three dimensions of porosity, uranium and potassium (Figure 22).

The regression coefficients associated with the independent variables show a consistent pattern of increasing permeability with increasing porosity, but decreasing permeability with greater concentrations of uranium and potassium. Both of these elements are statistically significant contributors to the regression model and so must be correlated with features of pore geometry. The potassium content appears to reflect small concentrations of illite which adversely affect permeability. The explanation for the role of uranium is more speculative, but may be linked with preferential leaching and improvement of transmissibility within the pore networks.

The porosity (Φ), uranium (Ur) and potassium (K)

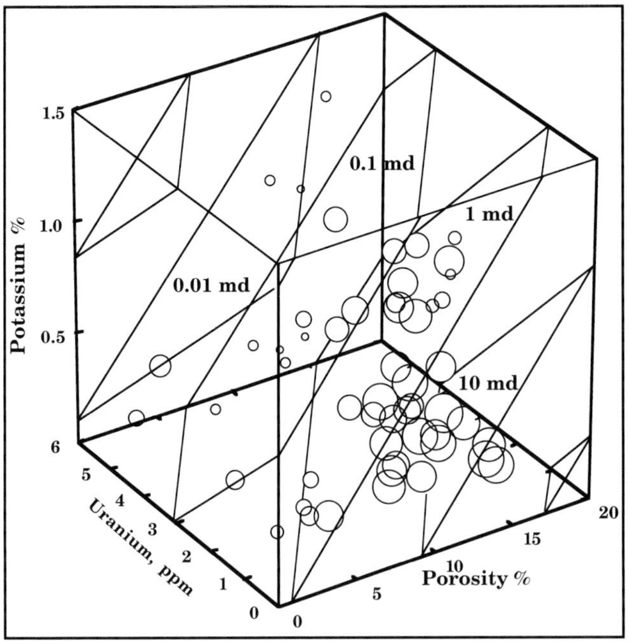

Figure 22. Contour planes of predicted permeability from multiple regression of logarithmic permeability on porosity, uranium, and potassium in the Chase Group example. The diameters of the bubbles are proportional to the measured permeabilities.

logs were transformed into a continuous profile of permeabilities through the application of the multiple regression equation:

$$\log K = -0.07 + 0.12\ \Phi - 0.30\ \mathrm{Ur} - 1.35\ \mathrm{K}$$

Permeability predictions outside the range of data used for the regression were discarded in order to screen out unwarranted extrapolations beyond reasonable prediction limits. The intervals eliminated by this procedure consisted of shales and shaly carbonate zones. The log is shown together with the core measurements of permeability in Figure 23. The match between them appears to be reasonable, although the regression accounts for only 55% of the total variability. The basic characteristic of the multiple regression as a method that tends to estimate the mean can be seen in the pattern of underestimates at higher values and overestimates of at lower values of permeability.

In this chapter we have focused on the role of simple statistical methods as important aids in reservoir property characterization using logs. The power of statistical inference is that it can be used for systematic prediction under conditions of uncertainty. Consequently, probability statements can be made about predicted values and associated error ranges can be calculated. These concepts have a direct bearing on standard log-analysis procedures and the economic consequences of their results. The increasing use of digital data has been a major factor in stimulating

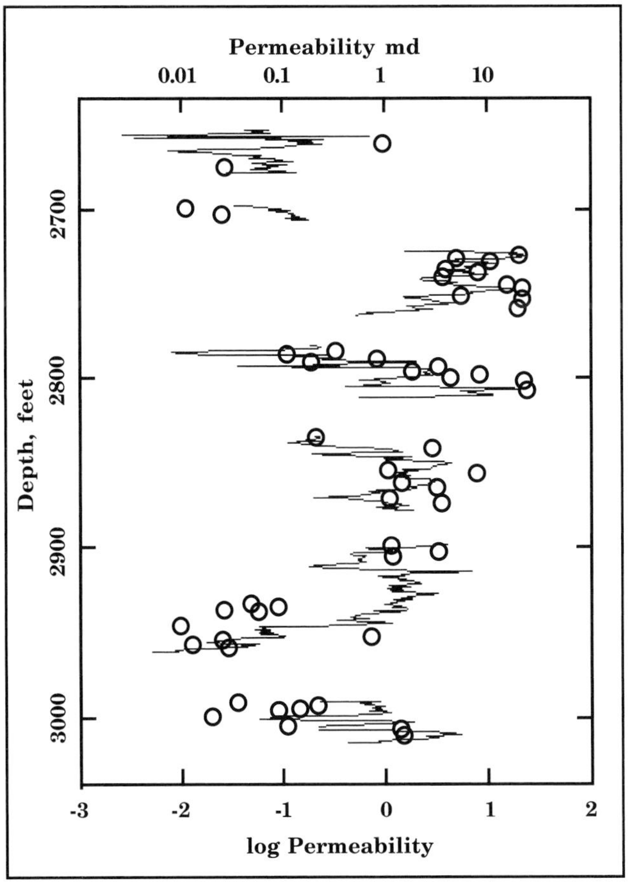

Figure 23. Predicted permeability log of the Chase Group section calculated from multiple regression on porosity, uranium, and potassium, and contrasted with core measurements of permeability (open circles).

interest in these methods among log analysts. However, considerably more remains to be done, particularly in the wider use of error analysis based on repeat sections and comparisons between wells.

In the narrative and examples of this chapter, statistical inference has been mostly restricted to procedures which attempt to predict reservoir properties on the basis of log measurements. This introductory treatment has not discussed the important ancillary techniques of statistical hypothesis tests which are used to judge the significance of conclusions. The interested reader is encouraged to refer to Davis (1986) as a text that expands on these themes, and is particularly useful because of its focus on geological data. In Chapter 3, we will return to statistical methods applied to multivariate data from logs and used for purposes of pattern recognition and discrimination.

REFERENCES CITED

Archie, G. E., 1942, The electrical resistivity log as an aid in determining some reservoir characteristics: Transactions of the AIME, v. 146, no. 1, p. 54-62.

Baker, P. E., 1957, Density logging with gamma rays: Petroleum Transactions of the AIME, v. 210, no. 3, p. 289-294.

Bebout, D. G., F. J. Lucia, C. R. Hocott, G. E. Fogg, and G. W. Vander Stoep, 1987, Characterization of the Grayburg Reservoir, University Lands Dune Field, Crane County, Texas: University of Texas Bureau of Economic Geology, Report of Investigations 168, 98 p.

Campbell, D. T., and J. C. Stanley, 1966, Experimental and quasi-experimental designs for research: Skokie, IL, Rand McNally, 78 p.

Carman, P.C., 1937, Fluid flow through a granular bed: Transactions of the Institute of Chemical Engineers, London, v. 15, no. 2, p. 150-156.

Chen, H. C., and J. H. Fang, 1986, Sensitivity analysis of the parameters in Archie's water saturation equation: The Log Analyst, v. 27, no. 5, p. 39-44.

Choquette, P. W., and J. D. Traut, 1963, Pennsylvanian carbonate reservoirs, Ismay Field, Utah and Colorado, in A symposium, shelf carbonates of the Paradox Basin: Four Corners Geological Society, 4th Field Conference, p. 157-184.

Coates, G. R., and J. L. Dumanoir, 1974, A new approach to improved log-derived permeability: The Log Analyst, v. 15, no. 1, p. 17-31.

Collins, H. N., 1984, Regression analysis—some loose ends: Canadian Well Logging Society Journal, v. 13, no. 1, p. 61-64.

Collins, H. N., and D. Pilles, 1979, Some uses of functional analysis in petrophysics: Canadian Well Logging Society 7th Annual Symposium, Paper E, 17 p.

Davis, J. C., 1986, Statistics and data analysis in geology: New York, John Wiley & Sons, 646 p.

Dent, B. M., 1935, On observations of points connected by a linear relation: Proceedings of the Physical Society of London, v. 47, pt. 1, p. 92-108.

Dorfman, M. H., 1984, Discussion of reservoir description using well logs: Journal of Petroleum Technology, v. 36, no. 13, p. 2195-2196.

Edmundson, H. N., 1988, Archie II: Electrical conductivity in hydrocarbon-bearing rock: The Technical Review, v. 36, no. 4, p. 12-21.

Etnyre, L. M., 1984a, Practical application of weighted least squares method to formation evaluation, Part I—The logarithmic transformation of non-linear data and selection of dependent variable: The Log Analyst, v. 25, no. 1, p. 11-21.

Etnyre, L. M., 1984b, Practical application of weighted least squares method to formation evaluation, Part II—Evaluating the uncertainty in least squares results: The Log Analyst, v. 25, no. 3, p. 11-20.

Farnan, R. A., and C. M. McHattie, 1984, Use of digital overlays and crossplots for log quality evaluation: The Log Analyst, v. 25, no. 1, p. 3-10.

Focke, J. W., and D. Munn, 1987, Cementation exponents in Middle Eastern carbonate reservoirs: SPE Formation Evaluation, v. 2, no. 2, p. 155-167.

Herron, M. M., 1987, Estimating the intrinsic permeability of clastic sediments from geochemical data: Transactions of the SPWLA 28th Annual Logging

Symposium, Paper HH, 22 p.
Heseldin, G. M., 1968, The use of error ratio in least square fitting of data: The Log Analyst, v. 9, no. 3, p. 22-25.
Keller, G. V., 1953, Effect of wetability on the electrical resistivity of sand: Oil and Gas Journal, v. 51, no. 1, p. 62-65.
Kendall, M. K., and A. Stuart, 1979, The advanced theory of statistics: Volume 2, Inference and relationship (4th Edition): London, Griffin and Co. Ltd., 748 p.
Krumbein, W. C., and G. D. Monk, 1942, Permeability as a function of the size parameters of sedimentary particles: AIME Technical Publication 1942, p. 153-162.
Lindley, D. V., 1947, Regression lines and the linear functional relationship: Journal of the Royal Statistical Society Supplement, v. 9, p. 218-244.
Looyestijn, W. J., 1982, Deconvolution of petrophysical logs: Applications and limitations: Transactions of the SPWLA 23rd Annual Logging Symposium, Paper W, 20 p.
Mark, D. M., and M. Church, 1977, On the misuse of regression in Earth science: Mathematical Geology, v. 9, no. 1, p. 63-75.
Ransom, R. C., 1974, Discussion of "A new approach to improved log-derived permeability": The Log Analyst, v. 15, no. 1, p. 30-31.
Raymer, L. L., E. R. Hunt, and J. S. Gardner, 1980, An improved sonic transit time-to-porosity transform: Transactions of the SPWLA 21st Annual Logging Symposium, Paper P, 12 p.
Rosepiler, M. J., 1981, Calculation and significance of water saturation in low porosity shaly gas sands: Oil and Gas Journal, v. 79, no. 28, p. 180-87.
Runge, R. J., and N. J. Powell, 1967, The effect of sampling on sonic log span adjustment: Transactions of the SPWLA 8th Annual Logging Symposium, Paper D, 14 p.
Schulze, R. P., G. L. Ives, E. A. Smalley, and W. E. Smith, 1985, Evaluation of low-resistivity Simpson Series of formations: SPE Paper 14282, 60th Annual Fall Meeting, 8 p.
Taylor, T. R., 1990, The influence of calcite dissolution on reservoir porosity in Miocene sandstones, Picaroon Field, Offshore Texas Gulf Coast: Journal of Sedimentary Petrology, v. 60, no. 3, p. 322-334.
Thomas, D. C., and V. J. Pugh, 1989, A statistical analysis of the accuracy and reproducibility of standard core analysis: The Log Analyst, v. 30, no. 2, p. 71-77.
Timur, A., 1968, An investigation of permeability, porosity, and residual water saturation relationships: Transactions of the SPWLA 9th Annual Logging Symposium, Paper J, 18 p.
Von Gonten, W. D., and J. S. Osoba, 1969, A method of predicting saturation exponents in logging: SPE Paper 2530, 44th Annual Fall Meeting, 9 p.
Walstrom, J. E., T. D. Mueller, and R. C. McFarlane, 1967, Evaluating uncertainty in engineering calculations: Journal of Petroleum Technology, v. 19, no. 12, p. 1595-1599.
Wendt, W. A., S. Sakurai, and P. H. Nelson, 1986, Permeability prediction from well logs using multiple regression, in L.W. Lake, and H. B. Carroll, Jr., eds., Reservoir characterization: San Diego, Academic Press, p. 181-221.
Whitman, W. W., 1986, Pickett plots with approximate geologic conditions: The Log Analyst, v. 27, no. 5, p. 11-28.
Worthington, P. F., J. E. Toussaint-Jackson, and N. Paillat, 1988, Effect of sample preparation upon saturation exponent in the Magnus Field, U. K. North Sea: The Log Analyst, v. 29, no. 1, p. 48-53.
Wyllie, M. R. J., A. R. Gregory, and L. W. Gardner, 1956, Elastic wave velocities in heterogeneous and porous media: Geophysics, v. 21, no. 1, p. 41-70.

Chapter 2

Graphical Techniques for the Analysis and Display of Logging Information

Even today, the petrophysicist who is equipped with a computer continues to make extensive use of crossplots in routine log analysis. The variability in reservoir rock properties often requires pattern-recognition skills to determine fundamental parameter values, distinguish natural groupings, and recognize potential anomalies. Once these are established, numerical processing comes into its own through the computation of properties of interest. As a general rule, computer methods cannot match the human ability to perceive trends and discriminate groups and anomalies within noisy data. However, this superior talent is only truly effective when the information is presented in graphical rather than numerical form. For most people, patterns that are obvious on a standard graph may be completely obscured when the same data are presented as a numerical table.

A fundamental limitation of both the paper medium and the computer screen is that they portray two-dimensional objects. Any graphical representation of logging information drawn in a Cartesian world is defined by two axes, but reviewed by petrophysicists whose everyday skills are designed for a three-dimensional world and whose logging problems often demand examination and interpretation in many dimensions! In spite of the limitation of graphical representation, a variety of techniques has been devised that condense multivariate logging problems into two-dimensional crossplots. Many of these methods were introduced before the widespread deployment of computers. Because log calculations and the production of crossplots are so labor-intensive, plotting by computer is ideal for these techniques. At the same time, advances in computer-graphics programs offer novel opportunities for visual presentations of logging data that would be unthinkable if they were to be done by hand. In this chapter, we will review the tried-and-true graphical logging techniques, as well as some of these newer methods and enhancements that can be added in a computer environment.

RESISTIVITY–POROSITY: THE HINGLE PLOT

The Hingle plot was introduced by Tom Hingle of Mobil in the late '50s, and was perhaps the earliest attempt to solve the Archie equation as a graphical procedure. At that time, logging calculations were made by nomograms and slide rules. Consequently, the pattern-recognition and computational properties of the plot made it a powerful and popular new method for log analysis. The crossplot was first described by Hingle (1959) and its principles are outlined in most standard log-analysis textbooks, as well as by Fertl (1979).

The plotting convention follows from the transformation of the Archie equation:

$$S_w^n = \frac{aR_w}{\Phi^m R_t}$$

to:

$$\Phi = \left(\frac{aR_w}{S_w^n}\right)^{\frac{1}{m}} \cdot \left(\frac{1}{R_t}\right)^{\frac{1}{m}}$$

which describes a family of lines on a plot of porosity versus the mth root of the bulk resistivity (Figure 1). Each line passes through an origin of zero porosity and zero conductivity (infinite resistivity) with a slope determined by the water saturation and the Archie equation constants. Two variants of the Hingle plot are commonly printed in logging service company chartbooks. One uses the Humble version of the Archie equation, in which a=0.62 and m=2.15, and is suitable for sandstones. The other is based on a=1 and m=2, and so is applicable to carbonates with intercrystalline and/or intergranular porosity. Using

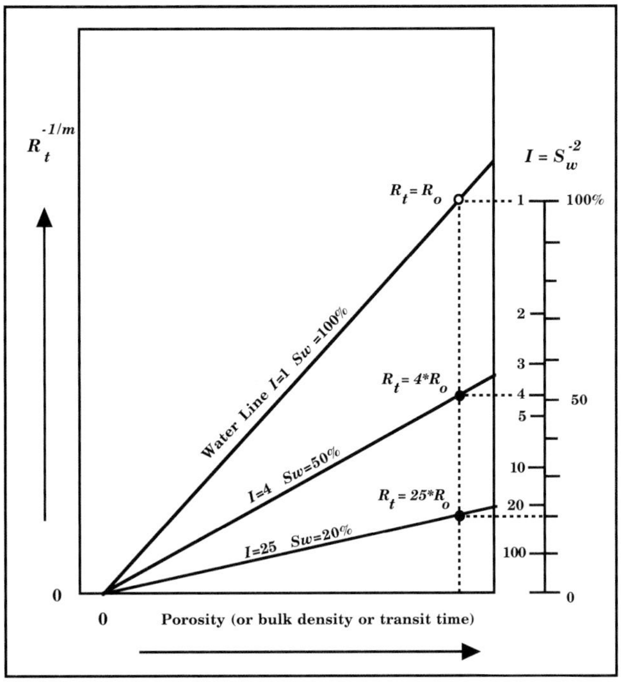

Figure 1. Principles and plotting convention of the Hingle plot.

the appropriate crossplot type, the graph can be used to solve implicitly the formation water resistivity and (where necessary) the calibration of the porosity log in porosity units. The broad methodology is shown by the following example.

Bulk densities and resistivities listed in Table 1 are for zones from a well in a limestone section in South Texas (Fertl, 1979). The paired values are crossplotted on a carbonate form of the Hingle plot as shown in Figure 2. Any zones that are completely water-saturated should lie to the northwest on a common line whose intercept is the matrix value that corresponds with zero porosity and a slope determined by water resistivity. The resistivity values on this water line are those of R_0 at the corresponding values of porosity. Once the water line is established, lines for any value of resistivity index, I, are readily drawn, because:

$$R_t = IR_o$$

for any given porosity. This procedure can be more clearly understood from a thoughtful inspection of Figure 1. The final step of estimating water-saturation values is made from a nomogram that converts resistivity-index values using a saturation exponent, n, of 2 (see Figure 2).

The Hingle plot is a powerful log-analysis tool considering its striking simplicity. Water zones do not have to be identified prior to the plot, but will become evident as points on a linear boundary. The intermediate steps of porosity conversion and water resistivity estimation can be ignored if water saturations are the goal, because the resistivity index lines can be established by simple drawing operations. In fact, the Hingle plot is an interesting example of a computer application in that it is itself a paper analog computer!

Table 1. Log Readings of Density, Resistivity, and Porosity (calculated from density) for Limestone Zones from a South Texas Well (data from Fertl, 1979).

Depth (ft)	Density (gm/cc)	Resistivity (ohm-m)	Porosity (%)
7160	2.57	28.0	8.2
7168	2.51	11.0	11.7
7174	2.62	16.0	5.3
7180	2.53	7.0	10.5
7185	2.58	4.5	7.6
7199	2.53	20.0	10.5
7212	2.60	19.0	6.4
7230	2.57	20.0	8.2
7241	2.53	15.0	10.5
7247	2.55	16.0	9.4
7264	2.52	6.3	11.1
7269	2.52	2.0	11.1
7275	2.62	9.0	5.3

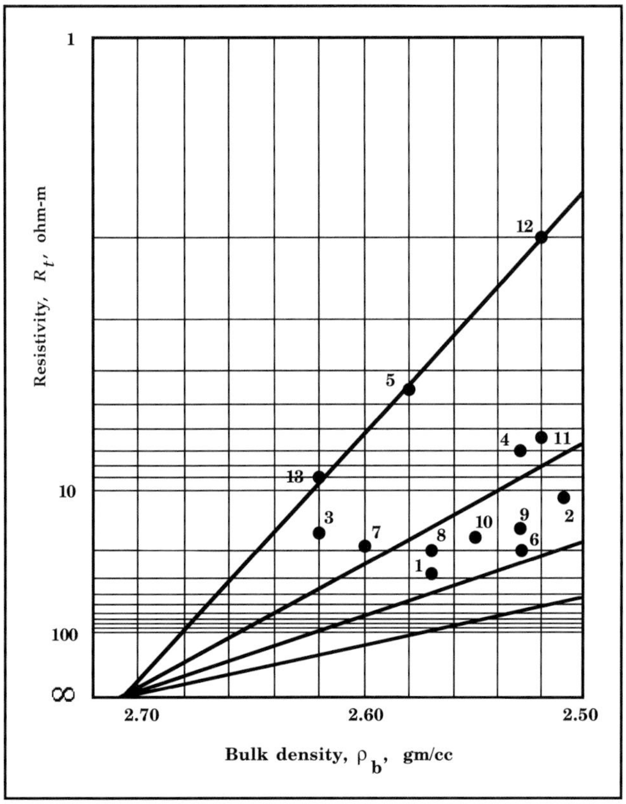

Figure 2. Hingle plot of limestone zone data from a well from South Texas (data from Fertl, 1979).

RESISTIVITY–POROSITY: THE PICKETT PLOT

The Hingle plot has several limitations that can hamper its usefulness. As most commonly used, a choice must be made between chart designs based on either the Humble or limestone forms of the Archie equation. In practice there will be instances when the cementation exponent is either unknown or is suspected to show some variability. This problem can be resolved by using another species of resistivity–porosity crossplot which is termed a "Pickett plot" and does not require prior knowledge of the constants in the Archie equation.

The basic method for the plot that bears his name was described by Pickett (1966). In a later paper, Pickett (1973) described in detail the pattern-recognition properties of the plot which make it a particularly powerful method for log interpretation. If anything, the mathematics of the Pickett plot are simpler than those of the Hingle plot; the plot is another transformation of the Archie equation,

$$S_w^n = \frac{aR_w}{\Phi^m R_t}$$

Rearranging the Archie equation and substituting the resistivity index, I, gives:

$$R_t = \frac{aR_w I}{\Phi^m}$$

Taking logarithms, the equation becomes:

$$\log R_t = \log\left(aR_w\right) + \log I - m \log \Phi$$

which describes a family of parallel lines for different resistivity-index values whose slope is the negative of the cementation exponent ($-m$). When the resistivity index, I, is unity, the line is the water line with an intercept equal to $A{\cdot}R_w$. Other resistivity lines are displaced to the northeast and are drawn easily as lines parallel to the water line and with resistivities which are the water-line resistivities multiplied by the index at common values of porosity. In common with most visual methods, these concepts are more obvious when sketched out graphically (Figure 3).

Pickett (1973) emphasized that this crossplot method was a powerful pattern-recognition technique. He demonstrated that the process involves the recognition and analysis of a complex variety of geometrical trends and patterns. These are the consequence of geological and reservoir engineering properties, as well as logging-tool characteristics. For a relatively homogeneous reservoir, water zones will be differentiated as a northwest–southeast limiting trend with hydrocarbon zones displaced to the northeast (see Figure 3). The water line can be established by eye or numerically by using the techniques of regression analysis that were described in the previous chapter, drawn from the work of Etnyre (1984) and Whitman (1986).

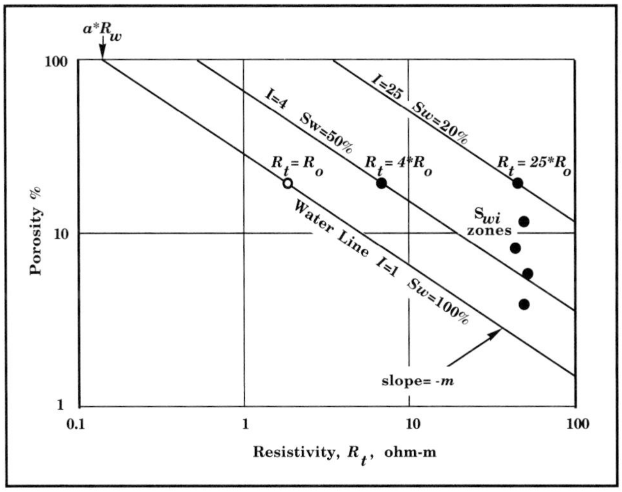

Figure 3. Principles and plotting convention of the Pickett plot. The S_{wi} points represent hypothetical reservoir zones with irreducible water saturation.

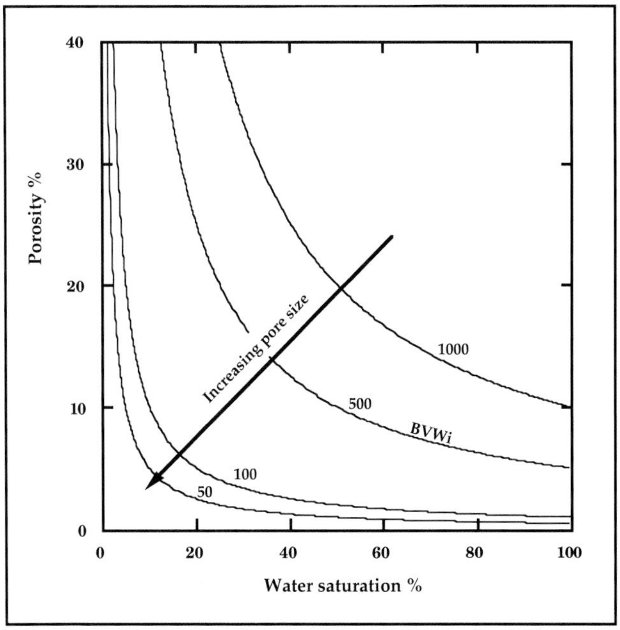

Figure 4. Trend of pore size and contours of irreducible bulk volume water (BVW_i) on graph of water saturation versus porosity.

Pickett (1973) also recognized that reservoir zones at irreducible water saturation should tend to lie on a steeper linear trend whose intercept with the water line reflected the grain or pore size. This is because of a general inverse relationship between irreducible water saturation and porosity that was noted by several early authors, including Archie (1952). They correctly attributed differences in the curvilinear trends to differences in reservoir pore sizes (Figure 4). This is because irreducible water saturation is controlled by surface tension at the internal surfaces and capillary pressure. While zones at irreducible water saturation in a moderately homogeneous reservoir should lie on a common curve, transition zones will be displaced to higher values of bulk water volume (ΦS_{wi}). The distinction is important because it determines which zones should produce water-free oil or gas and which should produce water or water-cut hydrocarbon. Computations of bulk volume water are therefore a critical additional step in log analysis for the assessment of producibility, as pointed out by Morris and Biggs (1967), Asquith (1985), and others.

Buckles (1965) made an extensive numerical analysis of reservoir measurements and concluded that the quadrilateral hyperbolic function:

$$\Phi S_{wi} = c$$

was a good first-order approximation to real field data. Low values of c reflect large average pore sizes; high values are linked with finer pores, as a direct consequence of control by internal surface area. Buckles (1965) reasoned that the form of the empirical relationship was to be expected, because the specific surface area was proportional to the reciprocal of the grain size, and that this was in turn, approximately linearly related to porosity.

The quantity c is simply the irreducible bulk volume water (BVW_i) which will be effectively a constant, provided that there is a limited range in pore size. Zones with comparable pore size that have higher values of bulk water volume should be water-cut or totally water-bearing. When computed for a field or reservoir, the characteristic value is often known as the "Buckles number" (especially if it is computed as the product of porosity and water saturation in percentages, rather than fractional amounts). Figure 5 shows ranges of field values plotted as cumulative frequency curves for sandstone (using data from Bond, 1978) and carbonate reservoirs with vugular and intercrystalline/ intergranular porosities (based on data from Chilingar et al., 1972). Carbonate reservoirs dominated by fracture porosities showed an erratic range of Buckles numbers. Morris and Biggs (1967) point out that is to be expected because of the very minor contribution of fracture porosity to bulk volume. Otherwise the data show systematic trends that reflect distinctions in internal surface area and provide values that are useful in the following analysis.

The hyperbolic relationship of:

$$\Phi S_{wi} = c$$

can be linearized to:

$$\log S_{wi} = \log c - \log \Phi$$

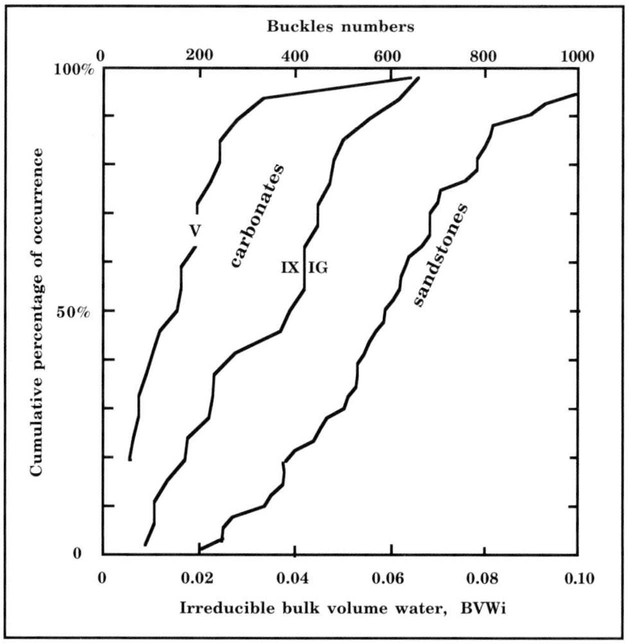

Figure 5. Cumulative frequency plots of irreducible bulk volume water (or Buckles number) for carbonate reservoirs with vugular porosity (V) and intercrystalline/intergranular porosity (IX–IG), as well as sandstone reservoirs. Carbonate data from Chilingar et al. (1972); sandstone data from Bond (1978).

Substituting the Archie equation solution for water saturation and rearranging, the relationship becomes:

$$\log R_t = \log(aR_w) - n \log c - (n - m)\log \Phi$$

which describes a line on the Pickett plot with a slope of (n–m) and an intersection with the water line at a porosity corresponding to the water line.

Greengold (1986) was the first to describe the systematic graphic properties of the irreducible bulk volume water on the Pickett plot. When the cementation and saturation exponents are equal, zones at irreducible water saturation should follow a line parallel to the porosity axis (Figure 6). Otherwise the line will be inclined according to whether the saturation exponent is greater or less than the cementation exponent. The parameters that determine the line give a powerful new means to extend the function of the Pickett plot beyond its traditional roles of cementation exponent and formation water resistivity. If the zones of irreducible saturation form a coherent trend, then the saturation exponent can be estimated directly from the plot for water-saturation calculations, while the producibility will be indicated for any zone. These ideas are explored in the following example.

The resistivity and porosity values from the South Texas limestone data set (Table 1) are shown located on a Pickett plot in Figure 7. The visual fit of a water line to the extreme points has a slope whose estimate

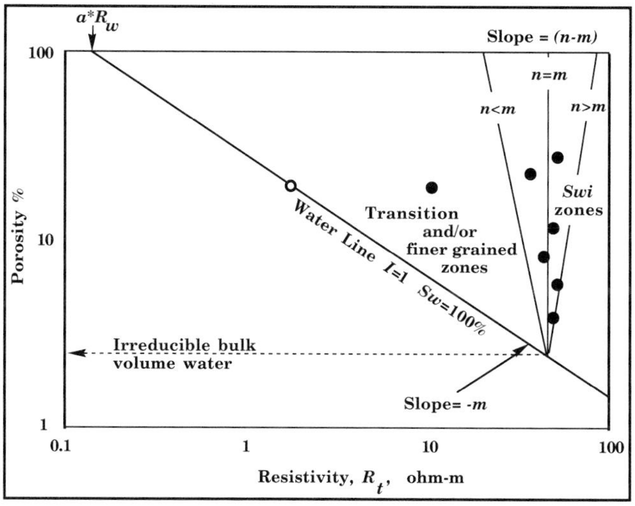

Figure 6. Location of irreducible bulk volume water trend on Pickett plot.

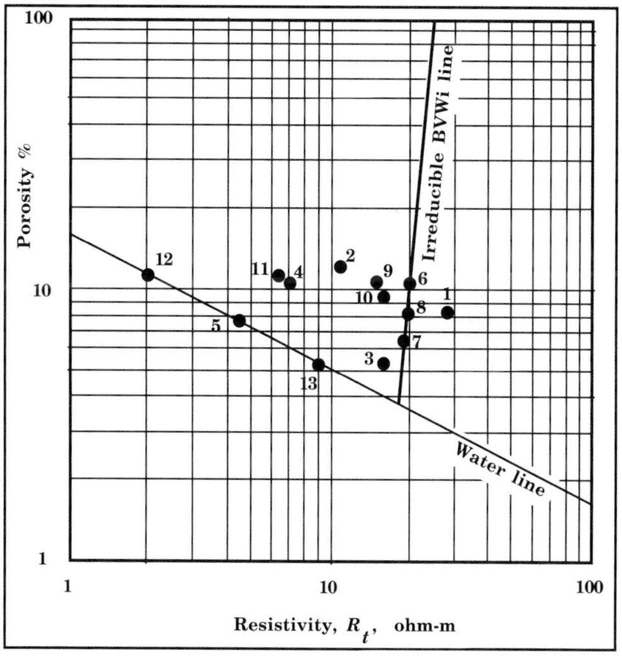

Figure 7. Pickett plot of limestone zone data from a well from South Texas indexed with water and irreducible bulk volume water lines (data from Fertl, 1979).

of the cementation exponent is close to a value of 2, and a water resistivity intercept value of 0.025 ohm-m (assuming the parameter "a" to be unity). If there is a single irreducible bulk water trend, it should be located as a high resistivity boundary of the potential hydrocarbon zones. The irreducible trend marked on Figure 7 gives a saturation exponent which is, again, close to 2, and a Buckles number of about 400 (or 0.04, in fractional terms). All of these parameter estimates are good functional numbers for a limestone reservoir with intergranular and/or intercrystalline porosities.

THE HOUGH TRANSFORM: A NEW RESISTIVITY–POROSITY CROSSPLOT METHOD

Although the Pickett plot has many useful properties for pattern recognition, there is still room for improvement. When fitting either the water line or a line of irreducible saturation, the resulting values of cementation and saturation exponents are not immediately obvious, but must be calculated from the slopes. The normal range of porosities also means that there is often a fair degree of uncertainty in the estimate of water resistivity when extrapolating to the intercept at 100% porosity. An alternative approach to this problem is through the use of the Hough transform, by plotting the data directly in parameter space.

The Hough transform was introduced by Hough (1962) as a means to detect patterns of points in binary image data. Illingworth and Kittler (1988) provided an excellent survey of the subsequent application of the Hough transform to a variety of scientific fields. Torres et al. (1990) reported their use of the Hough transform in petrophysics to extract sinusoidal features from borehole imaging logs, in terms of the dip and strike of the planar features that they represented. The basic concept is very simple, which is an immediate recommendation of its power. When searching for trends such as straight lines in a plot of X–Y coordinate data, the original points can be considered to be represented in "image space," although the purpose of the analysis is to extract the parameters of any systematic features. If parameters are the goal, then the problem can be simplified by replotting the data in "parameter space." The basic idea can be clarified by referring to Figure 8. Here the task is to locate any linear trend through the set of points plotted in X–Y Cartesian space. A line is specified by the equation:

$$Y = a + bX$$

where a is the intercept and b is the slope. The objective of the Hough transform then becomes to replot the data in a–b parameter space, rather than X–Y image space. Any single X–Y coordinate pair can be "back-projected" as a line on an a–b plot that gives the slopes and intercepts of all possible lines that could pass through this point. Alternatively, the slopes and intercepts of lines that link all pairs of points can be plotted as a cloud of parameter points, as shown in Figure 8. Using this method, a plot of n points in image space will be expanded to $(n^2–n)/2$ points in parameter space. Any clusterings of parameter points should reflect systematic linear trends within the original data.

Now that the basic concept of the Hough transform has been explained, we will follow its application to the limestone zone data from the South Texas well (Table 1). The parameters of the water line are

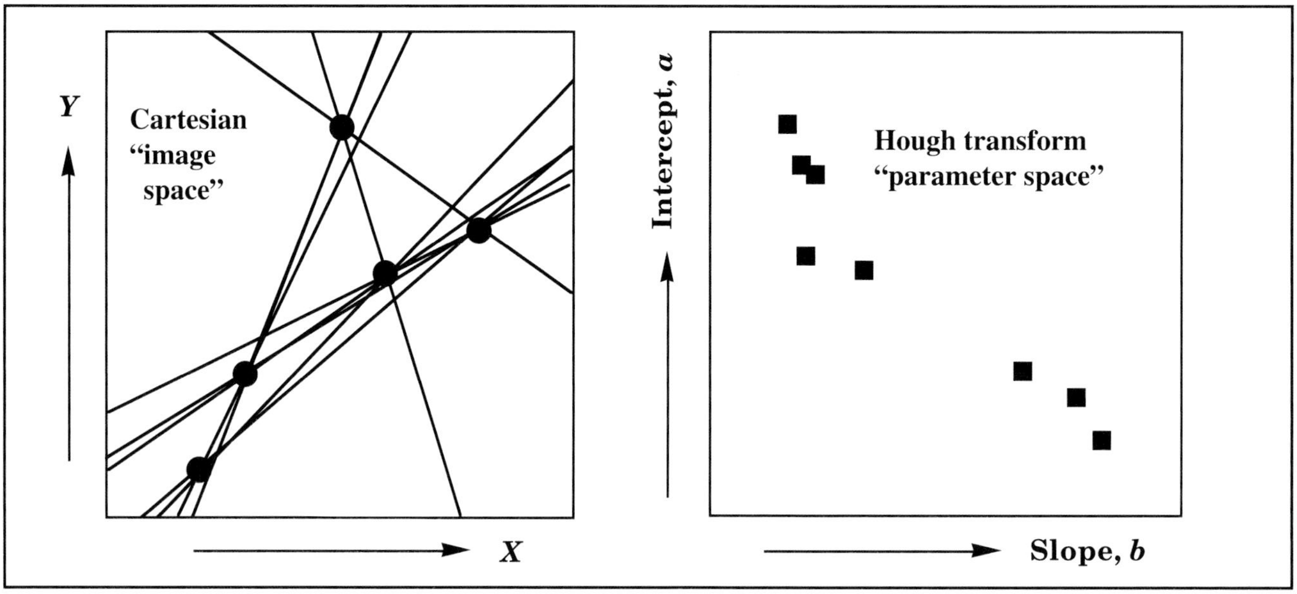

Figure 8. Simple example of Hough transform applied to the back-projection of lines linking pairs of points plotted in Cartesian coordinate space to a parameter space of slopes and intercepts.

the water resistivity intercept (assuming the constant "a" to be unity) and the cementation exponent slope. For any pair of zone readings of resistivity, R_t, and porosity, Φ, the estimate of cementation exponent is given by:

$$\hat{m} = \frac{\log R_{t1} - \log R_{t2}}{(\log \Phi_1 - \log \Phi_2)}$$

For pairs that are both water zones, the estimate is a crude guess of the cementation exponent; otherwise the estimate will generally be wildly off the mark. Similarly, the corresponding estimate of water resistivity is then calculated by:

$$\hat{R}_w = \exp\left(-\log R_{t1} - m \log \Phi_1\right)$$

The estimates of the parameters are then crossplotted as the water line Hough transform, as in Figure 9. In more typical field studies there would generally be a larger sample of water zones which would generate an elongate cluster of parameter choices, contrasted with an incoherent scatter generated by hydrocarbon and problem zones. The small data sample has only three water zones so that the number of parameter points caught in the range of the graph is fairly small. However, the principles of functional analysis outlined in the previous chapter are equally applicable here. The choice of water resistivity– cementation exponent coordinates should honor both the data and petrophysical expectations. Following this guideline, the best values are a cementation exponent of 2.03 and a formation water resistivity of 0.024 ohm-m. These values are a reasonable value for a limestone exponent, while Fertl (1979) reported a water resistivity of 0.025 ohm-m measured from a drill-stem test sample.

In a second application of the Hough transform we set the goal of locating a linear trend of irreducible water saturations, using the South Texas data once again. As derived previously, the equation of a theoretical irreducible bulk volume water line is given by:

$$\log R_t = \log\left(aR_w\right) - n \log c - (n - m) \log \Phi$$

For any pair of points, estimates of the parameters of the Buckles number and the saturation exponent can be calculated from this equation by rearranging the terms and inserting the global estimates of water resistivity and cementation exponent from the first Hough transform analysis:

$$\hat{n} = m + \frac{(\log R_{t1} - \log R_{t2})}{(\log \Phi_1 - \log \Phi_2)}$$

and:

$$\hat{c} = -\exp\left\{\frac{1}{n}\left[\log R_w - (n - m)\log \Phi - \log R_t\right]\right\}$$

The parameter estimates are plotted in Figure 10 as a Hough transform of Buckles number and saturation exponent. Wild estimates fall outside the range of the plot, whose axis bounds are restricted to reasonable ranges. Once again the notion of functional analysis is an important guide to an optimal selection. For this

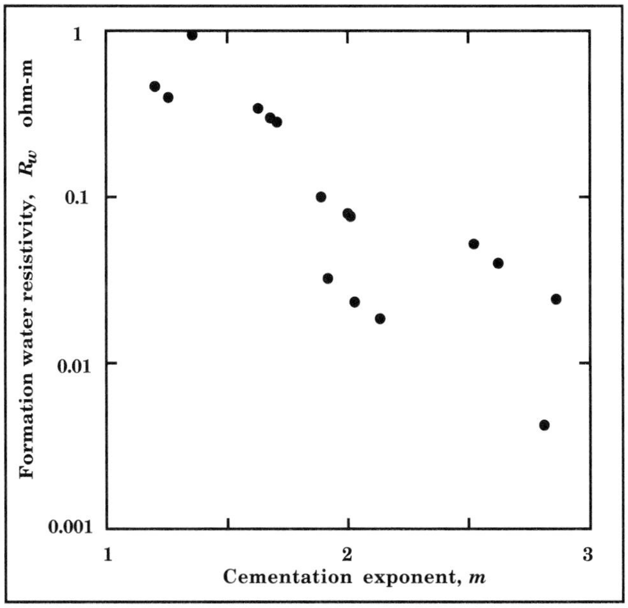

Figure 9. Water line Hough transform estimates of cementation exponent and formation water resistivity applied to limestone zones in a South Texas well. Logging data from Fertl (1979).

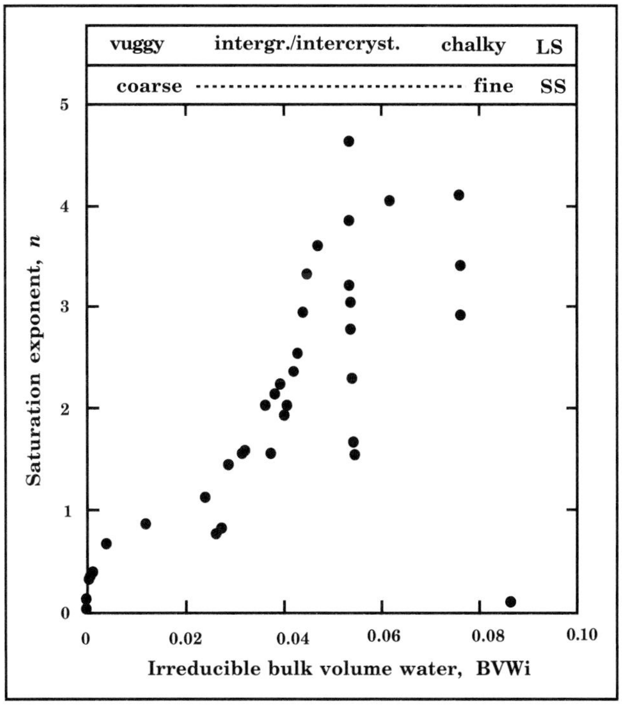

Figure 10. Hough transform of the estimates of saturation exponent and irreducible bulk volume water applied to limestone zones in a South Texas well. Logging data from Fertl (1979).

example, the choice of a saturation exponent of 2.02 and a Buckles number of 380 (or fractional 0.038) appears a good match with a limestone reservoir with intercrystalline porosity.

Clearly, because of the large number of calculations that would be needed in typical logging applications, the Hough transform would be unthinkable if it were not a computer application. However, the programming requirements are straightforward and the method plays off the ability of modern hardware to process massive amounts of data. At the same time, the methodology as outlined here capitalizes on the skills of the log analyst, both as a more adept pattern discriminator and as a judge of parameter estimates that make petrophysical sense. In future computer applications, the complementary abilities of log analyst and machine should be exploited in new ways that extend graphics beyond methods that were developed in the era of pencil and paper. Sometimes enhanced features can develop by serendipity. For example, the relative depth of zones is a valuable piece of information that is often lost on conventional Pickett plots. However, the depth ordering of zones is obvious in the generation of a computer display of the Pickett plot because it controls the order of their appearance on the screen. The analyst can then watch the data clouds and trends develop and follow the passage of zones through a reservoir at irreducible saturation, down though a transition zone, and along water-wet sequences.

LITHOLOGY DISTINCTIONS ON POROSITY LOG CROSSPLOTS

A knowledge of the porosity or, at least, an assumption of constant lithology is presumed in the application of the resistivity–porosity methods described up to this point. In reality, there is often a significant variability in mineral composition between reservoir zones. This variation impedes accurate porosity estimates from either density, sonic, or neutron logs, because the measurements are influenced by both fluid and mineral properties. The situation can be resolved to a large degree by crossplotting porosity logs. The most common choice for this purpose is a neutron–density plot, although neutron–sonic and sonic–density plots are often effective.

A neutron–density crossplot is shown in Figure 11 from a Middle Ordovician sequence of limestones, cherty dolomites, sandstones, and shales from northern Kansas. The trends and cloud patterns give good indications of all these lithologies; but, in fact, the plot can only show three compositional components, because it has only two dimensions. These components are two minerals plus porosity. The zones in the center of the plot are particularly ambiguous and could be either dolomitic limestones, cherty limestones or dolomites, or shales. However if porosity estimation is the major goal, then these multiple alternatives have only a minor bearing, because the lithology lines are orthogonal to the contours of true porosity.

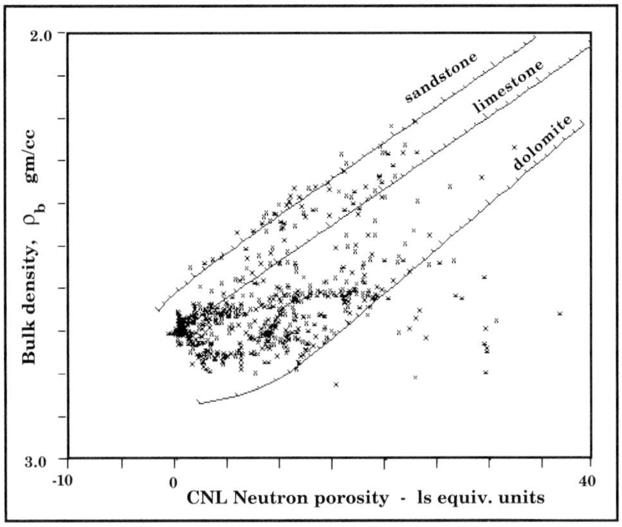

Figure 11. Neutron-density–porosity log crossplot of a Middle Ordovician carbonate–shale– sandstone sequence from northern Kansas.

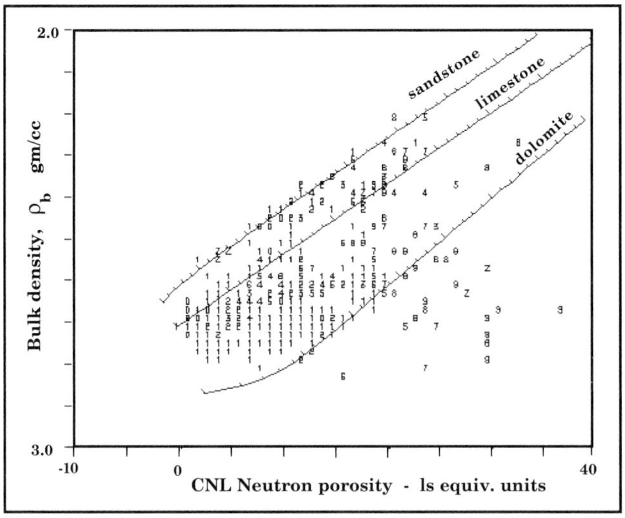

Figure 12. Neutron-density–Z crossplot of a Middle Ordovician carbonate–shale–sandstone sequence from northern Kansas. The integers on the plot represent gamma-ray log values as a third Z-axis normal to the plane of the crossplot.

The discrimination of shales is often made by a modification called the Z-plot (Figure 12). The Z-plot gives an illusion of three dimensions by substituting at each point an integer that represents its gamma-ray reading. The integer value can be considered to represent a height above the X–Y neutron–density plane, parallel to a gamma-ray axis, Z. A view of a three-dimensional representation of a Z-plot that differentiates a stream of shale points from the limestones, dolomites and sandstones in the basal plane is shown in Figure 13. Although interactive three-dimensional plotting could formerly be done only on large, dedicated computers, software to generate plots such as that shown in Figure 13 is now a standard option on many statistical packages for microcomputers. The techniques for display and analysis were originally developed at the Stanford Linear Accelerator Center (Fisherkeller et al., 1974). The Stanford PRIM-9 system was designed to search for graphic patterns of clusters, rods, and planes in the nine dimensions of measurements recorded from collisions between subatomic particles. This objective required the rotation of the hyperdimensional cloud in a search for views of the data in which interesting features appeared. Points in these clusters would then be isolated and identified to investigate relationships and meaning.

Huber (1987) has written a critical review of his experiences with three-dimensional scatterplots, in which he discussed both their strengths and limitations. He noted that observers typically and subconsciously refer data points to the axes of a two-dimensional plot. This is generally not the case for a three-dimensional plot, where axes are distracting and difficult to relate to the data. Instead, the viewer is much more sensitive to the internal structure of the cloud itself. Part of the problem is that the view is not truly three-dimensional, but is a two-dimensional plot for which the viewer provides the third dimension from visual cues. The depth cues are

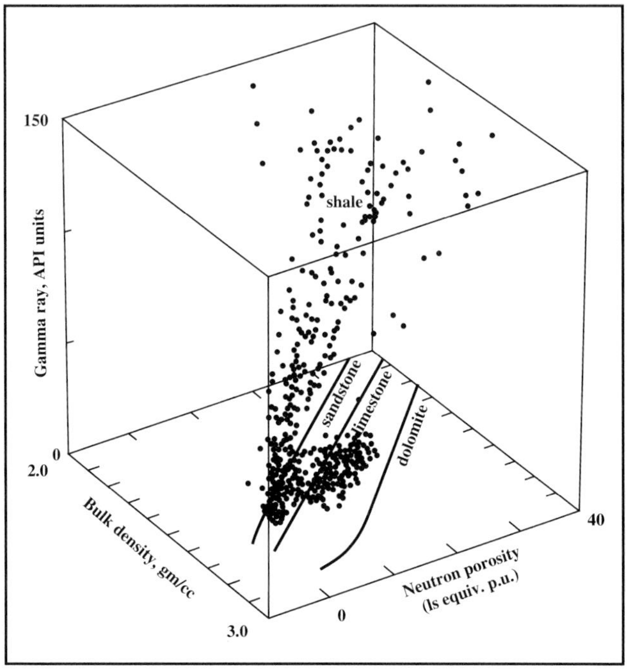

Figure 13. Three-dimensional neutron-density–gamma-ray log crossplot of a Middle Ordovician carbonate–shale–sandstone sequence from northern Kansas.

given either by real-time rotation or by the production of hard-copy stereo pairs. The best practical use for cloud rotation appears to be as a means to find the best plane for pattern discrimination.

These remarks suggest that three-dimensional scatterplots would be useful in the pattern recognition of electrofacies from multiple log data. This application is discussed in Chapter 4, together with other multivariate techniques. However, special-purpose two-dimensional projections are probably the better choice in cases where the aim is the explicit identification of distinctive minerals. When the logging properties of the minerals are known, their coordinates provide a reference framework to which zones of unknown composition can be related. Examples of the common projection plots used in log analysis are described in the following section.

MINERAL IDENTIFICATION ON PROJECTION PLOTS FROM POROSITY LOGS

When the density (ρ), neutron (Φ), and sonic (Δt) logs are matched with three orthogonal axes, log readings of zones are located in a three-dimensional space. Their composition can be expressed in terms of three minerals and a pore fluid volume. The problem of compressing three porosity logs into a plane can be solved by temporarily removing the variable of porosity from consideration. Algebraically, the operation is done by computing the following quantities:

$$M = \frac{(\Delta t_f - \Delta t)}{(\rho_b - \rho_f)}$$

and:

$$N = \left(\frac{\Phi_{nf} - \Phi_n}{\rho_b - \rho_f} \right)$$

which cause three logging measurements to be condensed into two variables.

The *M* and *N* values are the slopes of lines which connect the zone coordinates in three-dimensional log space with the hypothetical fluid point (see Figure 14). The data compression results from the fact that all zones with the same mineral composition will lie on a common lithology line, which has the constant slope (*M* and *N*), regardless of the porosity. Zone values of *M* and *N* can be plotted on an *M–N* plot and their composition deduced from their relation to mineral reference points (Figure 14). In addition to mineral identification, secondary porosity is often shown by anomalously high *M* values. This phenomenon is caused by the relative insensitivity of the sonic tool to vugs and fractures, as contrasted with the neutron and density tools which respond to all porosity types. Notice that the term "secondary porosity" is used here in the petrophysical sense of larger pore types. "Primary porosity" describes the finer-scaled intercrystalline and intergranular pore fraction. Shaly compositions cause zones to drift towards low values of *M*. The mathematics of the *M–N* plot produces a conical projection that would

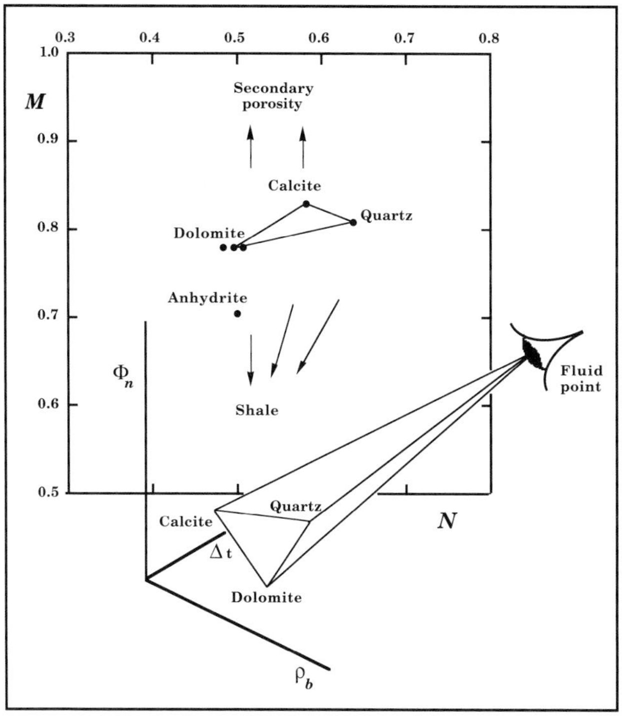

Figure 14. The *M-N* plot and its visualization as a conical projection of points in neutron-density–sonic log space, as would be seen by the eye located at the fluid point.

be seen if an observer's eye, placed at the fluid point, looked down through three-porosity log space at the points of zone log reading. The points inside a hypothetical calcite–dolomite–quartz–fluid tetrahedron would appear to be located on the mineral composition triangle at the base (see Figure 14).

The *M–N* plot was introduced by Burke et al. (1969) and has been widely used for mineral identification. However, although it is easy to generate by a simple computer program, it has a number of minor drawbacks. In reality, the isolithology lines of the projection are only approximate because both the neutron and sonic logs are not linear. Consequently, actual slope values will vary along their length as tangents to the log curves. The effect is most pronounced on the dolomite neutron response, and accounts for the use of three different values of *N* on the *M–N* plot, corresponding to different levels of porosity (Figure 14). In addition, the fluid point location will be variable to some degree, as regulated by the salinity of the fluid in the flushed zone, and will cause small changes in *M–N* values. Finally, the values of the *M* and *N* slopes have no intuitively obvious meaning to log analysts and tend to be seen as awkward, abstract numbers.

Clavier and Rust (1976) tackled all these problems in proposing a projection variation, which they called the MID plot. Rather than compute slopes of lines linking zone readings to the fluid point, they calculated the hypothetical intercepts of the density and sonic logs at an effective porosity of zero. These projected points are then estimates of the grain density and transit time of the rock framework mineral mixture. The nonlinearities of the tool responses were accommodated either through the use of specialized charts or by applying computer algorithms that were digital versions of the charts. The location of mineral points on the resulting MID plot has an overall similarity to the *M–N* plot. The MID plot method resolves many of the limitations of the *M–N* plot, but is more difficult to compute. Both plots are slightly flawed by the fact that the fair degree of correlation between acoustic velocity and density of minerals results in a tendency to colinearity of some mineral species. This factor can cause ambiguous or poorly resolved interpretations of composition. The situation was helped considerably through the introduction of the photoelectric absorption logging curve. This measurement can be used to substitute for the sonic log in an improved projection, which is a crossplot of *RHOmaa* (apparent matrix density) versus *Umaa* (apparent matrix volumetric photoelectric absorption).

The photoelectric logging curve records the absorption of low-energy gamma rays which is a direct function of aggregate atomic number of elements in the formation, and so is especially sensitive to mineralogy. The *RHOmaa–Umaa* crossplot was introduced to utilize measurements of the photoelectric index, neutron porosity, and bulk density for matrix mineral evaluation. The two dimensions of the crossplot require the three log variables to be condensed in some manner. This is done in an approach similar to the MID-plot methodology through the suppression of porosity as a variable and the calculation of the apparent properties of the rock matrix.

In a first step, the photoelectric index, *Pe*, must be converted to a volumetric measure, *U*. This is because *Pe* is measured in barns per electron, rather than barns per cc. The conversion is made by multiplying by the electron density, ρ_e:

$$U = Pe \cdot \rho_e = \frac{Pe\left(\rho_b + 0.1883\right)}{1.07}$$

or closely approximated by:

$$U = Pe \cdot \rho_b$$

The bulk density, ρ_b, and the volumetric photoelectric absorption, *U*, are the properties of the combined matrix and pore fluid. The elimination of the contribution of the pore fluid to these quantities will yield estimates of the apparent density (*RHOmaa*) and photoelectric absorption (*Umaa*) of the matrix. In order to do this, the true porosity, Φ_t, must first be estimated from a neutron–density crossplot (see following page) or calculated. The term "true porosity" refers to an estimate of the actual pore volume of a zone. It is differentiated from the apparent porosity units that are used to scale density, neutron, or sonic logs when referenced to an arbitrary mineral (usually calcite). *RHOmaa* and *Umaa* can be calculated as follows. Because:

$$\rho_b = \Phi_t\ \rho_f + \left(1 - \Phi_t\right) RHOmaa$$

then:

$$RHOmaa = \frac{\left(\rho_b - \Phi_t\ \rho_f\right)}{\left(1 - \Phi_t\right)}$$

and because:

$$U = \Phi_t\ U_f + \left(1 - \Phi_t\right) Umaa$$

then:

$$Umaa = \frac{\left(U - \Phi_t\ U_f\right)}{\left(1 - \Phi_t\right)}$$

The density of the fluid, ρ_f, can be taken to be that of the mud filtrate, which will be about 1 gm/cc, if fresh water. The fluid photoelectric absorption, U_f, will also reflect the fluid character of the flushed zone, which for mud filtrate is approximately 0.5 barns/cc.

Solutions for values of *RHOmaa* and *Umaa* for any zone are best calculated by a computer program, such as the one published by Elphick (1987). The zone values can then be located on a *RHOmaa–Umaa* crossplot (see Figure 15) and interpreted in relation to the standard reference points of common minerals. The calcite–quartz–dolomite triangle is the most common compositional template for application in most sedimentary successions, but gas effects and evaporitic minerals are easily seen, and useful distinctions can be made within shales. The *RHOmaa–Umaa* crossplot of Figure 15 shows Middle Ordovician limestone, dolomite, sandstone, and shale zones from a well in northern Kansas. The dominant pattern within the calcite–quartz–dolomite triangle contrasts slightly cherty limestones of the lower Viola Limestone with the more cherty dolomites of the upper Viola. Points clustered around the quartz vertex are zones within the St. Peter Sandstone.

The *RHOmaa–Umaa* plot is also useful for distinctions in clay mineral compositions within shales. However, any conclusions are interpretative regarding general trends and aspects because of the natural variability of clay minerals. The *RHOmaa–Umaa* crossplot of Figure 16 is drawn from logs from a Lower Cretaceous Dakota sandstone section in central Kansas. The unit consists of sandstones and shales deposited in a variety of deltaic and paralic environments. Shale compositions are often quite variable, with contents primarily of illite, smectite, and kaolin-

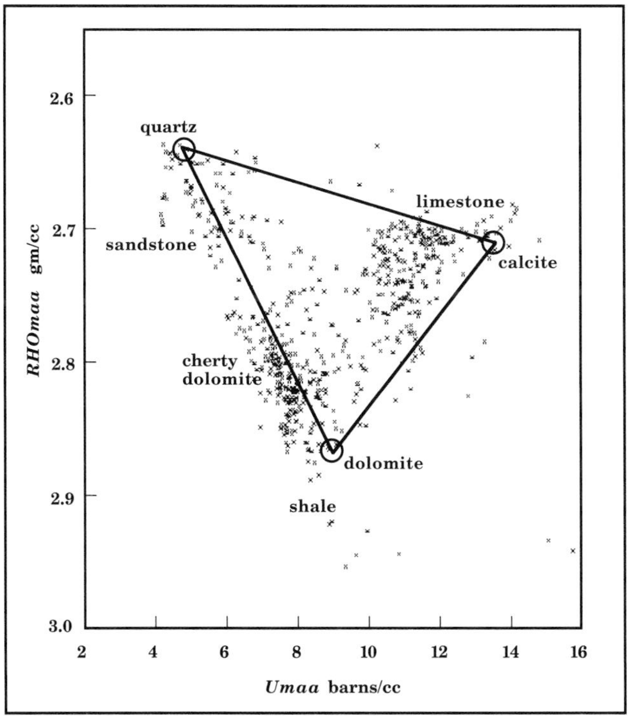

Figure 15. ***RHOmaa-Umaa*** **crossplot of a Middle Ordovician carbonate-shale-sandstone sequence from northern Kansas.**

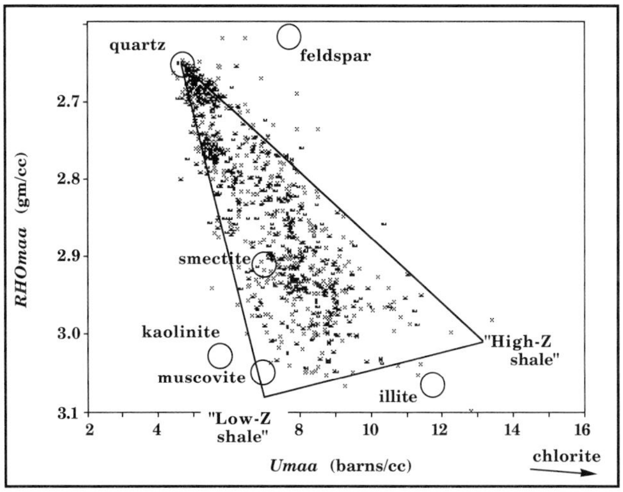

Figure 16. ***RHOmaa–Umaa*** **crossplot of a Lower Cretaceous Dakota sandstone sequence from central Kansas.**

ite. The crossplot shows a distinctive triangular wedge pattern whose upper vertex is highly focussed on the theoretical quartz point. The wedge diverges downwards to a shale facies represented by the base of the triangle. Although the shale has a restricted variation of apparent density, the wider range in *Umaa* value reflects compositional changes of clay minerals. The *Umaa* variable is a direct, but nonlinear function of the aggregate atomic number (Z) of the rock matrix. The wedge-shaped data cloud therefore represents a composition triangle whose vertex extremes are quartz, "low-Z shale," and "high-Z shale." The approximate locations of idealized clay minerals are marked on the figure, based on data from Schlumberger (1988). These serve as a general guide to the potential meaning of compositional trends, although the extremes of the clay mineral variation precludes more precise analysis.

According to Ellis (1987), changes in the photoelectric absorption of shales can be attributed almost entirely to iron content. This is because, in the absence of iron, the *Pe* of the aluminosilicate clays would be difficult to distinguish from quartz and feldspar. The shales with higher values of *Umaa* indicate the presence of significant amounts of iron-bearing clay minerals. These are most commonly illite and (especially) Fe-chlorite, in contrast with smectite and kaolinite, which contain little or no iron. The data of Figure 16 range between *Umaa* values for quartz and low-Z clays (kaolinite and smectite) and high-Z clays (illite and chlorite) and can be represented reasonably by a composition triangle.

GEOCHEMICAL INTERPRETATIONS OF SPECTRAL GAMMA-RAY PLOTS

The counts recorded on a standard gamma-ray log are the sum of all gamma rays, most of which are emitted by potassium-40 and isotopes in the uranium and thorium series. The energy levels of the gamma rays are characteristic of their source isotope and form distinctive spikes on a gamma-ray emission spectrum, although in practice these spikes are smeared by scattering events into a diffuse train of peaks. However, by subdividing the total energy range into restricted windows, estimates can be made of the separate contributions of potassium-40, uranium, and thorium using computer-processing methods. Spectral gamma-ray logs are commonly used for estimation of clay mineral volumes (and types) and the recognition of fractures in which uranium salts have been precipitated by ground-water systems.

Crossplots of potassium versus thorium have proved useful for recognition of clay minerals and distinction of micas and K-feldspars. The lines that radiate from the origin of the plot shown in Figure 17 have gradients matched with values of the Th/K ratio. The ratio is a relative measure of potassium richness as related to thorium. Illite has a higher potassium content than mixed-layer clays or smectite, while kaolinite has little or no potassium. The generalized fields of expectation of mineral responses on a thorium-potassium crossplot should only be used as a general guide to compositional aspects. Hurst (1990) cautioned against the "bland generalizations" that can be drawn from such crossplots and reviewed the values measured from laboratory samples. Clay minerals show wide variabilities in composition, and most shales contain mixtures of different clay mineral species.

The example of a thorium-potassium crossplot shown in Figure 17 is for a Permian–Cretaceous

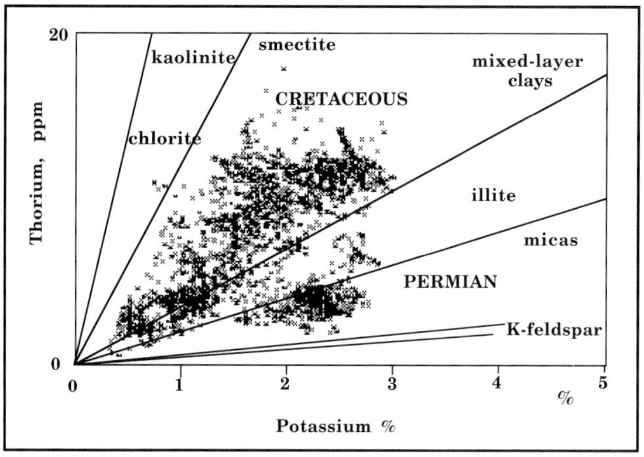

Figure 17. Thorium–potassium crossplot of a Permian to Cretaceous sandstone–shale sequence from central Kansas.

sequence from central Kansas. The Permian lithic subarkoses are easily distinguished as a separate cluster from the overlying Cretaceous sandstones and shales because of the higher levels of potassium linked with their K-feldspar fractions. The Cretaceous zones plot as a broader and more diffuse cloud; this characteristic reflects the contributions of illite, kaolinite, and smectite as volumetrically significant components.

The thorium-uranium ratio (Th/U) has also proved to be useful in the recognition of "geochemical facies." The Th/U ratio is an indicator of redox-potential. Uranium has an insoluble tetravalent state that is fixed under reducing conditions. However, under oxidizing conditions it is transformed to the soluble hexavalent state which may be mobilized into solution. In contrast, thorium has a single insoluble tetravalent state which is geochemically associated with uranium and is therefore a useful standard for comparison purposes. In a pioneer paper based on laboratory analysis of many samples of differing lithologies, Adams and Weaver (1958) concluded that the Th/U ratio was often strongly linked with depositional environment. They suggested that when the ratio was computed to be less than two (i.e., uranium-rich), the depositional environment had promoted uranium fixation, under probable reducing conditions, and was most commonly marine. At the other extreme, ratio values of greater than seven (uranium-poor), implied uranium mobilization through weathering and/or leaching,, thereby indicating an oxidizing, possibly terrestrial environment.

The Th/K and Th/U ratios from the Permian–Cretaceous sequence were plotted in the form of a crossplot (Figure 18), and as logs together with the gamma-ray trace and a graphic lithology log based on drill cuttings (Figure 19). The simultaneous consideration of these data throughout the sequence reveals striking and readily interpretable patterns, which were described by Macfarlane et al. (1989). An abrupt shift occurs at the Cretaceous–Permian contact and highlights clearly the major basal Cretaceous unconformity. The illite-feldspar signature of the Lower Permian Cedar Hills Sandstone changes to a Lower Cretaceous trace which oscillates between illitic and kaolinitic clay mineral facies, possibly linked with marine and deltaic fresh-water environments, respectively. The high amplitude variations in the Graneros Shale and Greenhorn Limestone may reflect the occurrence of bentonites (observed in the drill cuttings) interbedded with normal illitic marine shales. These bentonites represent altered ash deposits generated by explosive events from volcanoes in the Idaho–Montana and New Mexico–Arizona regions.

The Th/U ratio log was indexed with the diagnostic values of 2 and 7 suggested by Adams and Weaver (1958) to aid depositional-environment interpretation through its use as an oxidation-potential indicator. The ratio indicates an oxidizing environment for much of the Cedar Hills Sandstone, which would be expected from its postulated origin as eolian sands. Stacked repetitions of high and medium Th/U ratios characterize the Dakota Formation. These probably reflect high lateral variability in clastic facies and interplay between mostly brackish and fresh-water regimes of distributary channels, bays, and marginal marine deposits, which would be expected to typify a delta complex.

The relatively smooth, long-term cyclic pattern of the Th/U ratio in the marine sequence of the Upper Cretaceous is an excellent indicator of a broad transgression/regression couplet on an open-marine shelf. The broad sine-wave feature conforms precisely with the outcrop interpretation of the Greenhorn Cycle as a classic example of a symmetric, third-order tectono-eustatic cycle (Glenister and Kauffman, 1985). The transgressive phase of the cycle started in the uppermost part of the Dakota Formation, continued through the Graneros Shale, and reached maximum development in the Greenhorn Limestone. The regressive hemicyclothem was initiated at the top of the Greenhorn and continued through the Fairport Chalk and Blue Hill Shale to terminate in the Codell Sandstone. There is an abrupt break in the Th/U ratio log at the boundary between the Codell Sandstone and the overlying Fort Hays Limestone. This contact is thought to represent a long period of nondeposition followed by a major transgression (Hattin and Siemers, 1987).

CONTOUR MAPPING OF POINT DENSITIES ON CROSSPLOTS BY KERNEL ESTIMATION

Clusters of points can often be seen on log crossplots. They represent distinctive associations of petrophysical variables that have geological significance. Sometimes the clusters are tightly focussed and clearly separated from the remaining data cloud; at other times, the clusters are broad and diffuse, or have margins that blend with those of neighboring clusters. Methods that contour the areal density of data points across crossplot space are therefore useful both in the

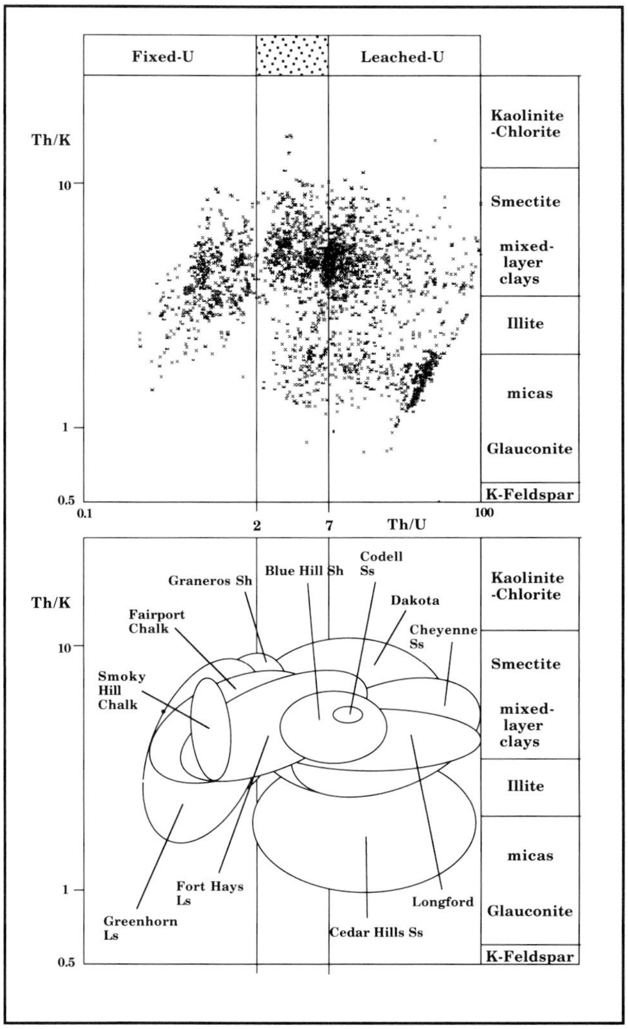

Figure 18. Th/K–Th/U ratio crossplot of a Permian–Cretaceous sequence from central Kansas.

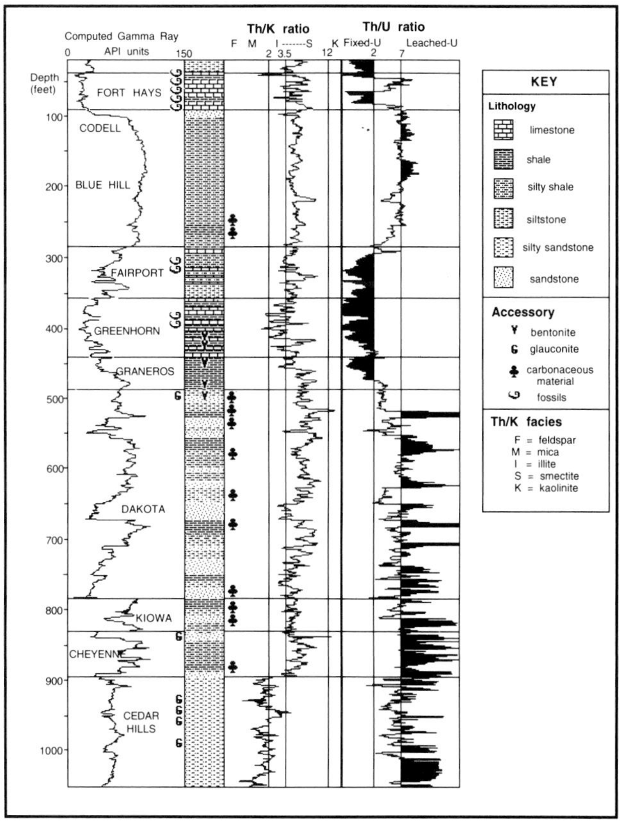

Figure 19. Gamma-ray, Th/K, Th/U ratio logs and cuttings lithology log of a Permian–Cretaceous sequence from central Kansas.

recognition of multiple modes within the total cloud and the definition of the locational coordinates of each cluster. However, any density-contouring method requires the stipulation of a counting-window size—and the form of the map will be controlled by this decision. Large windows tend to cause relatively smooth and featureless maps; windows that are too small result in maps with excessive detail and no useful local generalization.

The kernel-density method of contouring is widely used by data analysts who work with statistical graphics (Silverman, 1981) and has been applied successfully to logging data by Mwenifumbo (1993). The kernel function determines the shape of the "bump" that is centered on each data point and could take a triangular, rectangular, Gaussian, or other form. The Epanechnikov kernel is commonly used in graphical software packages and is a discontinuous parabola. The width of the kernel is set by the user; larger kernel widths smooth the data, narrower widths preserve finer detail.

Density estimation based on kernels can be applied to both univariate and bivariate data. Mwenifumbo (1993) presented a simple graphical example of the kernel method applied to a small data set (Figure 20), and also contrasted the results of alternative kernel widths on the distribution of values from a gamma-ray log (Figure 21). The kernel method is a powerful continuous alternative to the discontinuous step function of conventional histograms. Data-distribution representation by histograms can be adversely influenced by poor choices of both interval-boundary locations and interval widths. The continuity of the kernel summation process ensures that spurious "modes" will not be generated as artifacts of the interval-boundary locations.

The primary purpose of the density-estimation process is to locate systematic modes within any data-point distribution. When applied to logging data, the procedure has a two-fold purpose: the recognition of distinctive petrophysical associations and the estimation of the coordinates of their centroids. The choice of kernel width is important. Minor modes tend to be lost if the applied kernel is too broad; but, excessively narrow kernels will preserve the variability of the raw crossplot with little useful generalization. Mwenifumbo (1993) used an Epanechnikov kernel to contour data-point densities in a crossplot of gamma-ray and temperature gradient logs of a massive sulfide zone in New Brunswick (Figure 22). Although diffuse clusters can be seen on the raw crossplot, the

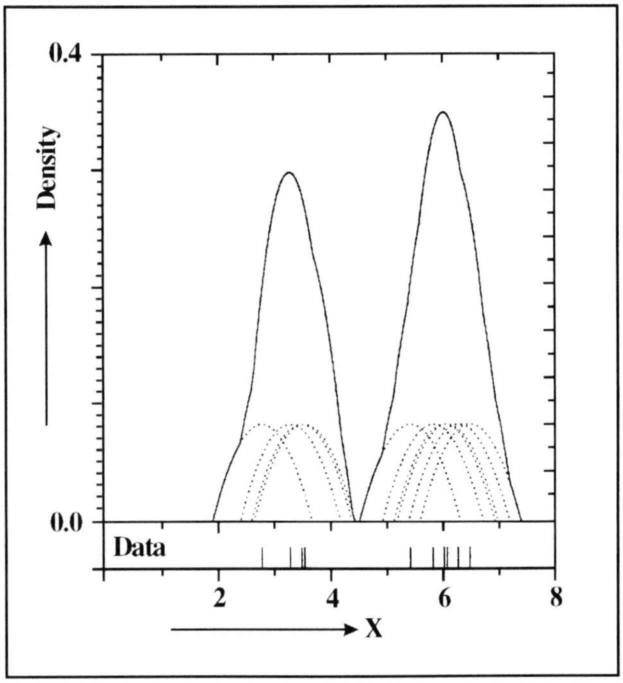

Figure 20. Epanechnikov kernel estimation of data-point density for a single variable, X. Data locations are shown by vertical bars; individual kernels correspond to the truncated parabolas. Adapted from Mwenifumbo (1993).

density estimation procedure locates the modes and clarifies the form of the overall cloud. The modes can be identified readily with the two types of volcanic tuffs, which are clearly distinguished from zones of gabbro and sulfides. Mafic dikes prove difficult to differentiate due to the wide spread in gamma-ray and temperature-gradient values within these units.

Kernels can be extended to the estimation of data-point densities in multivariate data sets. However, as previously discussed, a major problem must be addressed: How can such data be presented in a graphical form that can be comprehended easily? A three-dimensional example is provided by Mwenifumbo (1993), in which he contoured the first three principal components of resistivity from the New Brunswick sulfide zone: gamma-ray, density, and temperature-gradient logs (Figure 23). The application of principal component analysis reduced the dimensionality of the problem from four to the three major axes of the multidimensional cloud that absorb most of the data variability. The method of principal components is particularly powerful for the condensation of information in multivariate data; it is described at length in Chapter 4.

GRAPHICAL METHODS FOR ENCODING INFORMATION FROM MULTIPLE LOGS

One solution to the problem of representing multivariate logging data is to draw pictures that collectively summarize the variables. These pictures are known usually as "glyphs" or "icons." A good choice of glyph design can aid the recognition of complex distinctions or similarities between sets of observations that would be obscure if presented as tables of numbers. The glyph can also be designed to alert the viewer to zones with particularly favorable or unfavorable log characteristics.

The glyph that is probably most familiar to oil industry professionals is the "Stiff diagram," which graphs the compositions of natural brines. Stiff (1951) introduced the technique to show patterns of anion

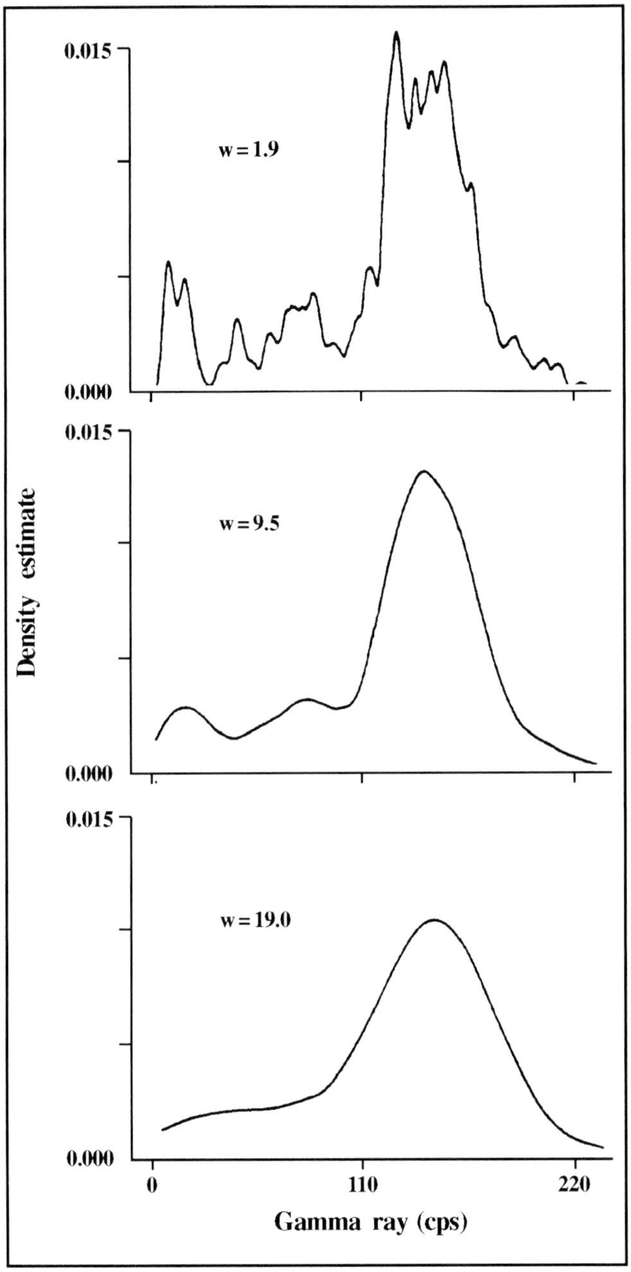

Figure 21. Alternative density estimates of 812 gamma-ray log values using Epanechnikov kernels with widths of 1.9, 9.5, and 19.0 cps. Adapted from Mwenifumbo (1993).

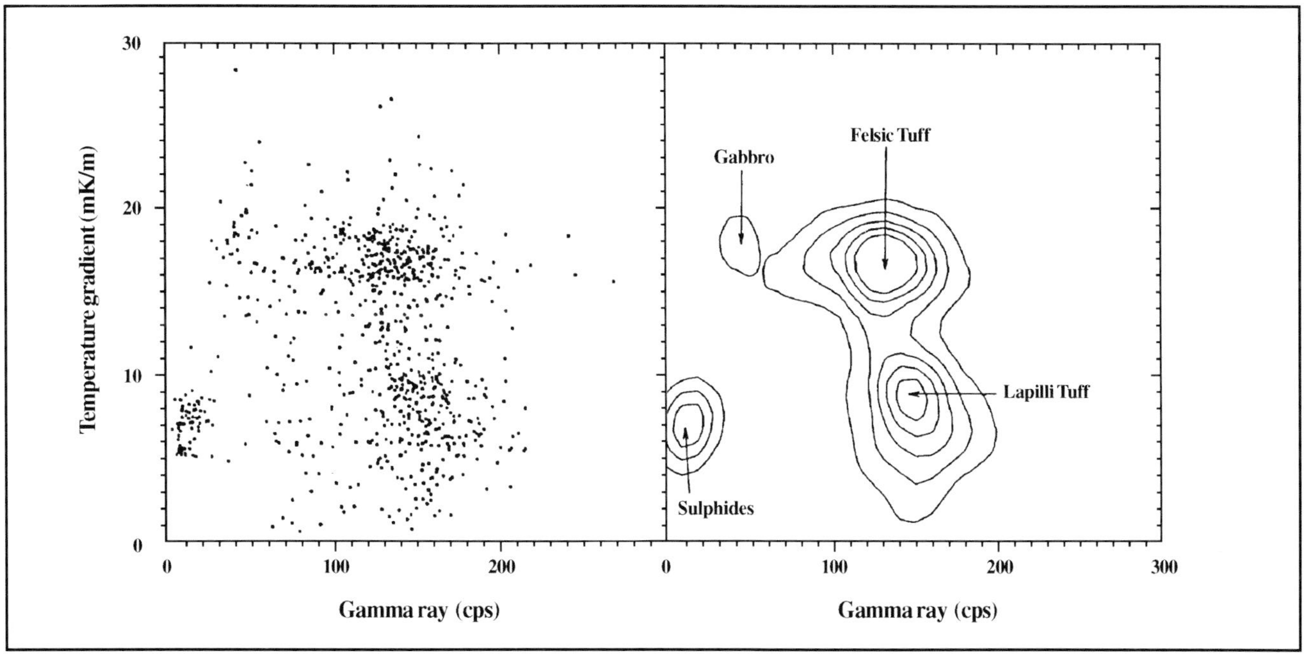

Figure 22. Scatter plot of gamma-ray and temperature gradient log readings (left) and the same data contour mapped by an Epanechnikov kernel-density estimate (right). Adapted from Mwenifumbo (1993).

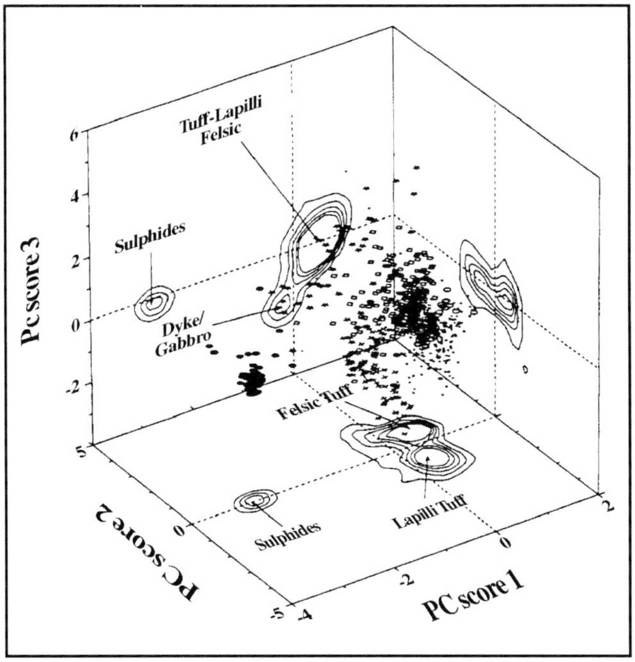

Figure 23. Three-dimensional scatter plot of three principal components of gamma-ray, temperature gradient, logarithmic resistivity, and density log readings contour mapped by an Epanechnikov kernel-density estimate. Adapted from Mwenifumbo (1993).

and cation associations that allow rapid comparisons between many analyses to be made easily. The shapes are particularly useful when plotted on a map to delineate aquifer zones with different water chemistries. As the first step in the construction of a Stiff diagram, the major anions (chloride, bicarbonate, and sulfate) and cations (sodium, calcium, and magnesium) are converted to equivalents per liter. The values are then plotted on horizontal axes and linked to form the boundary of an irregular hexagon. The basic idea is illustrated using water analyses tabulated by Lloyd (1986) from famous brewing areas in Europe (Figure 24). The data are drafted as Stiff diagrams and show a basic differentiation between waters best suited for ale (Burton, Birmingham, and Dortmund) and lager waters (Wrexham, Pilsen, and Munich). The major differences are the higher dissolved solids and elevated sulfate content in the waters used for ales. The ions important to the brewing process are calcium and bicarbonate. The major role of sulfate in ale waters is as a key taste factor, but it also functions as a good stabilizing agent that has allowed these ales to be exported around the world (Lloyd, 1986).

Serra and Abbott (1980) applied concepts similar to the Stiff diagram in their attempts to standardize simple graphical patterns to represent electrofacies. They defined an electrofacies as "the set of log responses that characterizes a sediment and permits the sediment to be distinguished from others." It is possible to describe any electrofacies in terms of a range of values on a set of different logs. The numerical form of this description has the potential to provide a more precise tool for classification of subsurface zones than the vaguer, verbal descriptions of lithofacies. However, most people find patterns within sets of numbers difficult to visualize; they are better equipped to discern similarities or differences when these same numbers are plotted as pictures. Serra and Abbott (1980) proposed two alternative glyphs, the spider web and

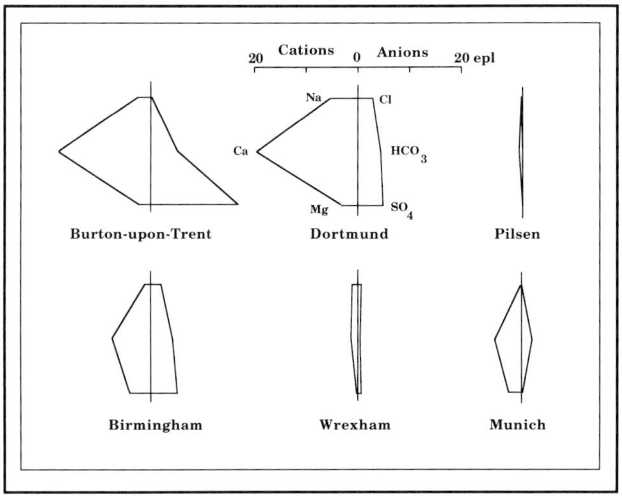

Figure 24. Stiff diagrams of typical waters from famous European brewing areas. Analytical data from Lloyd (1986).

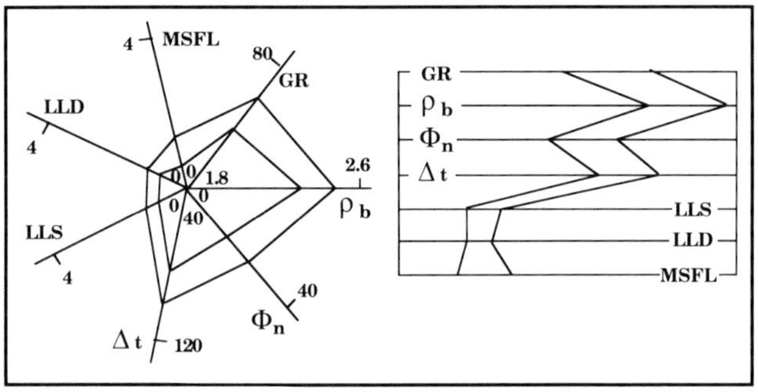

Figure 25. Spider web (left) and ladder (right) diagrams of an electrofacies based on ranges of gamma-ray, density, neutron porosity, acoustic transit time, and resistivity log readings of the electrofacies zones. Adapted from Serra and Abbott (1980). Copyright SPE.

the ladder (Figure 25), to express the ranges of variation of a single electrofacies, in terms of seven logs. Immediate comparisons could then be made between glyphs of multiple zones that would reflect both differences in mineralogy and textural changes linked with sedimentology (Figure 26).

A common problem with many designs of glyph is that while the encoding scheme used to absorb the multivariate information is obvious to the designer, for others it may be painfully difficult to learn. Impressions gained from glyphs should be immediate, intuitive, and should reflect their basic information content. Herman Chernoff, a statistics professor at Stanford University, proposed the use of cartoon faces to represent multivariate data (Chernoff, 1973). He argued that human beings are especially attuned to the subtle variations of facial expressions, since this is a lifetime study for all of us. Consequently, a large number of variables can be coded within a cartoon face, while trends and discriminations within sets of faces can be readily recognized. The method has been

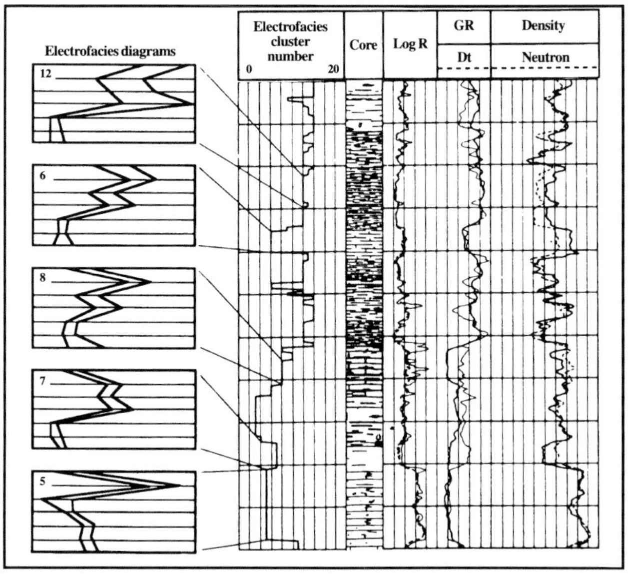

Figure 26. Example of different electrofacies referenced with lithofacies observed in core. On the electrofacies ladder diagrams, the logs and their scales are (from the top) gamma ray (0–100 API units), bulk density (1.7–2.7 gm/cc), neutron (0–50 porosity units), sonic (40–140 microseconds/foot), shallow laterolog, deep laterolog, and microspherically focused resistivity logs (0–5 ohm-m). These same scales are used on the logs at the right, with the exception of the resistivity logs that range between 0 and 2.5 ohm-m. From Serra and Abbott (1980). Copyright SPE.

applied to studies ranging from lunar craters to Soviet policy in Sub-Saharan Africa (Chernoff, 1978), and is commonly included in statistical computer packages that feature icon displays of multivariate data.

Doveton and Cable (1980) included a Chernoff face option within the reservoir analysis module of an interactive log-analysis computer system. The face was devised to incorporate the variables of porosity, water saturation, shaliness, and permeability. The following rules were used for face construction:

(i) The area of each eye is proportional to porosity. At a maximum theoretical porosity of 42% the eyes will touch.

(ii) The dark sector of each eye is proportional to hydrocarbon saturation; the blank sector is matched with water saturation.

(iii) The area of the nose is proportional to the shale content, using the same scaling convention as the eyes.

(iv) The curvature of the mouth is a function of the permeability. A neutral mouth set corresponds to a permeability of 5 md. Broader smiles represent progressively higher permeabilities; grimaces are generated by lower permeabilities.

An example of a computer display of Chernoff faces for zones from a Middle Pennsylvanian oil reservoir sandstone is shown in Figure 27. The faces were created from variables generated by log analysis

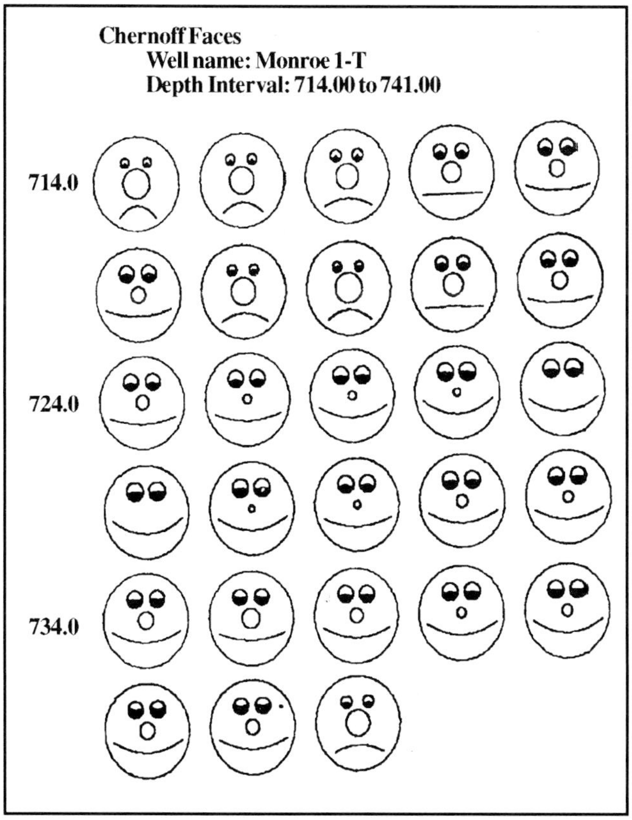

Figure 27. Chernoff faces of a zone succession in a Pennsylvanian oil-reservoir sandstone. Each zone represents 1 ft of section and the faces are ordered by depth, reading from left to right and top to bottom. Eye size reflects porosity; pupil size, oil saturation; nose size, shale proportion; mouth curvature, permeability. From Doveton and Cable (1980).

and interpretation using the rules described above. Each face represents a foot of section and the faces are arranged in order of increasing depth. Particularly favorable zones are matched by an aggregate of happy and harmonious facial features. Less favorable zones are relatively sad or glum. The example is very simple, in that it only encodes four petrophysical variables. There is plenty of room to incorporate additional variables in a more traditional and full-blown Chernoff face, which would add ears, eyebrows, and would allow for modifications in the shape of the head.

COLOR INFORMATION IMAGE TRANSFORMS OF WIRELINE LOGS

Glyphs and icons are comparatively exotic graphic methods for display of logging data which have had limited application to date. Almost all pattern recognition is still drawn either from multiple log traces or from crossplots of logging data. Log curves are graphed against depth, so that the stratigraphic framework and vertical changes in lithology and

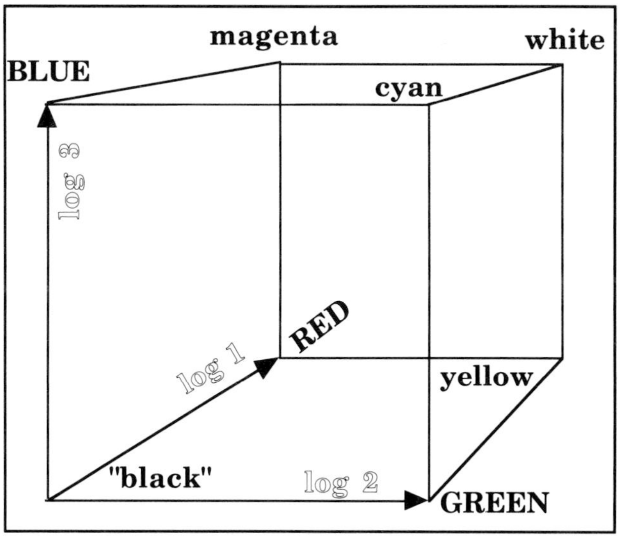

Figure 28. The Briggs color cube of log representation, in which zone readings can be matched with hues generated by the additive mixing of the primary colors of red, blue, and green.

reservoir quality are readily seen. However, in complex zones the increased number of logs can sometimes be difficult to interpret. By contrast, crossplotted data can be referenced to composition end members and trends and clusters are immediately apparent. However, the critical information on relative depth is lost in the simple crossplot. Ideally, the useful properties of the two conventions should be combined in a way that preserves immediacy, high information content, meaning, accessibility, and a rendition of depth. The use of color as a medium of information satisfies these criteria and is a tool that is ideally suited for a computer application.

Color monitors are increasingly common on all types of computers. However, color displays are often arbitrary and may even be distracting, "garish and content-empty colorjunk" (Tufte, 1990, p. 90). When used to graph information, all color systems have a structure of three dimensions. The Munsell classification expresses these in terms of hue, saturation, and value. Printing inks are controlled by a subtractive system with end members of cyan, magenta, and yellow. Finally, the operation of the color computer monitor is governed by an additive scheme that combines combinations of red, green, and blue.

Briggs (1985) introduced the concept of the color cube as a reference framework for the simultaneous display of several logs. When the three primary colors of red, green, and blue are made sources of light, then their mixing obeys additive rules. So, for example, red and green combined together generate yellow, as can be seen in stage lighting. This result contrasts with the brown made from the mixing of red and green paint pigments, which follows subtractive rules. By arranging red, green, and blue as three orthogonal axes, a spectrum of hues is generated in three-dimensional space (Figure 28). At the origin of the axes, the primary colors have no value and the result is darkness or "black." At the maximum value on all three axes the equal blending of the primaries causes white light. All other locations in the framework will be matched with distinctive hues which are unique and immediate visual information about their three coordinate values.

If the color axes are equated with logs, the color cube is the framework for a three-dimensional crossplot. As an illustrative example, Briggs (1985) chose sonic, gamma-ray, and resistivity logs as color-cube axes. When applied to a logged section, the crossplotted points formed clusters which were differentiated from one another by their change in hue. The hue assigned to each zone was then played back as a sequence ordered by depth to generate a color strip log. The result was a color image transformation, by which shale zones were differentiated and potential hydrocarbon zones highlighted in a striking manner.

Collins and Doveton (1986) named the color "Briggs cube" presentation of logs in honor of its originator and explored new ways of working with the cube that would incorporate standard log-analysis concepts. Rather than work with the Briggs cube as a continuum of color, they elected to subdivide it into a set of discrete cells. The planes between layers of cells were chosen to coincide with distinctive coordinate values that marked the boundaries between reference minerals, levels of porosity, or shale content. Used in this manner, the cube operates as an automatic classification device with explicit meanings assigned to cell hues. Also, the choice of axis designation is not restricted to raw logs, but can include variables that are themselves arithmetic combinations of several logs.

The use of blue to represent gamma-ray response, green for neutron porosity, and red for the difference between neutron and density porosities was applied as a color-cube transformation of logs from a variety of Kansas subsurface sequences. Although the design was geared to a basic discrimination of limestones, sandstones, dolomites, shales, and different levels of porosity, the cube was found to be a powerful medium of representation for a variety of lithologies. Distinctive and readily interpretable color-image logs were generated not only for carbonate, sandstone, and shale sequences, but also for evaporites, coal measures, and igneous and metamorphic rocks.

Collins and Doveton (1988) later extended the color-cube concept to new designs that condensed the information of the lithodensity–neutron–gamma-ray logs to colors equated with lithology and the spectral gamma-ray measurements in a color square to aid in clay mineral typing and interpretations of the geochemistry of deposition and diagenesis. The *RHOmaa–Umaa* crossplot was matched with red and green color axes (Figure 29). To aid interpretation, a square in this color coordinate plane was divided into nine color hue cells whose boundaries were chosen to subdivide the fundamental calcite–quartz–dolomite triangle into reasonably equant subareas. Sandstones are signified by red, limestones by chartreuse, and dolomites by dark green. Although the design is

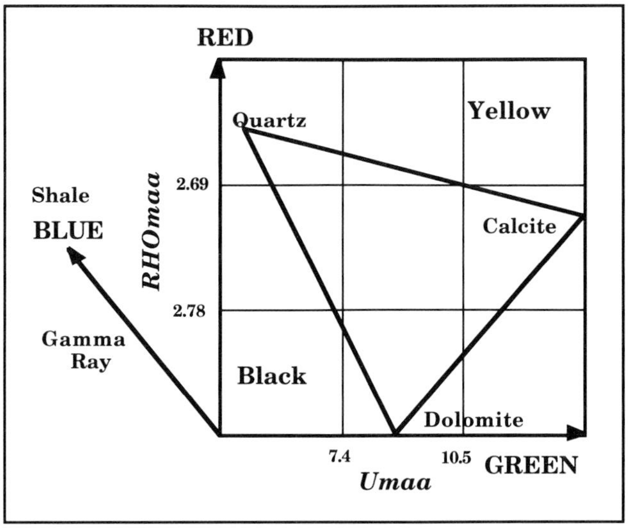

Figure 29. The Briggs color cube applied to apparent grain density (*RHOmaa*), apparent volumetric photoelectric absorption (*Umaa*), and gamma ray log values. From Collins and Doveton (1988).

geared to the most common non-shale lithologies, other minerals will have characteristic colors. So, for example, anhydrite beds will be bright green, while halite will be yellow. The introduction of the gamma-ray log as the blue axis completes the cube and equates the blue component with shale effects. The hues associated with shales are more variable and result from the interplay of several factors. In general, the degree of "blueness" reflects the proportion of clay minerals, allowing discrimination of shales, silty shales, and shaly sandstones. At the same time, the *Umaa* (green axis) variable is sensitive to variation in clay mineral types, ranging from kaolinite through smectite and mixed-layer clays to illite and chlorite. As a consequence, the hue variation within and between shales is often an immediate indication of both silt content and clay mineral facies.

Figure 18 depicted thorium/potassium and thorium/uranium ratios as the two axes of a crossplot—the former keyed to radioactive mineral composition and the latter as a geochemical facies indicator linked with reduction/oxidation potential. These crossplot axes were matched with the color primaries of green and red (Figure 30) and subdivided into nine cells to form a color square. The two cell boundaries on the Th/U ratio axis correspond to the critical values advocated by Adams and Weaver (1958). The column boundaries of the Th/K axis are more arbitrary, but were selected from gradients on the standard potassium–thorium crossplot to subdivide mineral associations into three approximately equant fields. The colors and resulting cell hues were carefully chosen to mimic "natural" colors in order to aid pattern recognition on color-transform logs. High values of Th/U (implied oxidation) correspond to the cell hues of yellow, orange and red. These are the colors of oxidized ferric iron which commonly colors rocks of this facies.

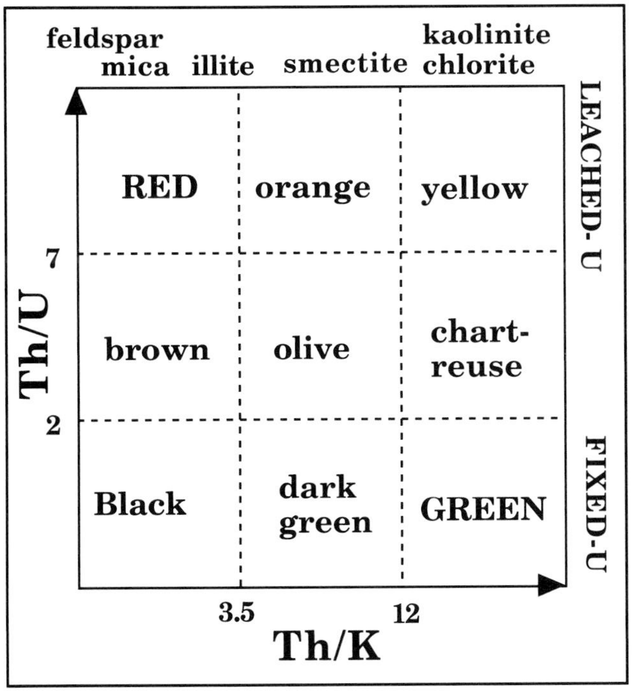

Figure 30. Green–red color square for hue generation matched with diagnostic values of Th/K and Th/U ratios from the spectral gamma-ray log. From Collins and Doveton (1988).

Low values of Th/U (implied reduction) generate cell hues ranging from black through dark green to bright green, which are the colors of reduced ferrous iron as seen in green and black shales. Color assignments with respect to the Th/K ratio are less clear-cut, but the convention used often matches rock color.

A procedure similar to that used on the lithologic color cube was applied to the generation of a color strip log from the spectral-ratio color square. Digitized values of the thorium, uranium, and potassium logs were condensed to Th/K and Th/U ratios at each depth increment. By reference to cell boundaries within the color square, a color hue was assigned to each increment and the cumulative results displayed as a color-image log. The style of the display is different than the lithologic color-image log so that it can accommodate the statistics of the spectral-log measurements. For zones with high counts for the individual radioactive sources, the ratios were not unreasonable estimates of their real values, especially when taking into account the coarse classification of the color square. However, at lower levels the ratios were poor estimates which fluctuated wildly as a consequence of the stochastic nature of the measurement and the poor statistics associated with the small count samples. Consequently, the color-image log was generated as a color fill within the trace of the gamma-ray log. By this means, the image was automatically and graphically scaled in strength by the total sample size traced out by the gamma-ray curve.

As an example, the color imaging method was used to transform lithodensity–neutron–spectral gamma-ray logs from the Permian–Cretaceous section used several times earlier in this chapter (see Figure 19). The result is shown in Figure 31, where the gamma-ray, lithodensity, and neutron logs are graphed in conventional format on the left and the color *RHOmaa–Umaa*-gamma-ray image is positioned as a strip log to the right. The Permian Cedar Hills Sandstone is magenta red and dark magenta with subsidiary brown and red bands and highlights sandstones which range from quartz arenites to subarkoses. At the basal Cretaceous unconformity there is an abrupt change to bandings of browns, dark blues, and dark cyan in the deltaic deposits of the Dakota Formation that reflect micaceous and shaly sandstones interbedded with illite-kaolinite shales. Above the Dakota Formation, the marine Greenhorn cyclothem is marked by chartreuse Greenhorn Limestone beds, green marls and dark cyan shales. At the top of the sequence, the Fort Hays Limestone of the Niobrara Formation is picked out by a thick chartreuse-colored feature.

The spectral gamma-ray log was transformed to the color-image log used as the infill to the gamma-ray curve to the left on Figure 31 by using the spectral ratio square. As before, the colors that are masked by this black and white rendition will be described and interpreted. The eolian feldspathic sandstones of the Permian Cedar Hills are dominantly red with subordinate brown bands. The sandstones and illite-kaolinitic shales of the Lower Cretaceous Dakota Formation deltaic deposits are colored orange and olive. The marine beds of Lower Cretaceous shales and limestones follow a color sequence which tracks with the transgressive-regressive history of the Greenhorn cyclothem. Initial transgressive shales are dark green, changing to black at maximum transgression. In the regressive phase, the color switches back to dark green, then olive, and finally orange at the peak of regression. The Niobrara, at the top of the succession, marks a new transgressive deepening cycle and is matched by dark green.

The image applications described are custom-designed variants of the color cube used for multiple log representation. The basic principle can be extended to other designs. These would be based on alternative logs and crossplot conventions which highlight properties such as hydrocarbon saturation. The design principle can be geared to either "supervised" (as in this paper) or "unsupervised" pattern recognition. Supervised pattern-recognition techniques are used when prior knowledge is available concerning the patterns to be classified. The locations of matrix minerals and approximate locations of clay minerals are known with respect to both *RHOmaa–Umaa* and spectral gamma-ray plots. Consequently, boundary planes within the color cube can be selected *a priori* to discriminate between common mineral associations. In an alternative approach, unsupervised statistical methods can be used in conjunction with the color cube to differentiate naturally occurring data clusters such as those generated by "electrofacies." In this case, the location of the color-hue planes would be

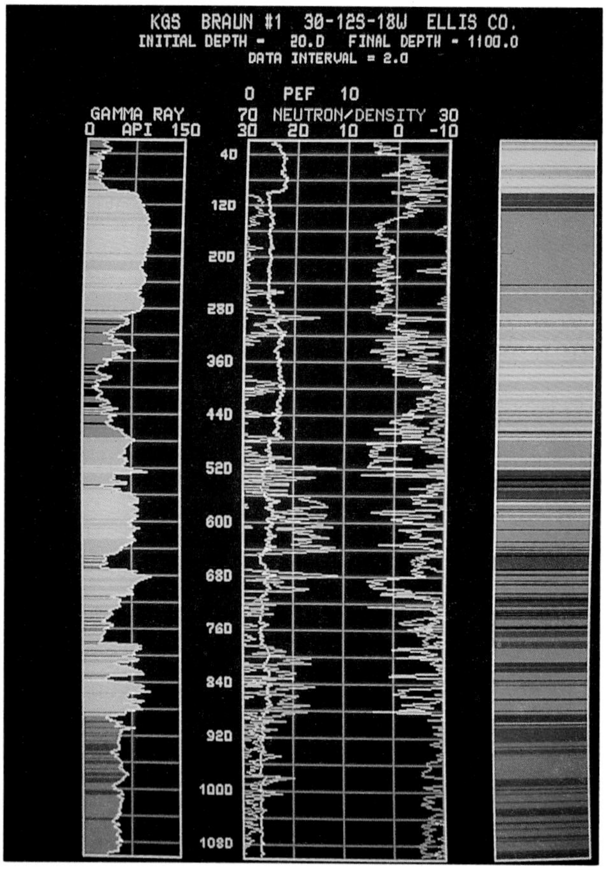

Figure 31. Color-cube transformation of gamma-ray–lithodensity–neutron logs as an image strip log (at right) and color-square rendition of spectral gamma-ray logs (as infill to gamma-ray log at left). The section is a Permian–Cretaceous sequence from central Kansas used earlier in this chapter (compare with Figure 19). From Collins and Doveton (1988).

dictated by the variability observed in the crossplotted logging data. Finally, there is no intrinsic reason why the color cube should be subdivided into discrete cells. The mixture of primary colors forms a continuum across triaxial color space. Logs are analog curves and their composite transformation to continuous color space would be a faithful color rendition, with no loss of information. Trends on logs would be preserved as gradations in hue, while sharp changes in value would appear as abrupt shifts of color.

GRAY-TONE AND COLOR-INTENSITY IMAGES OF REGIONAL CROSS SECTIONS FROM WIRELINE LOGS

The color-cube method is a particularly powerful application when lengthy log sequences are transformed to color images. At this scale it becomes increasingly difficult to disentangle the patterns of multiple curves, but a color image still has an immediate impact. A well-designed image transformation mimics subsurface geology, so that distinctively different lithologies have distinctive colors, while sharp contacts will show as breaks in color and are contrasted with transitional blends of gradational sections. The observation concerning long sequences is also true for both regional stratigraphic sections and profiles across large reservoirs based on logs. Here, structural variations must be traced visually between logs at the same time as the information on lateral changes in log responses are absorbed. Ideally, the result of a transect of color-image logs is a false-color image in which structure and stratigraphy are revealed automatically as the eye interpolates lateral changes between the separate images at their point locations.

The basic concept is explored in the following example from Collins et al. (1992) of a regional cross-section image based on a single type of log. In this case, a gray-level scale is adequate to capture the variation of the gamma-ray log. Because the gamma-ray log is primarily sensitive to volumetric content of shale, the obvious choice of gray-scale is to equate the lowest readings with white, the highest with black. Using this convention, the gray-level image from a gamma-ray log will often show a striking mimicry of interbedded shale sequences as would be seen in an outcrop. The approach is similar in principle to that used in the production of electrical borehole images, where multiple microresistivity traces are converted to gray-scale equivalents (Ekstrom et al., 1987).

Gamma-ray logs of closely spaced wells along a lengthy east–west traverse across central Kansas were digitized from just below the Stone Corral (Lower Permian) to the Fort Hays (Upper Cretaceous). In a conventional presentation of the regional geology based on gamma-ray logs, the individual traces would be hung in a single presentation, with correlation lines marked to aid visualization. Each digital gamma-ray log was transformed to a gray-level image strip, where the darkness intensity is a function of the natural gamma radioactivity of the logged formations. Under this system, sandstones and limestones appeared as white or pale gray, while shales registered as dark gray or black. As an enhancement of this method, the gray intensities were converted to an arbitrary color scale in order to accentuate differences between stratigraphic units. The methodology was then similar to that used on topographic maps, where the single variable of elevation is color-coded as a visual aid. When hung together on a common stratigraphic horizon and arranged in correct geographic order, a regional image of the subsurface geology of the Permian to Cretaceous sequence resulted. The major advantage of this approach is that the information is coded in a visual form that closely simulates how the geology would actually appear, making allowance for the vertical exaggeration.

The resulting regional image indexed with some major subsurface units is shown in Figure 32. The Stone Corral forms a distinct and continuous band at the base of the section. The Cedar Hills Sandstone (Lower Permian) is the most obvious feature in the overlying Permian, forming a thick wedge that sub-

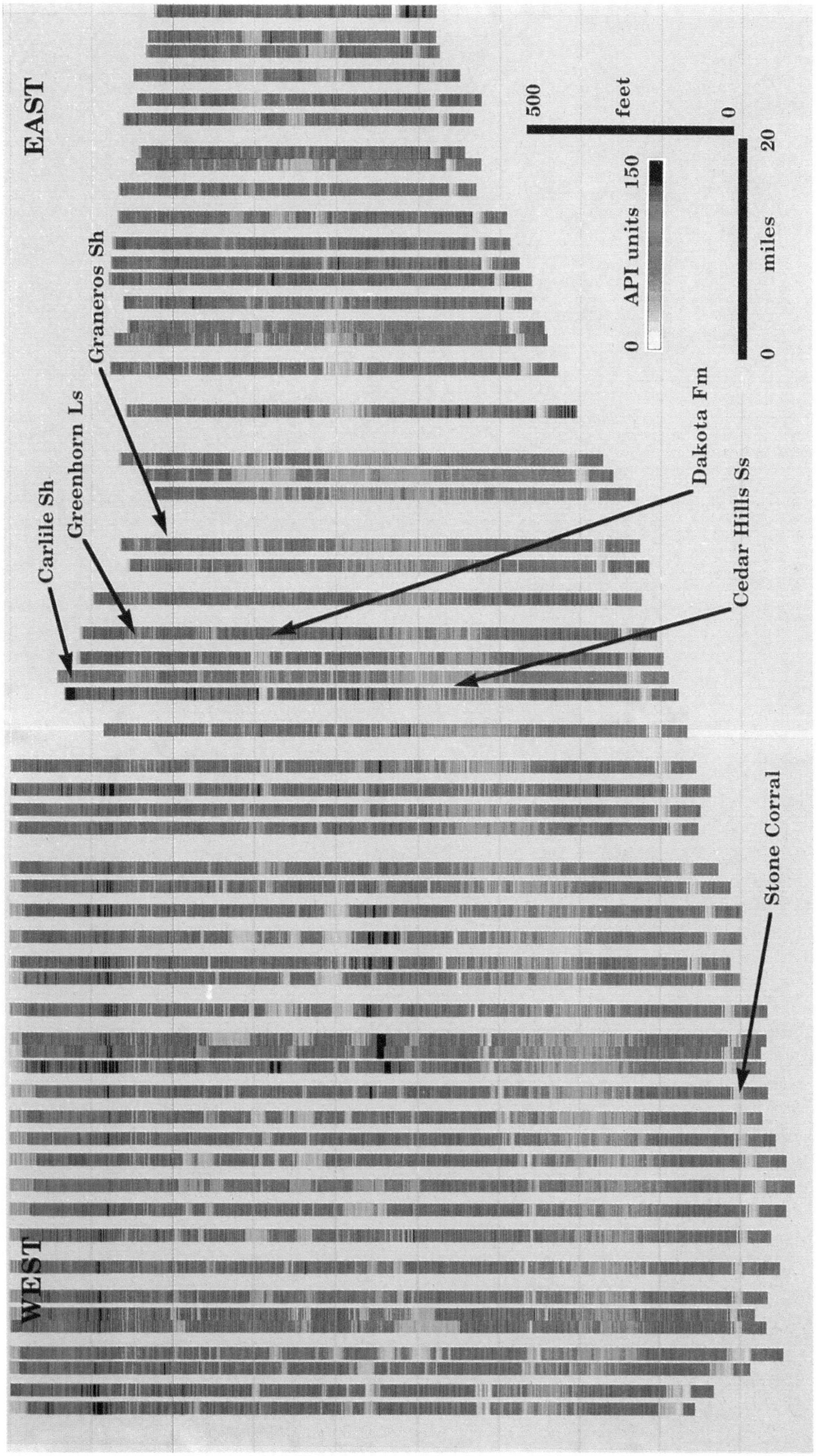

Figure 32. East–west regional gray-level image of the Permian–Cretaceous section of western Kansas based on gamma-ray logs. The gray levels have been converted to a false-color spectrum scaled to gamma-ray values (see key). From Collins et al. (1992).

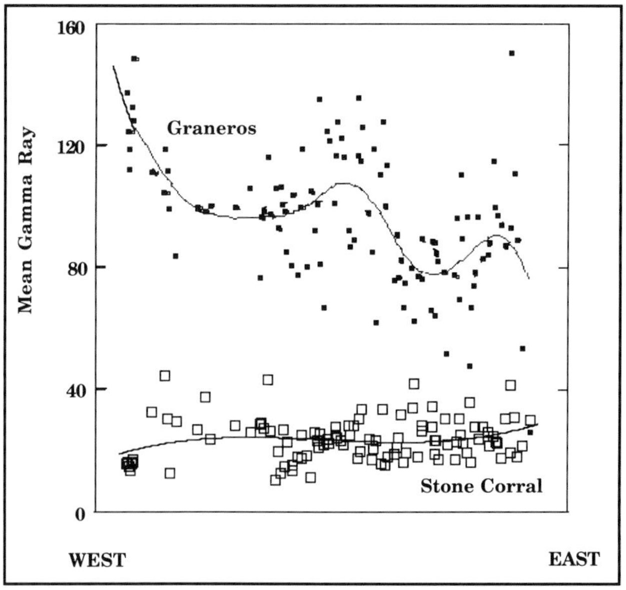

Figure 33. Plot of mean gamma-ray intensity of Graneros Shale and Stone Corral along an east–west regional transect of western Kansas. The data are fitted with a high-order spline (Graneros) and a quartic polynomial (Stone Corral).

crops at the base of the Cretaceous to the east. Sandstones at the base of the Jurassic Morrison Formation appear to be traceable for considerable distances, and contrast with the rather patchy sandstone developments within the Dakota Formation. These sporadic sandstones are the expected outcome of distributary channel networks and the complex facies mosaics of the delta and coastal plains which deposited the original sediments. The marine transgressive sequence of the Upper Cretaceous (Graneros Shale, Greenhorn Limestone, Carlile Shale) shows as a broad regional bands at the top of the section.

The original version of the regional image required remedial processing to remove errors associated with the recording of the gamma-ray logs on which the image is based. The use of raw measurements caused individual log strip images that appeared to be either distinctly darker or lighter than the strip images of adjacent wells. Most of these anomalies are small but are sufficiently distracting to hinder the ability of the observer's eye to trace lateral features. The cause of this phenomenon is directly attributable to random calibration errors linked with the original gamma-ray logs. This problem is common to all regional studies that involve log measurements, where systematic lateral changes must be distinguished from variations caused by tool measurement error. The remedial process is usually called normalization, for which simple, empirical methods have been described by Neinast and Knox (1973). Normalization is also an ideal subject for statistical processing because the goal is to differentiate systematic trends from random error. Doveton and Bornemann (1981) described the application of regression trend surface analysis to the estimation of tool measurement errors in a study based on areal variations in a limestone marker zone used as a calibration unit. A similar concept was used to normalize the gamma-ray logs on the regional cross-section.

Two stratigraphic units were selected to function as calibration standards: the Graneros Shale as the "hot" standard (high gamma-ray values); the Stone Corral as the "cold" standard. The use of two different standards allows correction of random errors that involve both shift and stretch of the log with respect to its true scale. The choice of each was based on a geological perception that major local changes (at a scale less than the distance between section wells) would be fairly minor. Instead, the gamma-ray response of these units would be dominated by a regional trend, but confounded by random measurement error. This working hypothesis leads to a numerical model in which curvilinear trends may be fitted to the raw gamma-ray values of the units to extract regional changes. The appropriate choice of trend can be guided by the statistics of fit and the criterion that the residuals should represent measurement error and thus show no autocorrelation.

The procedure was applied to the Stone Corral by fitting a succession of progressively more complex polynomial regression curves. There is little variation in the unit's gamma-ray values when traced across the state, so that the significant quartic polynomial trend shows only minor fluctuations from a constant (Figure 33). The residuals from this trend are essentially random, and so were equated with measurement error. By contrast, the Graneros Shale shows systematic, but complex regional changes that could not be absorbed satisfactorily by low-order polynomial trends. In an alternative strategy, the Graneros data were fitted with splines: short stretches of cubic polynomials with continuous derivatives at their links. The degree of complexity of the optimum spline fit was chosen as that which resulted in random residuals (Figure 33). More localized features on the curve can be attributed to the influence of Dakota deltaic paleogeography with respect to the line of the section. However, there is a systematic and generalized drift in gamma-ray values moving westwards. This aspect reflects the marine transgression of the Graneros from the west and the location of the Cretaceous Interior Seaway in that direction. Increasing gamma-ray values are commonly associated with deeper marine shales and are caused by higher concentrations of uranium fixed by organic matter under reducing conditions.

Measurement error estimates produced at each well have been used to correct digital gamma-ray records in the processing of revised versions of the image section. The final product provides a unique insight into the regional geometry of units within the Cretaceous and Permian of Kansas and is a closer representation of reality than standard, interpretative cross-sections. This prototype study was restricted to the production of a regional gray-level image based on a single log. However, there is no reason why a regional color image should not be produced as an extension of the color-cube processing described earlier. The most irksome aspect would be the increased load of normalizing several logs simultaneously, although the procedure could be automated satisfactorily. The most obvious candidates for the generation of these images are larger oil and gas fields that have been extensively logged with more recent tools. The field images would show not only the gross features of structure and trends in lithology, but also the finer scaled detail that is diagnostic of internal heterogeneity and flow units within the reservoir.

REFERENCES CITED

Adams, J. A. S., and C. E. Weaver, 1958, Thorium to uranium ratios as indications of sedimentary processes: Example of concept of geochemical facies: AAPG Bulletin, v. 42, no. 2, p. 387-430.

Archie, G. E., 1952, Classification of reservoir rocks and petrophysical considerations: AAPG Bulletin, v. 36, no. 2, p. 278-298.

Asquith, G. B., 1985, Handbook of log evaluation techniques for carbonate reservoirs: Tulsa, OK, AAPG, Methods in Exploration Series No. 5, 47 p.

Bond, D. C., 1978, Determination of residual oil saturation: Oklahoma City, Interstate Oil Compact Commission Report, 217 p.

Briggs, P. L., 1985, Color display of well logs [Abstract]: Mathematical Geology, v. 17, no. 4, p. 481.

Buckles, R. S., 1965, Correlating and averaging connate water saturation data: Journal of Canadian Petroleum Technology, v. 9, no. 1, p. 42-52.

Burke, J. A., R. L. Campbell, Jr., and A. W. Schmidt, 1969, The lithoporosity crossplot: Transactions of the SPWLA 10th Annual Logging Symposium, Paper Y, 29 p.

Chernoff, H., 1973, The use of faces to represent points in k-dimensional space graphically: Journal of the American Statistical Association, v. 68, no. 342, p. 361-368.

Chernoff, H., 1978, Graphical representations as a discipline, *in* P. C. C. Wang, ed., Graphical representation of multivariate data: New York, Academic Press, p. 1-11.

Chilingar, G. V., R. W. Mannon, and H. H. Rieke, III, eds., 1972, Oil and gas production from carbonate rocks: New York, Elsevier, 408 p.

Clavier, C., and D. H. Rust, 1976, MID plot: A new lithology technique: The Log Analyst, v. 17, no. 6, p. 16-24.

Collins, D. R., and J. H. Doveton, 1986, Color images of Kansas subsurface geology from well logs: Computers & Geosciences, v. 12, no. 4B, p. 519-516.

Collins, D. R., and J. H. Doveton, 1988, Application of color information theory to the display of lithologic images from wireline logs: Transactions of the SPWLA 29th Annual Logging Symposium, Paper II, 15 p. [reprinted in Geobyte, 1989, v. 4, no. 1, p. 16-24].

Collins, D. R., J. H. Doveton, and P. A. Macfarlane, 1992, Regional gamma-ray gray-tone intensity images of the Permian-Cretaceous sequence of western Kansas [Abstract]: SEPM 1992 Theme Meeting, Fort Collins, CO, p. 19.

Doveton, J. H., and E. Bornemann, 1981, Log normalization by trend surface analysis: The Log Analyst, v. 22, no. 4, p. 3-8.

Doveton, J. H., and H. W. Cable, 1980, KOALA: Kansas on-line automated log analysis system abecedarium and flight manual: Kansas Geological Survey, Petrophysics Series 2, 190 p.

Ekstrom, M. P., C. A. Dahan, M. Y. Chen, P. M. Lloyd, and D. J. Rossi, 1987, Formation imaging with microelectrical scanning arrays: The Log Analyst, v. 28, no. 3, p. 294-306.

Ellis, D. V., 1987, Well logging for Earth scientists: New York, Elsevier, 532 p.

Elphick, R. Y., 1987, Nuclear log interpretation in hard rock formations: Geobyte, v. 2, no. 3, p. 44-47.

Etnyre, L. M., 1984a, Practical application of weighted least squares method to formation evaluation, Part I—The logarithmic transformation of non-linear data and selection of dependent variable: The Log Analyst, v. 25, no. 1, p. 11-21.

Etnyre, L. M., 1984b, Practical application of weighted least squares method to formation evaluation, Part II—Evaluating the uncertainty in least squares results: The Log Analyst, v. 25, no. 3, p. 11-20.

Fertl, W. H., 1979, Hingle crossplot speeds long-interval evaluation: Oil and Gas Journal, v. 77, no. 3, p. 115-118.

Fisherkeller, M. A., J. H. Friedman, and J. W. Tukey, 1974, PRIM-9: An interactive multidimensional data display and analysis system: Stanford, CA, Stanford Linear Accelerator Center, SLAC Publication 1408, 78 p.

Glenister, L. M., and E. G. Kauffman, 1985, High-resolution stratigraphy and depositional history of the Greenhorn regressive hemicyclothem, Rock Canyon anticline, Pueblo, Colorado: Society of Economic Paleontologists and Mineralogists Field Trip Guidebook No. 4, p. 170-183.

Greengold, G. E., 1986, The graphical representation of bulk volume water on the Pickett crossplot: The Log Analyst, v. 27, no. 3, p. 21-25.

Hattin, D. E., and C. T. Siemers, 1987, Guidebook—Upper Cretaceous stratigraphy and depositional environments of western Kansas: Kansas Geological Survey, Guidebook Series 3, 55 p.

Hingle, A. T., 1959, The use of logs in exploration problems: Society of Exploration Geophysicists 29th Annual Meeting, Abstracts with Program, p. 7.

Hough, P. V. C., 1962, A method and means for recognizing complex patterns: U. S. Patent 3, 069, 654.

Huber, P. J., 1987, Experiences with three-dimensional scatterplots: Journal of the American Statistical Association, v. 82, no. 398, p. 448-453.

Hurst, A., 1990, Natural gamma-ray spectrometry in hydrocarbon-bearing sandstones from the Norwegian continental shelf, *in* A. Hurst, M. A. Lovell, and A. C. Morton, eds., Geological applications of wireline logs: Geological Society of London, Special Publication 48, p. 211-222.

Illingworth, J., and J. Kittler, 1988, A survey of the Hough transform: Computer Vision, Graphics, and Image Processing, v. 44, p. 87-116.

Lloyd, J. W., 1986, Hydrogeology and beer: Proceedings of the Geological Association, London, v. 97, no. 3, p. 213-219.

Macfarlane, P. A., J. H. Doveton, and G. Coble, 1989, Interpretation of lithologies and depositional environments of Cretaceous and Lower Permian rocks by using a diverse suite of logs from a borehole in central Kansas: Geology, v. 17, p. 303-306.

Morris, R. L., and W. P. Biggs, 1967, Using log-derived values of water saturation and porosity: Transactions of the SPWLA 8th Annual Logging Symposium, Paper X, 26 p.

Mwenifumbo, C. J., 1993, Kernel density estimation in the analysis and presentation of borehole geophysical data: The Log Analyst, v. 34, no. 5, p. 34-45.

Neinast, G. S., and C. C. Knox, 1973, Normalization of well log data: Transactions of the SPWLA 14th Annual Logging Symposium, Paper I, 19 p.

Pickett, G. R., 1966, A review of current techniques for determination of water saturation from logs: Journal of Petroleum Technology, v. 18, no. 11, p. 1425-33.

Pickett, G. R., 1973, Pattern recognition as a means of formation evaluation: The Log Analyst, v. 14, no. 4, p. 3-11.

Schlumberger, 1988, Log interpretation charts: Schlumberger Educational Services, Houston, Texas, 150 p.

Serra, O., and H. T. Abbott, 1980, The contribution of logging data to sedimentology and stratigraphy: SPE 9270, 55th Annual Fall Technical Conference and Exhibition, Dallas, Texas, 19 p.

Silverman, B. W., 1981, Density estimation for univariate and bivariate data, *in* V. Barnett, ed., Interpreting multivariate data: Chichester, John Wiley & Sons, p. 37-53.

Stiff, H. A., Jr., 1951, The interpretation of chemical water analysis by means of patterns: Journal of Petroleum Technology, v. 3, no. 10, p. 15-16.

Torres, D., R. Stickland, and M. Gianzero, 1990, A new approach to determining dip and strike using borehole images: Transactions of the SPWLA 31st Annual Logging Symposium, Paper K, 20 p.

Tufte, E. R., 1990, Envisioning information: Cheshire, CT, Graphics Press, 126 p.

Whitman, W. W., 1986, Pickett plots with approximate geologic conditions: The Log Analyst, v. 27, no. 5, p. 11-28.

Chapter 3

Compositional Analysis of Lithologies from Wireline Logs

Crossplots are an excellent medium for the pattern-recognition and problem-solving tasks that are at the core of most log analysis. As we have seen in the last chapter, even the limitation of the computer screen to the display of two dimensions can be overcome to some degree. Projections, glyphs, and the use of color are alternative strategies for accommodating the extra dimensions of information from multiple log traces. Ultimately, all these graphical methods have their limits, prescribed both by the capacity of the technique to represent complex data and the ability of the human mind to understand the results.

The most common problem of traditional log analysis is to establish the composition of the logged rocks in terms of the volumes of fluids and gases contained within the pore space and, secondarily, to determine their mineral content. The use of crossplots allows the identification of the mineral phases and crude volumetric estimations of mineral and fluid proportions. For rocks with simple compositions, the estimates of porosity are often sufficiently accurate for routine reservoir analysis. For more complex rock types, compositional analysis by traditional crossplot methods is more challenging because of the increased number of components that must be incorporated.

The compositional methods described in this chapter are algebraic solutions of data patterns that can be seen on log crossplots. Operations of matrix algebra cause the coordinates of data that are located in "log space" to be remapped as coordinates in "composition space." This duality of geometry and algebra is an important concept, particularly in the log analysis of sections with complex lithologies. Crossplots and other graphic techniques are valuable aids both in the design of the composition model and in the review of the model results. In reality, the number and compositional variability of rock minerals can challenge the most ingenious mathematical methods, especially in instances of restricted log suites. However, the methods of this chapter can often generate surprisingly good representations of subsurface rock compositions when monitored and guided by pattern recognition skills and geological insight.

COMPONENT ESTIMATION FROM POROSITY LOGS

In an alternative strategy, the proportions of multiple components can be estimated directly from a set of simultaneous equations, written in matrix algebra form as:

$$\mathbf{CV} = \mathbf{L}$$

where **C** is a matrix of the component log properties, **V** is a vector of the component unknown proportions, and **L** is a vector of the log responses of the zone to be evaluated. This equation specifies a linear model that describes the link between log measurements and rock properties. The situation is one of an "inverse problem," because the rock composition must be deduced from its physical properties. As such, it contrasts with a "forward problem," in which the task would be to predict the logging properties of a rock with known composition.

Solving the inverse problem was one of the earliest applications of computer processing to log analysis that went beyond simple machine adaptations of routine manual procedures. As described by Savre (1963), the method proved particularly effective in the Permian carbonates of West Texas as a means of improving the log estimates of porosity calculated by extant methods. Most commonly, porosities had been evaluated from neutron logs, but the values often proved to be unduly optimistic in zones with significant gypsum contents. Hydrogen within the water of crystallization of gypsum results in an apparent porosity in excess of 50% for pure gypsum, so that even moderate amounts cause large porosity errors. The grain density and transit time of gypsum are also markedly dissimilar from other common matrix minerals, so that the substitution of an alternative porosity log is not particularly helpful. Finally, the use of a conventional crossplot is difficult because the lithologies are a mixture of three minerals: dolomite, gypsum, and anhydrite. The total system actually consists of four , components when pore fluid is considered, and is difficult to represent graphically in a useful manner.

An effective log-analysis solution to this problem is one that computes the contents of gypsum and the other minerals and corrects apparent porosities recorded by the logs to true volumetric porosities. The inverse solution of a log-response equation set achieves this objective very neatly by computing mineral proportions and true porosities simultaneously. Because the unknowns are the fractions of four components, four equations are required for a unique solution. For this application, the log-response equations are:

Neutron:

$$\Phi_n = n_g \cdot G + n_a \cdot A + n_d \cdot D + n_f \cdot \Phi$$

Sonic:

$$\Delta t = \Delta t_g \cdot G + \Delta t_a \cdot A + \Delta t_d \cdot D + \Delta t_f \cdot \Phi$$

Density:

$$\rho_b = \rho_g \cdot G + \rho_a \cdot A + \rho_d \cdot D + \rho_f \cdot \Phi$$

In these first three equations, Φ_n, Δt, and ρ_b are the neutron porosity, sonic transit time, and bulk density log readings of a zone of interest. In each equation, the zone log reading is equated with the sum of the proportions of gypsum (G), anhydrite (A), dolomite (D), and fractional porosity (Φ), multiplied by the log response of the pure mineral or fluid.

The model is completed by a fourth "unity" equation which reflects the fact that the unknown proportions of gypsum (G), anhydrite (A), dolomite (D), and true fractional porosity (Φ) form a closed system:

Unity: $1 = G + A + D + \Phi$

Rewritten in the more compact matrix algebra convention, **CV=L**, **C** is the matrix of neutron porosities, transit times, and grain densities of gypsum, anhydrite, dolomite, and pore fluid, augmented by a line of unit values; **V** is the vector of their unknown proportions in the zone; and **L** is a vector of the zone log readings of neutron porosity, transit time, and bulk density, together with a unit value.

Defined in this manner, the equation set is fully determined and the solution for the unknown vector, **V** is:

$$\mathbf{V} = \mathbf{C}^{-1}\mathbf{L}$$

where $\mathbf{C}^{-1}$ is the inverse of the **C** matrix. This procedure can be coded in a computer program and executes quickly. The most time-consuming task is the inversion operation, which is made only once on any computer run. After this is done, the composition of each zone is calculated by a single matrix multiplication. A more expansive discussion of the basic matrix algebra concepts is given by Doveton (1986, p. 154-161) and code for KIWI, a FORTRAN computer program for compositional analysis, is included in the same source (p. 264-266).

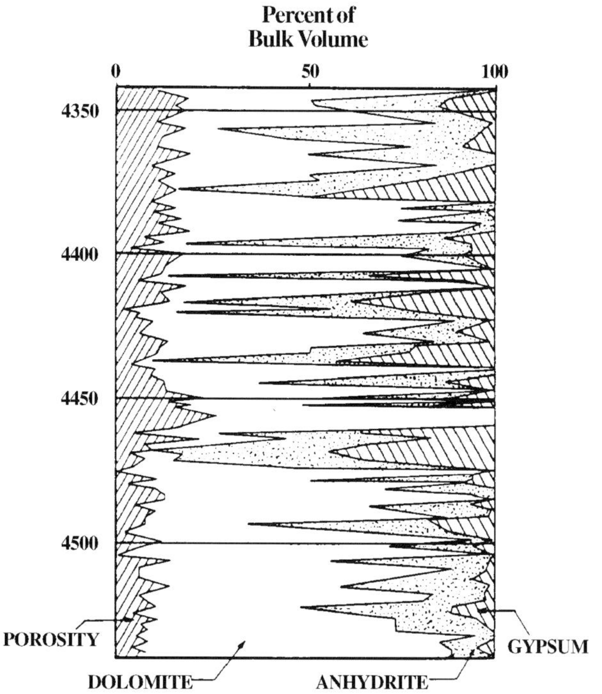

Figure 1. Early example of a computer-processed transformation of three porosity logs into mineral and pore fluid volumes in the San Andres Formation, Texas. From Alger et al. (1963). Copyright SPE.

A particularly useful property of the inverse solution procedure is that the results can be displayed as a function of depth, in contrast to the results of classical crossplots. An example of the graphical output drafted from one of the earliest computer runs is given by Figure 1 (Alger et al., 1963). Here, profiles of porosity, dolomite, anhydrite, and gypsum are shown from a Permian San Andres Formation section in West Texas. The form of the graphics is a useful reminder that although the model is designed to solve for mineral compositions, the result is bound by the limitations of the original logs, particularly with regard to effective vertical resolution. Put another way, the compositional profiles are simply numerical transformations of the original log traces which have been processed by the inverse matrix operator.

Savre (1963) carefully evaluated his computer analysis as an effective prediction device for true porosity by comparing the results with porosities measured from core. In an interesting twist to the story, it turned out that the core measurements, themselves, were often highly erroneous because of the

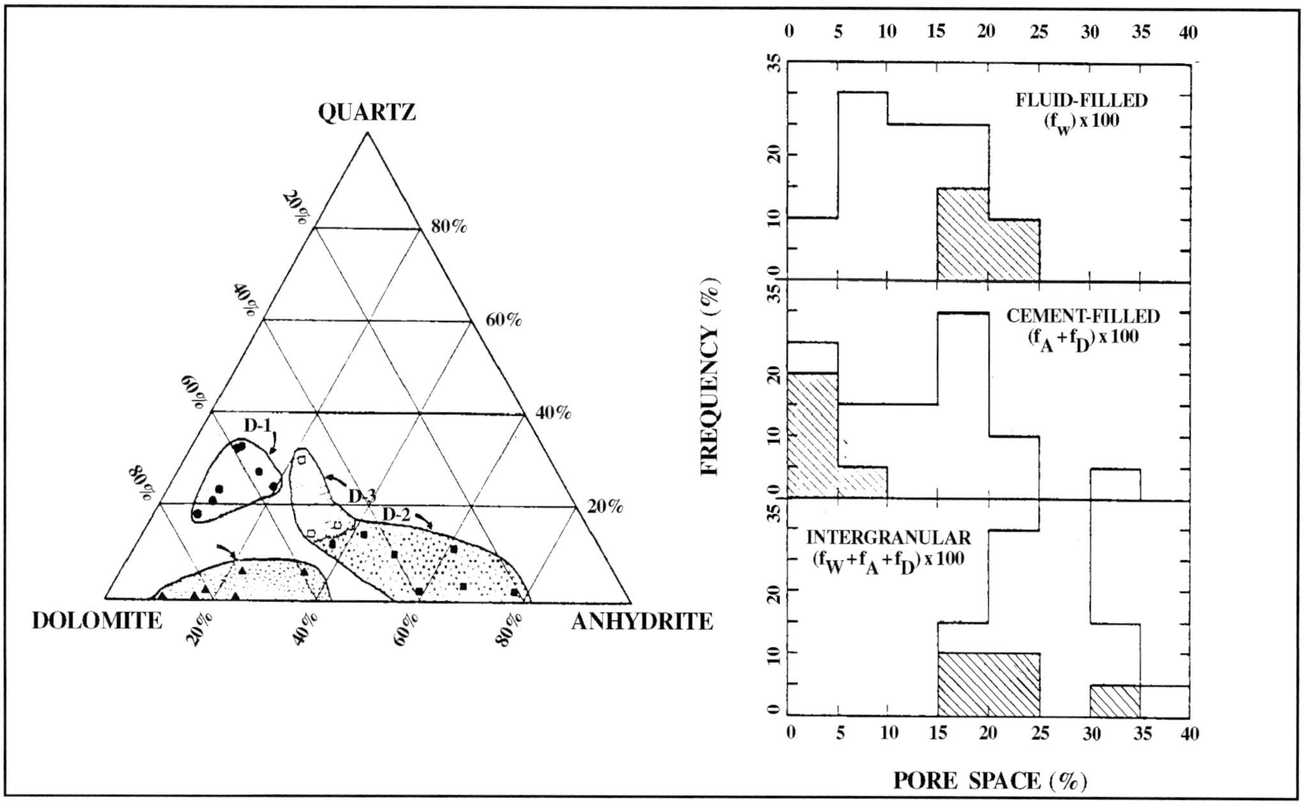

Figure 2. Summaries of log compositional analysis of the Permian upper Minnelusa Formation in the Powder River basin of Wyoming. At the left, the average composition of Minnelusa dolomites are plotted, where each point represents a dolomite unit in one well. At the right, log-calculated distributions of fluid-filled, cement-filled, and intergranular porosities are shown for the Minnelusa sandstones. Data from the eolian dune complex sandstones are hachured. From Schmoker and Schenk (1988).

dehydration of gypsum during core preparation. The systematic and significant bias towards porosity overestimation in gypsiferous samples by older core-analysis methods has been described by Hurd and Fitch (1959). There was a reasonable match between computed log estimates and core values when precautions were taken against gypsum dehydration. However, the crossplotted results indicated that a systematic bias to core overestimates of porosity in samples with high gypsum contents still existed. A moral to be drawn is that, wherever possible, log estimates of rock properties should be cross-validated with core or cuttings data, but also, that the investigator should be constantly vigilant for errors and bias in sampling and laboratory procedures.

The "tri-porosity" solution for minerals and porosity by the inverse solution computer method is perfectly general and can be applied to other mineral assemblages. The most accurate results will generally occur when there are pronounced differences in the logged properties of the individual components, so that the inevitable tool errors will tend to be small by comparison. So, for example, the technique was found to be very useful for tonnage estimates of sulfur extracted in boreholes by the Frasch process from salt dome caprocks in Louisiana. In this case the three minerals, anhydrite, calcite, and sulfur, together with pore volume, were computed from density, neutron, and sonic logs (Tixier and Alger, 1967). The distinctive properties of sulfur caused good volumetric estimates that were an improvement on core evaluations in cases where the sulfur tended to be selectively sloughed away in the coring process.

In a more recent study, Schmoker and Schenk (1988) described the use of this same method for the compositional analysis of the Permian upper part of the Minnelusa Formation in the Powder River basin, Wyoming. They applied sonic, neutron, and density logs to derive estimates of quartz, dolomite, anhydrite, and fluid-filled pore space within beds of dolomite and sandstone. Clear compositional distinctions could be made between the four upper Minnelusa dolomites representing three depositional cycles (see Figure 2). The differentiation helped to establish subsurface correlations from logs keyed to specific compositional information, rather than vague generalities drawn from log character.

The calculated mineralogies and pore volumes of the Minnelusa sandstones were used in the interpretation of their diagenetic histories. Petrographic studies indicated that the porosity was initially occluded by anhydrite cement, part of which was later dissolved in the creation of a secondary porosity. Pressure solution of quartz and some dolomite precipitation then caused some additional loss of pore volume. The mineral proportions computed from the logs were allocated between cement-filled (anhydrite and dolomite) and fluid-filled porosity. Schmoker and Schenk (1988) showed that eolian dune sandstones had significantly higher fluid-filled pore space and significantly lower cement than sandstones without dune buildups (see Figure 2). They concluded that the better reservoir porosity of the dune sandstones was caused more by the paucity of occluding cement than the volume of intergranular porosity.

When applied to cherty, dolomitic limestones, the computed mineral compositions are less precise because of the closer similarity between the porosity-log responses of calcite, dolomite, and quartz. However, the radically different properties of the pore fluid tend to ensure that porosity estimates are much more accurate. In recent years the introduction of the photoelectric absorption measurement has generally resulted in the substitution of this curve for the sonic log. There are several reasons for this. First, there is a strong degree of correlation between the speed of sound and the density of common minerals. The effect can be seen on service company charts of sonic–density crossplots, where the common lithology lines plot closely together. Effectively, this means that the acoustic velocity often does not provide a major source of new information when the density log is already considered. In addition, the sonic measurement responds primarily to intercrystalline and intergranular porosity, rather than the larger features of vugs and fractures. By contrast, the neutron and density tools measure the total porosity, regardless of pore size. This discrepancy can cause solution errors in zones with significant vug or fracture porosity, although it can be turned to advantage in an expanded evaluation of two porosity components.

NEGATIVE COMPONENTS: PROBLEMS AND DIAGNOSTICS

The inverse solution procedure is a simple and powerful model for compositional analysis, but its simplicity carries certain assumptions that must be considered carefully. In particular, the basic model contains no built-in constraint to prevent negative estimates of compositional proportions. The algebra of the material balance unity equation guarantees that the proportions collectively sum to unity. However, the equation does not safeguard against any individual proportion finding a value that is either negative or that exceeds unity. This shortcoming often comes as an unpleasant surprise and may cause misunderstanding and poor choices for a remedial strategy to arrive at a rational solution. The generation of negative proportions is a perfectly natural consequence of the model and contains useful feedback information.

The seeming paradox of negative components can be understood in the context of a simple example. Overlaid neutron–density porosity logs are commonly scaled in limestone equivalent units with a typical primary range between –10 and 30 porosity units. What are the alternative meanings that we could assign to a zone with a density porosity reading of –5 units? Clearly, a negative pore volume of 5% is impossible. Two possible explanations are that either the tool malfunctioned or that the mud program contained a heavy mineral such as barite. The potential role of such logging and borehole environmental problems should be reviewed first, before evaluating geological explanations. The inverse compositional analysis would calculate that the zone consisted of –5% fluid and 105% calcite. The solution is predicated on the assumption that the system is composed only of calcite and water. Clearly it is not. As the old Chinese proverb tells us: "The beginning of wisdom is calling a thing by its right name." A revised model run should be keyed to another mineral, and this mineral must have a density greater than calcite. Dolomite (density of about 2.87 gm/cc) would be a possible candidate; quartz (density of 2.65 gm/cc) would not.

This simple example is the basis for understanding the implications of negative components in more complex situations. Consider the compositional analysis of a cherty, dolomitic limestone by a neutron–density–photoelectric factor log combination. As an initial step, the photoelectric factor in units of barns per electron (P_e) would be converted to a bulk volumetric absorption measurement with units of barns per cubic centimeter (U) through multiplication by bulk density (ρ_e):

$$U = P_e \cdot \rho_e$$

The coding of the component log-response matrix is the algebraic equivalent of a geometric construction. In this case, a three-dimensional space is defined, whose reference axes are density, neutron, and photoelectric absorption. The log responses of the components identify four locations in this space which become the vertices of a tetrahedron when they are linked by straight lines (Figure 3). All rational mixed compositions must lie within this tetrahedron. If the log responses of a zone result in coordinates that lie outside the tetrahedron, then a solution will result with at least one negative component. As stated before, the diagnostic procedure should first evaluate operational causes such as tool error, adverse borehole environment, or poor choices of component log-response values. In addition, allowances must be made for measurement error, particularly in cases of low mineral abundances or poor mineral resolution. If these possibilities can be discounted, then a redefinition of the component system may be necessary. Some clues to the nature of the additional components are implied by the identity of the negative components and revealed by the appropriate crossplot.

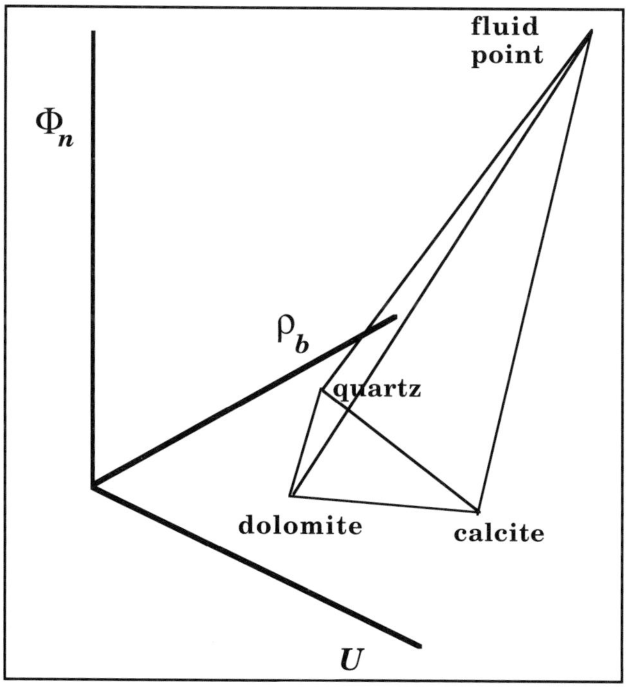

Figure 3. Schematic representation of a cherty, dolomitic limestone system in a three-dimensional log space with axes of the volumetric photoelectric cross-section *(U)*, bulk density (ρ_b) and neutron porosity, (Φ_n) logs.

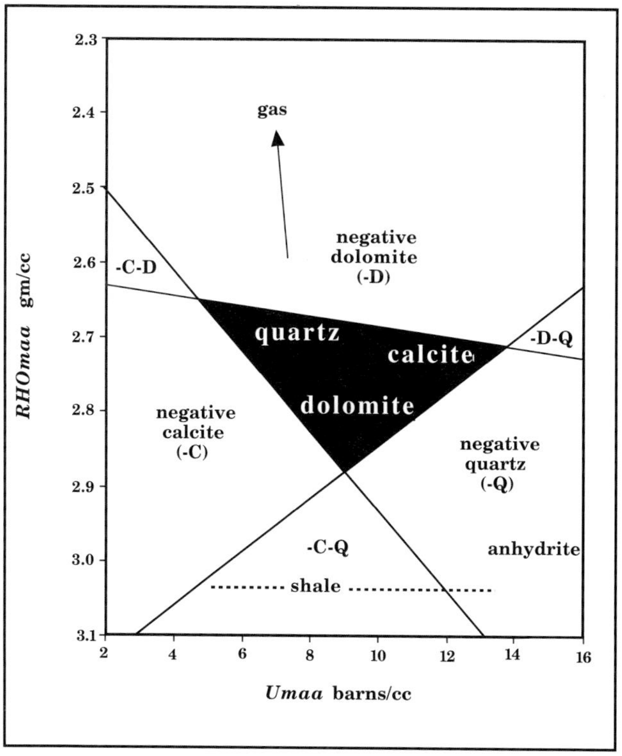

Figure 4. Negative component domains on a *RHOmaa–Umaa* crossplot projection of a quartz–calcite–dolomite fluid system.

The location of the quartz–calcite–dolomite triangle on the *RHOmaa–Umaa* crossplot is essentially a representation of the base of the trimineral–fluid tetrahedron. Zone responses that plot outside the tetrahedron also plot outside the triangle. The domains of negative component space are shown in Figure 4 and geological meanings are easy to assign to most of them. Some negative solutions are almost inevitable in the compositional analysis of sequences with relatively pure lithologies. So, for example, zones of pure limestone should be located close to the calcite–fluid line, but some will fall outside the tetrahedron. If a small negative value is corrected to zero, the corresponding geometric operation is equivalent to a movement of the zone over a short distance to the nearest tetrahedron face.

Displacement to larger distances within the negative component domain suggests the influence of systematic additional components. Negative dolomite may reflect uncorrected gas saturation in the flushed zone or a lighter mineral such as halite. Conversely, negative quartz indicates the possibility of some heavier mineral such as anhydrite. If shales are significant and unaccounted for, they will typically register their presence by either negative calcite, negative quartz, or both. These diagnostic remarks have been made for a neutron–density–photoelectric factor log combination, but very similar conclusions can be drawn for a neutron–density–sonic log combination (see Figure 14 in Chapter 2).

So, although negative components are unacceptable as an end result, they do have a role as a fundamental check on the potential validity of the model used for compositional analysis. They should be viewed as diagnostic errors with an information content useful for guiding the analysis to a better solution. The distinction between errors that are acceptable as minor, random measurement noise and systematic deviations is best made by a comparison between the original logs and the logs predicted by the model solution. The predictions are given by:

$$\hat{\mathbf{L}} = \mathbf{CV}$$

If the solution results in compositional proportions that are all positive, then there will be an exact match between the logs and model predictions. This equivalence does not imply that the result is geologically correct; it simply means that the solution is rational and consistent with the choice of components and their properties. There may be other satisfactory solutions based on alternative mineral suites. Strategies that evaluate alternative compositional models will be discussed later in this chapter.

If the inverse procedure has generated zone solutions with proportions that are negative or exceed unity, then the adjustment to rational proportions will result in log predictions that will deviate from the original logs. The deviations between measurements and predictions can then be examined to differentiate minor measurement error from systemat-

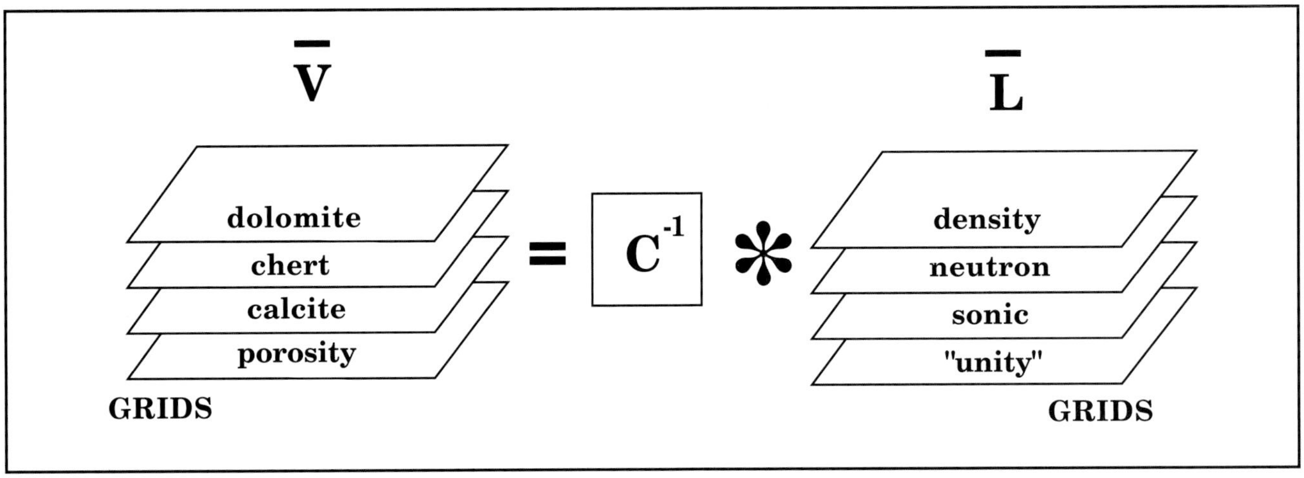

Figure 5. Schematic representation of the transformation of three porosity-log spatial grids into four component-composition grids through a matrix inversion solution applied as a grid-to-grid operation in a computer contouring package.

ic perturbations that require intervention and correction. In the more sophisticated models which we will review, tool-response errors are actively incorporated into the solution algorithm, together with constraints that preclude illogical compositional proportions.

INVERSION MAPPING OF COMPOSITIONS

The conventional result of a compositional analysis from simple inversion is equivalent to a transformation of log curves into a compositional profile graphed as a function of depth. The input vectors of log responses represent depth zones or digitally sampled increments of depth. However, instead of vectors sampled vertically along a depth axis, vectors of log responses can be input from geographic locations across a stratigraphic unit and the compositional results interpolated laterally to produce a lithofacies map.

Bornemann and Doveton (1983) described a case study of the application of this mapping paradigm to the lithofacies of the Middle Ordovician Viola Limestone in south-central Kansas. Density, neutron, and sonic logs were used to estimate the proportions of calcite, dolomite, chert, and pore volume. The logs were first normalized, based on the results of trend surface analysis applied to a calibration unit (Doveton and Bornemann, 1981). Average normalized log values of the Viola were then interpolated between well control, using a standard automated contouring package. The result of this step was the generation of three grids of average log values (Figure 5). A particular advantage of this approach is that all three logs are not required in any of the wells, because the inverse transformation is applied to the gridded values from interpolation, rather than the values at individual well locations.

The matrix algebra inversion procedure was then applied to the three grids supplemented by a grid of unit values. Each cell log-response vector was processed by the grid-to-grid operation in the production of four solution grids of compositional proportions (Figure 5). The grid values were combined in a single map that is a compositional expression of lithofacies (Figure 6). Each cell was assigned a symbol according to whether the dominant component was calcite, dolomite, or chert. The map was carefully validated, using standard lithological information available from drill-cuttings and core records. Three additional lithofacies were identified as a result of negative proportional solutions at a number of the cell locations. Negative dolomite and negative quartz were found to have an excellent match with areas of residual chert and karstically weathered sections. The association is caused by the insensitivity of the sonic measurement to larger pores in these facies. Negative calcite solutions reflect significant occurrences of shale as an additional component in a shaly carbonate facies.

LINEAR ASSUMPTIONS AND THE INCORPORATION OF THE RESISTIVITY LOG

The equations of the simple inversion procedure define a linear model that is an approximation of reality. Both neutron and sonic porosity ranges are curvilinear. In practice, the inaccuracies introduced by this assumption tend to be minor, especially when compared with errors that can be caused by poor parameter choices (Quirein et al., 1986). The nonlinearities can often be handled in more detailed work by choosing mineral log properties such that the line linking them with the fluid point tends to be tangential to the curve segment that is broadly representative of the porosity of the logged zones.

The resistivity log has a markedly nonlinear response and must be linearized to some kind of porosity function if it is to be included as part of an

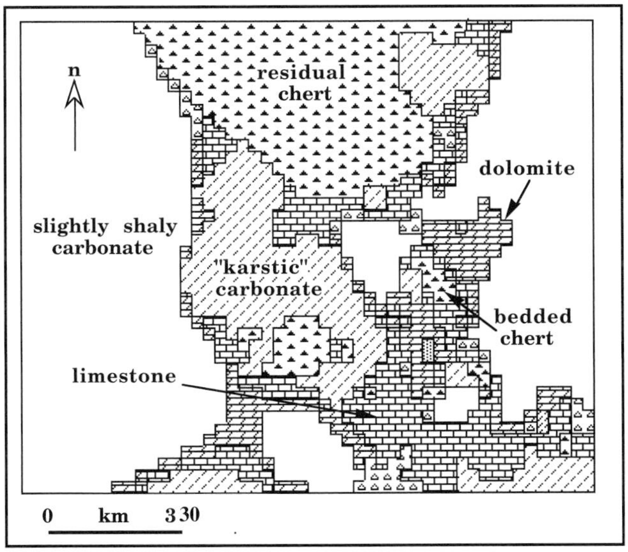

Figure 6. Lithofacies map of the Viola Limestone in south-central Kansas computed from the inversion of grids of average neutron, density, and transit times. Adapted from Bornemann and Doveton (1983).

expanded equation set. The necessary operation can be deduced from the Archie equation and is very simple if the cementation and saturation exponents can both be considered to have a value of two. The resistivity is controlled effectively by the bulk volume of water, since both the hydrocarbon phases and reservoir-rock minerals are insulators (assuming no clay or conductive metallic minerals are present). Then, from the Archie equation, the volume of water is given by:

$$W = \Phi \cdot S_w = \sqrt{\frac{R_w}{R_t}}$$

where Φ is fractional porosity, S_w is fractional water saturation, R_t is the formation resistivity, and R_w is the resistivity of the formation water. So, the resistivity log can be incorporated after a transformation to a square-root conductivity form. The log response for the water phase is then the square root of the formation-water conductivity, and the values for all the insulator components will be zero. The insertion of a new equation allows the solution of an additional variable, and this is provided by splitting the pore-volume term into two constituent volumes: a conducting phase (water) and a non-conducting phase (oil or gas).

Obviously there are some practical constraints to this expanded resistivity model. A major problem is introduced by the drastic differences in radial investigation by the separate tools. The porosity tools have shallow depths of investigation and their readings are drawn mostly from the flushed zone. The pore space of the flushed zone is filled with mud filtrate and any residual hydrocarbons. The resistivity tool is usually selected to have a deeper investigation so that it can be used to estimate the hydrocarbon saturation of the undisturbed formation. Consequently, the tools will be influenced by varying contents of hydrocarbon as dictated by their investigation characteristics. However, it would be possible to adapt the model to an iterative operation, with progressive refinement in pore-phase estimates—in which case it would emulate some of the classical procedures used to correct logs for hydrocarbon effects.

COMPOSITIONAL ANALYSIS OF UNDERDETERMINED SYSTEMS

The simple inversion procedure requires a precise match between the number of knowns and unknowns. Such a situation is called a "determined system." The alternative possibilities are that the number of logs is not sufficient to resolve all the components (an underdetermined system) or that the number of logs exceeds the number of components (an overdetermined system). In reality, most formations are likely to present underdetermined compositional problems if all the constituents are counted and matched against the number of logs run in a typical borehole. McCammon (1970) and Harris and McCammon (1971) considered alternative systematic solutions to the estimation of mineral compositions from logs in underdetermined cases. Although their algorithms have been superseded by more recent optimization procedures, their approach contains a number of useful lessons.

McCammon (1970) considered the underdetermined system from the viewpoint of classical information theory, which holds that the least biased solution is the one that maximizes the entropy function:

$$\mathrm{E} = \sum v_i \log v_i$$

where v_i is the proportion of the ith component. The expression for entropy is closely approximated by that of proportional variance:

$$\mathrm{P} = \left(\frac{n}{n-1}\right)\sum v_i(1-v_i) = \left(\frac{n}{n-1}\right)\left(1-\sum v_i^2\right)$$

where n is the number of components. A maximization of the variance function, P, is close to the condition of maximum entropy, and the resulting optimal solution is easier to compute from the matrix algebra equation:

$$\mathbf{V} = \mathbf{C}^{\mathrm{T}}\left(\mathbf{C}\mathbf{C}^{\mathrm{T}}\right)^{-1}\mathbf{L}$$

where **V** is the vector of unknown proportions, **C** is the matrix of component log properties, **T** signifies a matrix transpose, and **L** is the vector of zone log responses (Doveton and Cable, 1979).

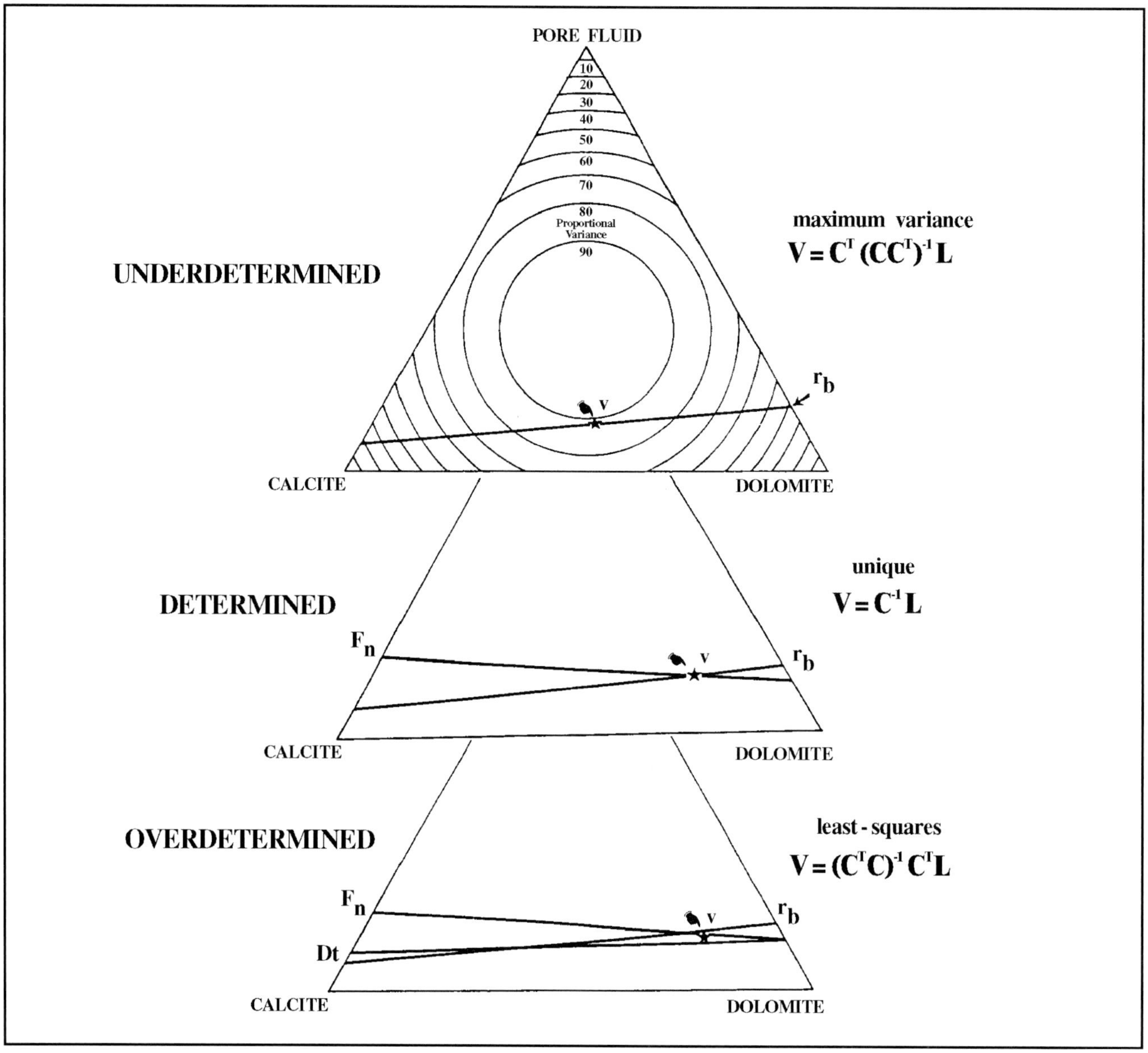

Figure 7. Graphic examples of solutions to the compositional analysis of a single zone in a calcareous dolomite based on either one, two, or three logs from density, neutron, and sonic measurements. All possible situations of determinancy are shown together with the appropriate matrix algebra solution in which the vector V represents the unknown component proportions, C is a matrix of the components' log properties, and L is a vector of the zone log readings. Adapted from Doveton (1986). Reprinted by permission of John Wiley & Sons, Inc., copyright© 1986.

The compositional solution provided by proportional variance algorithm is optimal from a classical statistical viewpoint: the average squared errors between the estimates and real compositions should be the minimum possible. This is a conservative philosophy that aims to be "least wrong," rather than "most right." The difference in the two positions can be illustrated by an analogy with coin tossing. The optimal prediction is 0.5 head/0.5 tail. This result never occurs, so the prediction is always wrong. However, the average squared error for either heads or tails is the minimum possible. The comparison of rock composition with coin-tossing trials is not as far-fetched as it might at first appear. Mineral proportions are frequently distributed in a highly unequal manner. Therefore the real rock composition will often be one of several extreme possibilities, rather than the less likely, seemingly homogeneous composition that can result from a minimum-variance solution. The correct interpretation of a bland compositional solution is that it represents the average of a range of possibilities. As such, it is a good estimate of the average, but may be a very poor prediction of the particular: the composition of the zone in question. Such a result is a useful diagnostic suggesting that several extreme alternatives should be reviewed and that extra information is required. The information can take a variety of forms, such as explicit geological knowledge of the range of actual compositions or additional constraints that preclude impossible solutions.

A graphic example of the general problem of finding a "correct" solution is shown in the upper composition triangle of Figure 7. Here, all the possible alternative rational solutions of calcite, dolomite, and fluid proportions are shown by a line for a density log reading of 2.6 gm/cc. The hand points to the location of the maximum variance solution. If all possible solutions are equally likely, then the maximum variance solution should give a minimum error *on average* when compared with the real composition.

If real rock compositions are made up of components that tend to the extremes of maximum or minimum proportions, then the real solution might be one of minimum entropy. The minimum entropy estimate would correspond to the most uneven possible distribution of proportions. Again, this would be a mathematically optimal solution, albeit using the information theoretic criterion turned on its head. On the basis of their experience with real examples, Harris and McCammon (1971) reported that errors were always less when using a minimum, rather than a maximum entropy measure. This conclusion should be considered cautiously. While it seems to be true that real compositions are likely to have low entropy values, they will often be matched with a local entropy minimum rather than the global entropy minimum. A true local minimum entropy solution is likely to be radically different from the global minimum because it will be located at a different extreme location within compositional space. In the final analysis, mathematical optimality does not necessarily dictate geological optimality.

COMPOSITIONAL ANALYSIS OF OVERDETERMINED SYSTEMS

The observation that compositional proportions are often distributed in a highly unequal manner means that many rocks are dominated by a relatively small number of components. Consequently, the number of logging-tool measurements may exceed the number of significant lithological components. The situation becomes overdetermined when the number of log-response equations is greater than the number of components. The appropriate solution is then one that most accurately reproduces the original logs when predicted logs are calculated from the compositional solutions. Using conventional statistical theory, this solution is the one that minimizes the sums of squares of the deviations between the original logs and their predictions. The least-squares solution is given readily by the matrix algebra equation:

$$\mathbf{V} = \left(\mathbf{C}^{\mathrm{T}}\mathbf{C}\right)^{-1}\mathbf{C}^{\mathrm{T}}\mathbf{L}$$

where the terms are the same as those in both the determined and underdetermined matrix algorithms written earlier. A geometrical illustration of the problem is shown by the lower triangle of Figure 7. Here, three log responses from a hypothetical zone are plotted on a calcite, dolomite, fluid composition triangle. Each log-response line locates all the possible solutions that would be satisfied by that value taken by itself as an underdetermined solution. The intersections of the three log-response lines, taken a pair at a time, are the three alternative determined solutions. The hand points to the least-squares solution as a compromise that minimizes the difference between the input responses and their values that are calculated from the solution.

The matrix formulation requires some additional weighting function to allow for the fact that the logging measurements are recorded in radically different units. Without any weighting, the error minimization is predicated on equal units and results in a solution which preferentially honors logs with the highest data ranges. The modified least-squares algorithm is then:

$$\mathbf{V} = \left(\mathbf{C}^{\mathrm{T}}\mathbf{W}\mathbf{C}\right)^{-1}\mathbf{C}^{\mathrm{T}}\mathbf{W}\mathbf{L}$$

where **W** is a diagonal matrix that contains the elements of a weight vector (Harvey et al., 1990). The weights may be input based on physical first principles or by a standardization scheme, such as transformation from the original measurement to a scale anchored to the mean and counted in standard deviation units.

For any given zone, the sum-of-squares error is given by:

$$e = \left(\mathbf{L} - \hat{\mathbf{L}}\right)^{\mathrm{T}}\left(\mathbf{L} - \hat{\mathbf{L}}\right)$$

where **L** is the vector of log responses associated with the least-squares solution. The error term can be plotted as a monitor log to highlight zones where there are striking inconsistencies between the model and the log responses. The overall performance of an algorithm may be judged from the standard error, computed from the summed zone errors as:

$$s_e \sqrt{\frac{\sum e}{(n - m - 1)}}$$

OPTIMIZATION MODELS FOR COMPOSITIONAL SOLUTIONS

A second generation of compositional analysis procedures has moved beyond the somewhat rigid inversion algorithms described above. In a more generalized approach, constraints and tool error functions have been incorporated as part of the solution process The methodology was first developed by Mayer and Sibbit (1980) in which modified steepest descent strategies were used to hunt for an optimal solution that minimized the "incoherence" between the logs and their predicted values. For any given log, the incoherence function is defined as:

$$I_A = \frac{(a - \hat{a})^2}{(\sigma_A^2 + \tau_A^2)}$$

where I_A is the incoherence for log A, a is the log response for the zone and $\hat{a}$ is its prediction, and σ^2 and τ^2 are the uncertainties associated with the log measurement and the response equation, respectively.

The uncertainty term for each log measurement is compounded from the sources of sensor error, data acquisition, and the dispersions associated with environmental corrections. Response equation dispersion represents the uncertainties introduced by linear approximations, erroneous choices of component log responses, and hidden factors such as the influence of textural parameters. It seems reasonable to suppose that these two types of uncertainty are independent, so that they can be summed as one total error term for each tool:

$$u_A^2 = \sigma_A^2 + \tau_A^2$$

The total log incoherence for any particular depth zone is the sum of the separate log incoherences:

$$I_1 = I_A = I_B + I_C + \ldots$$

The form of the equations shows that the solution tends to be most strongly influenced by the logs to which the most confidence can be attributed. Logs with large errors will have smaller incoherences and will contribute less to the total incoherence term.

Constraints are also included and take the general form of:

$$g_i(v_i) \geq 0$$

where g_i is some function that constrains the value of the unknown proportion of the ith component. Rigid mathematical constraints are those that preclude the occurrence of negative proportions or those that exceed unity. Geological and local constraints incorporate relations that conform to general geological principles or prior knowledge of local geology. These geological constraints are more generalized, so that appropriate uncertainties are assigned to them. The constraint dispersions generate additional incoherence terms to be considered. A combined incoherence function is then the sum of the log and constraint incoherences:

$$I_t = \sum \frac{(a_i - \hat{a}_i)^2}{\sigma_i^2 + \tau_i^2} + \sum \frac{g_j(v_j)^2}{\tau_j^2}$$

Notice that if the system is fully determined, then the total incoherence will be zero, provided that no constraints are violated. This special situation is the limiting case of applications which are otherwise presumed to be overdetermined. In a routine application of the optimization algorithm, the number of logs would be expected to exceed the number of components. In part this is feasible because the bulk of rock compositions tend to be dominated by relatively few components. In addition, the range of wireline measurements used today typically extends beyond the traditional porosity log to resistivity, spectral gamma-ray, and geochemical logs.

The optimization method of Mayer and Sibbit (1980) is an iterative search procedure. The system model of input logs and output components are first defined. The incoherence values associated with each log type are entered, together with the constraints to be met. For each zone an initial composition is estimated by an approximate method and used as the starting point for a sequence of intermediate solutions. At each step the incoherence between the input log responses and those predicted from the solution is calculated. A gradient is also computed as the means to generate the next solution, using a steepest descent technique. The process terminates when it is determined that convergence has been satisfied, at which time there is no appreciable difference between successive solutions. The final solution is approximate, but the total incoherence between the logs and the compositional estimate will be the minimum possible. The combined display of real and theoretical logs is invaluable as a quality-control mechanism to alert the user to problem zones which may be optimal but are flatly wrong. The generality of the approach allows alternative and remedial attempts to be made without major difficulty.

In further refinements, Gysen et al. (1987) described an extension of the method to the simultaneous optimization of component proportions and response parameters. Moss and Harrison (1985) also reported a technique to solve for the uncertainty multipliers which contain the total error associated with each tool. Although the errors cannot be solved for every depth zone, they can at least be estimated for selected intervals and assumed to be effectively constant between zones.

MULTIPLE MODEL SOLUTIONS OF ROCK COMPOSITION

The iterative search mechanism of error-minimization methods makes their operation slow, especially when compared with the speed of "classical" inversion algorithms described earlier. Faster solutions

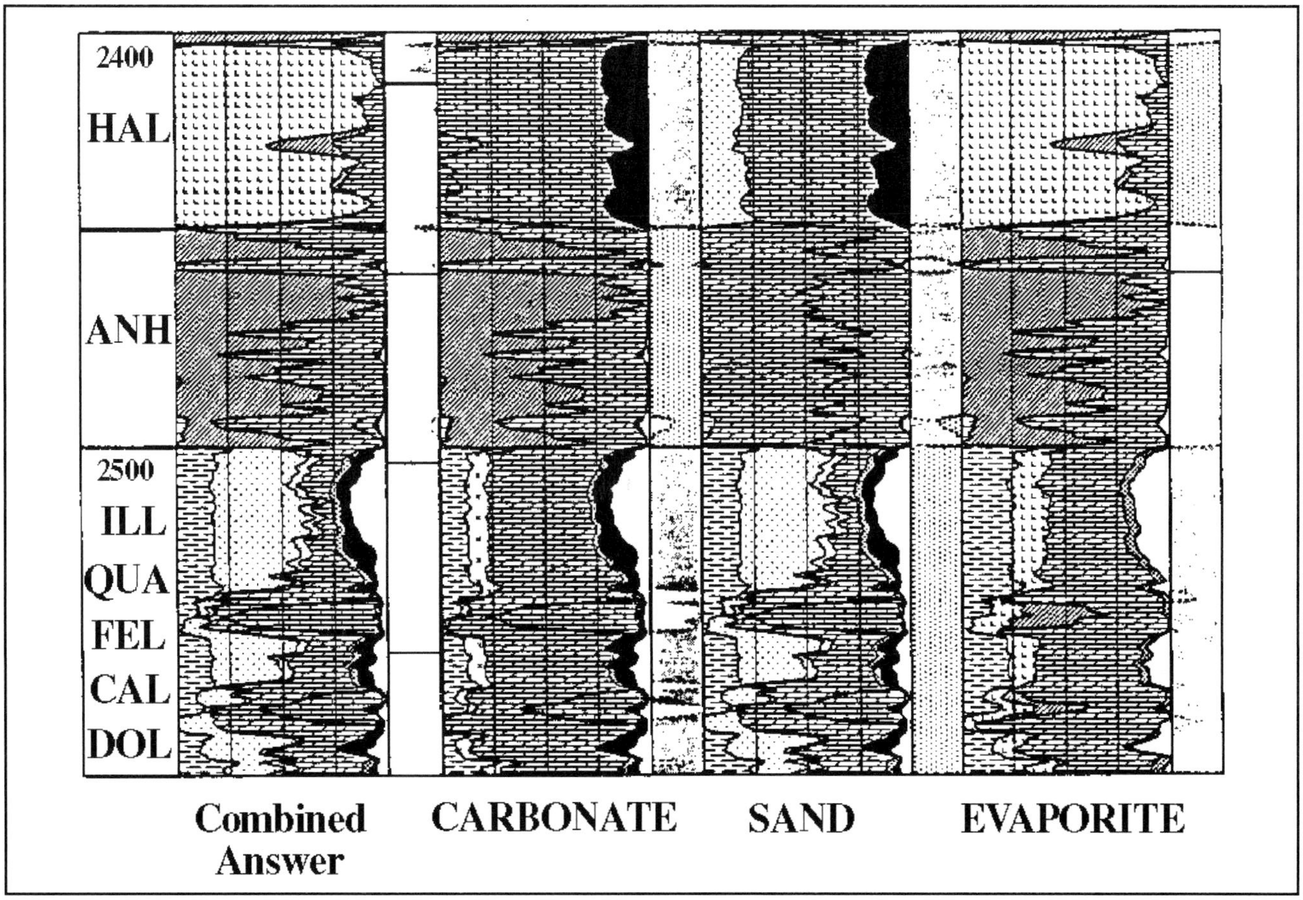

Figure 8. Multiple formation evaluation models applied to a Yates Formation section together with composite result generated by switching between models as determined by their probability of occurrence. The lithology symbols represent illite, quartz, feldspar, calcite, dolomite, anhydrite, halite, and pore fluid. The total horizontal scale of 100% bulk volume is subdivided between these components. From Quirein et al. (1986).

make workstations a practical environment for geologists to use interactive graphics to evaluate and improve on compositional analysis. The potential benefits of optimal minimum-error solutions are worthless if the component model is incorrectly specified. By contrast, patient and intelligent interaction with intermediate solutions and diagnostics from faster methods can often lead to more geologically meaningful results. Recent developments in minimization procedures have benefited from increased computation speeds and strategies that evaluate alternative geological models.

Quirein et al. (1986) reported a major reduction in computer time using quadratic programming techniques and linearized response equations, an improvement on the penalty-constraint approach used by earlier methods. In addition, they incorporated a program to solve for poorly known log responses of a component subset as an optimization procedure to be applied to specific depths that could be used for calibration. These calibration intervals, where both logs and compositions are known, are most typically those that have been cored. Knowledge of composition can also be utilized from other sources. Not all component log responses need to be estimated, since their properties are restricted to a limited range. However, the properties of a subset of mineral components are ambiguous and locally variable. The most notorious example of such components is clay minerals; these will be discussed more fully in the following section.

In common with earlier optimization methodologies, the system that links logs with components is assumed to be either determined or overdetermined. The use of multiple alternative models allows a more realistic treatment of this assumption because common associations can be modelled in parallel and a final selection made between them at any depth. Wherever possible, in an attempt to find a good match and to sidestep problems associated with the estimates of log and equation dispersions (Marett and Kimminau, 1990), each separate model is designed to be close to fully determined. The appropriate logs for each model are clearly those that discriminate well between the separate components. If a poor choice of logs is made, then the model is ill-conditioned. The model structure can be checked through the computation of the condition number of:

$$C^TDC$$

where **C** is the matrix of component log responses and **D** is a matrix of uncertainty values. The condition number is higher for ill-conditioned models and gives a measure of the sensitivity of proportion estimates to small changes in component log responses (Quirein et al., 1986).

The choice between alternative models for any zone can be made by the user based on an assessment of the relative incoherence of the solutions and their feasibility as reasonable geological descriptions. Alternatively, the decision can be made on the basis of probability established either from electrofacies assignation or from the volumetrics of the solution. An example of the system in operation is shown in Figure 8. Here, three alternative model solutions of carbonate, sandstone, and evaporite associations were calculated for a Yates Formation section from the Permian basin. The final solution is a composite of the associations in which the switch between each alternative at various depths was determined by their respective probabilities.

In a more expansive case study, King and Quirein (1987) described the systematic incorporation of geological data to arbitrate the choice of model association. The initial step was the creation of a database of electrofacies calculated from lithofacies types. Means and standard deviations were estimated as summaries of fully quantitative volumetric data from X-ray diffraction or thin-section analysis, or from semi-quantitative assessments of core and sample descriptions. The conversion of these lithofacies characteristics to electrofacies involved forward modelling in which the volumetric data were convolved with the tool and mineral responses, coupled with the uncertainties of the tool measurements. Each electrofacies was then represented by a hyperellipsoid in the multidimensional space of the log variables. Two-dimensional projections of electrofacies show as ellipses (see Figure 9). The log crossplots are useful as a means for displaying the relationship between electrofacies, the ambiguities in their distinction, and the need for multidimensional characterization.

The electrofacies database is then used as a classification device to make a prediction for each zone based on a Bayesian decision procedure. Bayes' theorem was found in the papers of the Reverend Thomas Bayes (1702-61) after his death. The theorem leads to the relationship:

$$P(B|A) = \frac{P(A|B)P(B)}{P(A)}$$

where P(B|A) is the conditional probability that event B occurs given that A has occurred, P(A|B) is the conditional probability that event A occurs given that B has occurred, and P(A) and P(B) are the unconditional probabilities of the occurrences of events A and B. The philosophy of Bayesian statistics differs from that of classical statistics, but is widely used as a means to revise initial estimates of probabilities in the light of additional information (Anscombe, 1961).

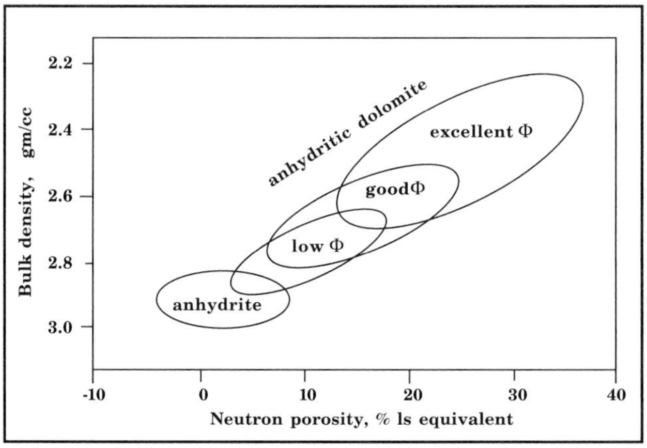

Figure 9. Example of projection of electrofacies ellipsoids in two-log space from an electrofacies database. Modified from King and Quirein (1987).

The electrofacies with the highest posterior probability is displayed as the selection at each depth level. The electrofacies choice determines which composition model should be used in the estimation of volumetric proportions. A graphical example of this procedure is shown in Figure 10, which shows actual core description, predicted lithology, and volumetric estimation for a complex lithology sequence from the Midland basin of West Texas (from King and Quirein, 1987). The section contains evaporite beds which overlie fine-grained, shaly sandstones interbedded with tight dolomites. The lithology classification was referenced to a database of 125 electrofacies types compiled to represent lithologies of the Permian basin. The choice of electrofacies determined which of four alternatives would be used for compositional analysis: "shaly sand," "carbonate," "shale," or "evaporite" models.

It should be noted that the thrust of improvements in the application of minimization procedures has been towards a closer control by geological constraints. So, for example, decisions about model selection are determined by electrofacies classifications, themselves calculated from known lithologies. Procedures have also been incorporated to estimate log responses for difficult components in calibration zones where component volumetric proportions are considered to be known. Although generally still applied to an overdetermined system, the multiple models are not far removed from determined matches of components and logs. In the case of a determined model, the solution is that of a simple and fast matrix inversion with zero incoherence, provided that the non-negative constraint is not violated. The analysis of the relative conditioning of the model system is a valuable mathematical contribution to the determi-

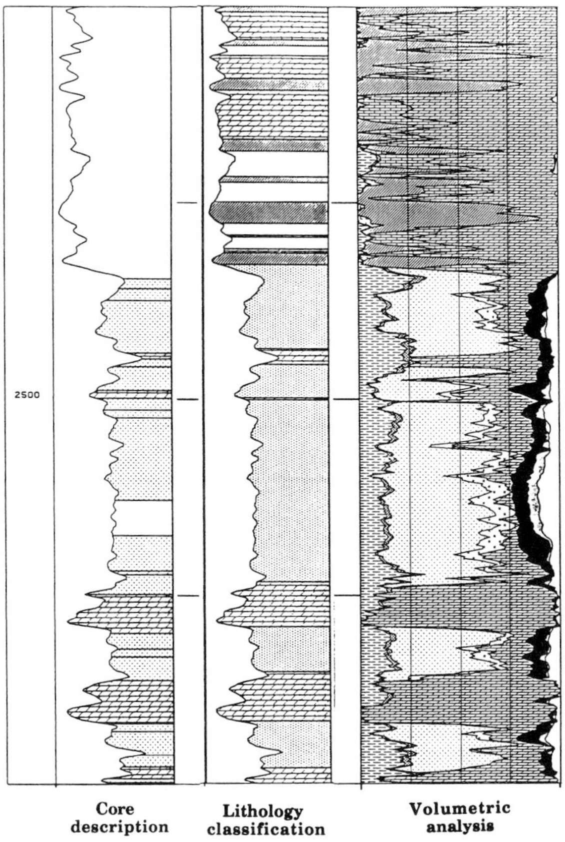

Figure 10. Core description, lithology classification from application of electrofacies database to wireline logs, and volumetric solution calculated for model selected by the classification. The section gamma ray log is shown superimposed on the core description and lithology classification tracks. The volumetric analysis is scaled such that the individual components sum to 100% bulk volume on the horizontal axis. The components are illite, chlorite, quartz, feldspar, dolomite, calcite, anhydrite, halite, water, and oil. Modified from King and Quirein (1987).

nation of which logs provide the maximum discrimination of model components that will lead to the most stable estimates of volumetric proportions. Finally, the reduction in computer time has made generalized compositional analysis a viable interactive procedure for workstations. This is the appropriate computational environment in which geologists can guide analysis toward realistic lithological solutions.

VOLUMETRIC ESTIMATES OF SHALES AND CLAYS

The compositional analysis of many sedimentary successions involves the estimation of volumetric proportions of shales. Shales typically are composed of a mixture of clay minerals, quartz, carbonates, and iron minerals, as well as other accessory components. Clay minerals are markedly different from other rock-forming minerals both in terms of their complexity and their variability. Shales present special problems for log interpretation and the meaning and limitations of the results of the many algorithms that have been designed for their volumetric estimation should be understood.

Both the spontaneous potential and gamma-ray logs are especially sensitive to clay mineral content to the degree that they are sometimes known as "shale logs." The deflection of the SP log between the extremes of a "shale baseline" and a "clean baseline" has often been used as a measure of shale content. However, Griffiths (1952) showed that the relationship is nonlinear, while Bacon (1948) demonstrated the varying influence of different clay minerals. Estimates of shale content from the SP log are therefore poorly semi-quantitative, with an inherent tendency to overestimation.

The gamma-ray measurement is a direct function of the weight concentration of radioactive isotopes. The log value can be transformed to a volumetric measure by multiplying by the bulk density, although this is usually a minor correction. The conversion of the log from API units to shale proportions is empirical. A zero-shale baseline is established as representative of shale-free lithologies in the analytical section. The selection of a gamma-ray value to typify a shale end member is a more pragmatic exercise. Extremely radioactive shales should be discounted, because the high values are caused largely by uranium contents fixed by organic matter. Even "normal" shales are likely to show some significant variability due, in part, to clay mineral changes, but controlled mainly by differences in silt content.

In fact, shale as a component of rock composition is radically different from the rock-forming minerals and pore fluids discussed up to this point. Part of the problem has been caused by the confusion between "clay" and "shale" that is common in the older log-analysis literature. The majority of shales are mixtures of clay minerals and a silt fraction that is typically composed of quartz, calcite, feldspar, iron oxides, and other materials. Most, but not all, of the radioactivity is associated with the clay mineral content. Yaalon (1962) found an average silt content of 41% from his study of thousands of shales. From this it follows that a gamma-ray value selected from a log to typify a total shale will most likely represent a shale with a moderate, but significant silt content. Zones of pure clay are uncommon in sedimentary successions. Further useful discussion is given by Heslop (1974) on the terminology and estimation of shales and clay mineral contents from the gamma-ray log.

The introduction of neutron–density crossplots provided a third approach to the characterization of shale content. The composition of lithologies in clastic successions can be represented by a simplified model of quartz, pore fluid, and shale. These three end members plot as a composition triangle on a neu-

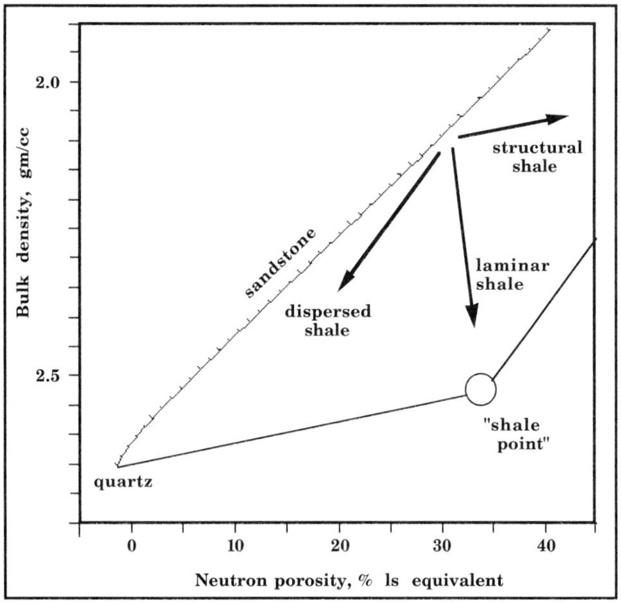

Figure 11. Generalized composition triangle of quartz, pore fluid, and shale on a neutron-density crossplot.

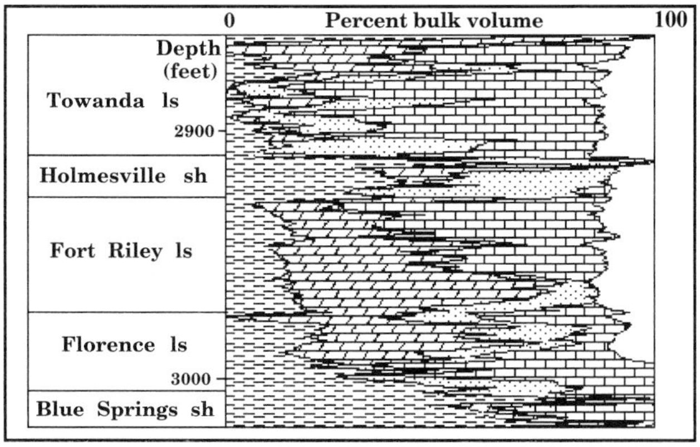

Figure 12. Compositional profile of part of the Chase Group (Permian) in southwestern Kansas based on inversion of gamma-ray, photoelectric-absorption, density, and neutron logs. Standard lithology symbols; pore fluid shown as white.

tron–density crossplot (Figure 11). The quartz and fluid points are predefined from their physical properties, but the shale point is selected as an extreme location that probably represents the least silty shale. Estimates of the proportions of quartz, shale, and pore fluid can then be solved by the inversion techniques described earlier. An implicit assumption of this model is that the clay mineral composition of the shale member is effectively constant. This is unlikely to be true, particularly when contrasting the clay minerals that are likely to occur within the pore networks of shaly sandstones with those of the shales used for the end member. Therefore, shale volumes from either the neutron–density crossplot, or the gamma-ray or spontaneous potential logs should be viewed as generalized estimates.

Compositional analysis by matrix inversion allows the simultaneous use of several shale indicators in a single shale estimate. The log properties of an idealized shale component must be defined either by inspection of the original logs or through pattern recognition applied to log crossplots. The range of shale variability generally makes these properties a matter for interpretation, in contrast with the predefined values for standard minerals and fluids. In practice, the results provide a serviceable, first-order profile of what is a vaguely defined lithological component (e.g. Figure 12). In the next chapter, we will review the use of principal component analysis as a statistical method to combine these separate shale indicators in a single estimator that is keyed to intrinsic variability.

In more detailed work, the older and broader methods of shale evaluation have been expanded to the quantitative assessment of clay mineral species. These relatively novel techniques have been made possible by the latest generation of nuclear logging tools. Clay minerals play a significant role in the producibility of many clastic reservoirs through their differing effects on permeability within pore networks. The cation exchange capacity of clay minerals is the central term in shaly sandstone modifications of the Archie equation. These and other considerations have provided the impetus for developments in the log estimation of clay mineral content.

Clay minerals show differing degrees of variability, but are generally subdivided between four major types: illite, smectite, kaolinite, and chlorite. Clay mineral typing is based on several log criteria which must be considered carefully and collectively. Ellis (1987, p. 460-461) noted that the four principal clay mineral types could be combined into two types, based on their hydroxyl content. Kaolinite and chlorite have eight hydroxyls, as contrasted with four for smectite and illite. The neutron log is sensitive to this difference, which can be used as one diagnostic guide through comparison of the neutron and density porosities when they are both scaled with respect to a quartz matrix. The photoelectric factor is also a useful clay discriminator because of its control by the aggregate atomic number. Ellis (1987, p. 451-454) pointed out that iron-free aluminosilicate clays would have photoelectric absorption characteristics virtually the same as those of quartz. Therefore variations in the photoelectric factor within shales are primarily a reflection of iron content. There is an overall tendency toward progressive increase in iron, ranging from low values in kaolinite through smectite and illite to high values for iron-bearing chlorite.

These general indicators have been supplemented by measurements from the spectral gamma-ray log. The subdivision of the total gamma-ray flux into its potassium, uranium, and thorium sources has found wide application in shale studies. The elimination of the uranium contribution in a computed gamma-ray (CGR) log results in a trace that is a much more reliable indicator of overall shale content. Quirein et al.

(1982) attempted to use the thorium and potassium measures in a quantitative model to estimate several clay mineral types (Figure 13). They recognized four principal component end members characterized by low potassium–low thorium (non-radioactive matrix minerals), high potassium–low thorium (K-feldspar), moderate potassium–high thorium (illite), and low potassium–high thorium (smectite, kaolinite, chlorite).

The quantitative estimation of clay mineral abundances from the neutron, density, photoelectric factor, and spectral gamma-ray measurements is fraught with difficulties. Wide compositional changes within clay mineral groups pose special problems. Useful quantitative models are not easy to define and their interpretation is frequently ambiguous. The most realistic approach would be to coordinate log measurements with laboratory analyses of core samples. The core values may be idealized as a calibration standard in the development of a statistical prediction model for clay minerals from logs. Even this strategy must be considered thoughtfully and honestly. The most widely used laboratory method for estimating clay mineral quantity is that of X-ray diffraction. Even with careful sample preparation procedures, the error associated with clay mineral content estimated by X-ray diffraction can routinely be expected to be 50% or more of the reported value (Eslinger and Pevear, 1988, p. A-24). Nevertheless, an important result of X-ray diffraction is that at least the appropriate mineral subset can be identified with some confidence. This ensures that the correct components will be selected for compositional analysis from logs. Reconciliation of the log estimates with X-ray diffraction analyses should then be made within a model that attributes appropriate error magnitudes to both data sources.

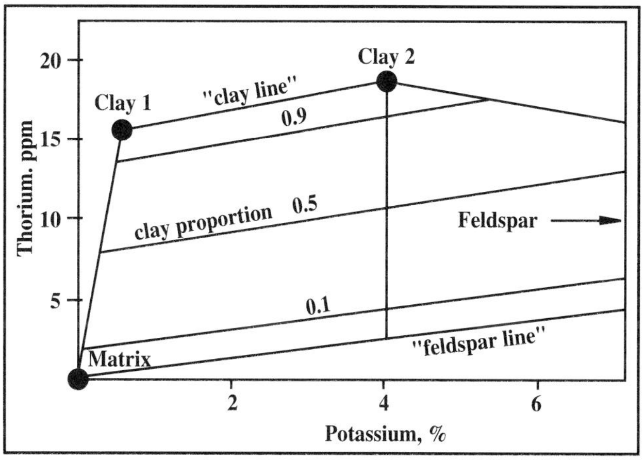

Figure 13. Potassium–thorium model for a four-component system of a low-potassium clay (Clay1), a high-potassium clay (Clay2), potassium feldspar, and matrix (non-radioactive minerals). Adapted from Quirein et al. (1982). Copyright SPE.

COMPOSITIONAL ANALYSIS FROM GEOCHEMICAL LOGS

The geochemical logging tool uses three types of nuclear measurement on a single string to estimate concentrations of ten elements: potassium, thorium, uranium (from the natural gamma-ray spectrum); aluminum (by delayed neutron activation analysis); and silicon, calcium, iron, sulfur, titanium, and gadolinium (from the prompt capture gamma-ray spectrum measured after a 14 Mev neutron burst). A major thrust of research connected with these logs has been aimed at the production of realistic mineral transforms.

"Normative" minerals calculated from oxide analyses have been widely used in igneous petrology since the CIPW (Cross-Iddings-Pirsson-Washington) norm was introduced by Cross et al. (1902). These normative minerals are contrasted with modal compositions, which are those mineral phases actually observed in the rock. The normative concept can be extended to sedimentary sequences in attempts to compute mineral assemblages. Krumbein and Pettijohn (1938, p. 490-492) explained the molecular ratio method, which computes the probable mineral composition of a rock based on chemical analyses of oxide percentages. As a first step, the minerals to be estimated are first identified from thin sections or other sources. The molecular ratios are then assigned in a piecemeal fashion to the minerals. The process is a logical order of steps that first accommodates unique associations between oxides and certain minerals and then allocates the remainder to other components. Imbrie and Poldervaart (1959) described a specific and routine method of sedimentary normative analysis and compared the results with modal estimates of mineralogy. From a detailed study of the Permian Florena Shale Member, they concluded that the estimates of chert, calcite, dolomite, and clay had errors of less than 5%. However, there was little agreement between computed clay mineral proportions and those produced by X-ray diffraction analysis. This discrepancy was not found to be surprising and was attributed to the known high variability of clay mineral compositions caused by isomorphous substitution.

Essentially the same problems facing petrologists are tackled in the computation of sedimentary normative minerals based on the ten elements currently measured on the geochemical logging-tool string (Herron, 1986). However, modern procedures capitalize on the availability of computers in an attempt to improve on the results from the older normative methods. The classical norm calculation is subtractive, deterministic, and rigidly leveraged. As Harvey et al. (1990) pointed out, the method is useful when certain elements can be assigned totally to single individual minerals. The assignations can be used in an ordered protocol of analysis partition between mineral species. Otherwise, simultaneous equations to link mineral compositions with elemental measures provide a much more general and powerful method.

The model that links minerals with elements can

be set up as a fully determined system and solved by standard matrix inversion using methods described earlier. Whenever the components are computed as positive proportions, the compositional solution is rational and honors the analysis perfectly. However, in common with the normative model, any apparent precision read into the result is illusory, because the determined system makes no allowance for analytical error. It is usually practical to model a rock with a set of minerals that is fewer in number than the set of elements available from geochemical logging. The system is then overdetermined and can be resolved by one or other of a variety of optimization techniques. The additional complexity in computation is offset by several distinctive advantages. The overdetermination allows constraints and error functions to be incorporated, both for optimal solution control and diagnostic evaluation of sources of analytical error. The choice of an overdetermined system also provides better assurance of a stable solution in situations where the mineral response matrix becomes sparse or there are potential compositional collinearities that link some of the mineral subsets (Harvey et al., 1990).

Strictly speaking, there will almost always be more minerals than elements, so the problem is usually underdetermined. However, as Herron and Herron (1990) note, the overwhelming majority of sedimentary minerals can be numbered as ten: quartz, four clays, three feldspars, and two carbonates. In practice, reasonable compositional solutions can be generated using relatively small mineral subsets, provided that the minerals have been identified correctly and that the compositions used are both fairly accurate and constant.

Mineral solutions may be calculated by two alternative strategies. In the first, the average chemical compositions of minerals drawn from a large database are used as end member responses and resolved by standard matrix inversion procedures. This result is normative and generic in the sense that it is based on a sample drawn from a universal mineral reference set and applied to a specific sequence in which local mineral compositions may deviate from the global average. The result is hypothetical, but it has the particular advantage that comparisons can be made between a variety of locations and do not require expensive ancillary core measurements. New methods of classification may also be needed in terms both of core and geochemical log data, as discussed by Herron (1988) in his study of terrigenous sands and shales.

In the second strategy, the mineral solution is calibrated to core data and laboratory determinations of mineralogy and elemental geochemistry are analyzed by multiple regression techniques to determine local mineral compositions. The result of this approach is linked to petrography, thus is philosophically closer to an estimated modal solution rather than the more hypothetical normative model. As mentioned earlier, realistic statistical calibration models should incorporate error terms from all sources of measurement.

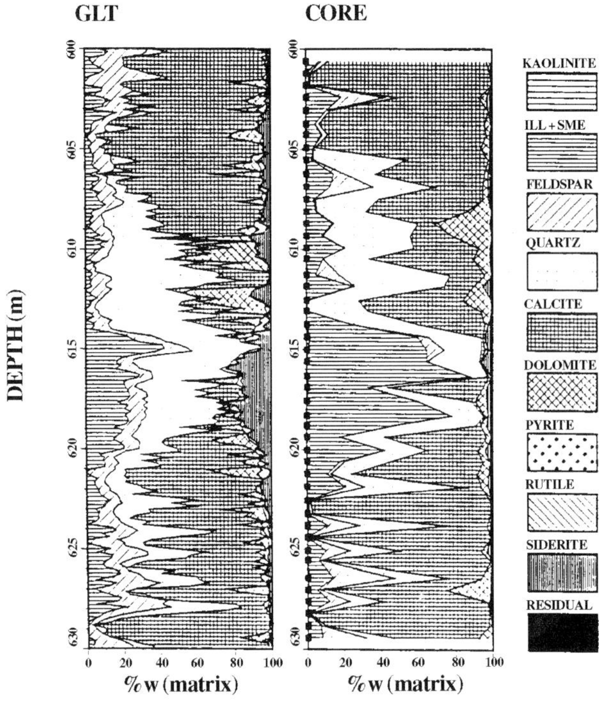

Figure 14. Comparison of mineralogy computed from geochemical logs with core analysis estimation (from van den Oord, 1990).

Several detailed studies have been made to assess the strengths and shortcomings of geochemical logging by exhaustive comparisons of borehole data and core elemental and mineralogical analyses. Examples include comparisons in the Conoco research well in Ponca City, Oklahoma, by Hertzog et al. (1987); the discussion by Wendlandt and Bhuyan (1990) of the results from an Exxon research well which penetrated Upper Cretaceous siliciclastic rocks in Utah; and an assessment of data from three Shell wells in the Netherlands, Oman, and the U.S. by van den Oord (1990). A typical comparative example of core and geochemical log estimates of mineralogy from van den Oord (1990) is shown in Figure 14.

In general, the prognosis is quite good for the infant technology of geochemical logging, particularly with regard to the relatively good match between borehole measurements and elements measured from core in the laboratory. "Teething problems" seem to be confined primarily to the appropriate choice of mineral transform strategies to obtain useful results. Most of the foregoing authors agree that local core calibration is the necessary step, rather than to resort to a generic normative solution. At the same time, it has been recognized that the precise resolution of sedimentary mineral assemblages is an inherently complex problem and that in some respects the technology is ahead of our understanding of the distribu-

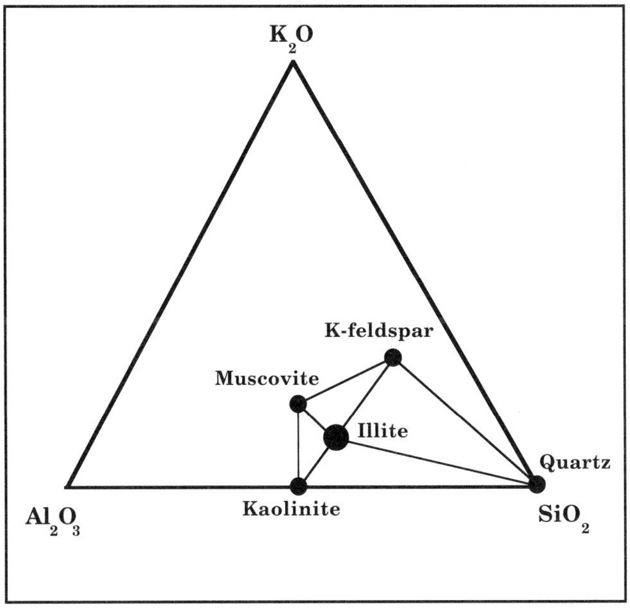

Figure 15. Location of some common sedimentary silicate minerals in the oxide system SiO_2–Al_2O_3–K_2O illustrating problems of compositional collinearity. From Harvey et al. (1992).

tion of elements in sedimentary sequences. So, for example, Wendlandt and Bhuyan (1990) point out that some knowledge concerning the controls on distribution patterns of gadolinium and titanium would be a very useful aid in future work.

A major obstacle to determining unique mineral transformations from element concentrations has been the problem of compositional collinearity. Ambiguities in the separate resolution of illite, mica, kaolinite, and potassium-feldspar by silicon, aluminum, and potassium can be understood when these minerals are plotted on a ternary diagram (Figure 15). Illite is located at a position intermediate between K-feldspar and kaolinite. If it is precisely collinear, then an infinite range of solutions is possible, causing matrix singularity and a breakdown of the inversion procedure. If average mineral compositions are used, a solution becomes possible, but it may be unstable (Harvey et al., 1992). Wendlandt and Bhuyan (1990) found that the use of silicon, potassium, and aluminum tended to result in overestimates of kaolinite; the use of iron to predict illite content caused underestimates of kaolinite. However, effective discrimination between illite and kaolinite contents was possible when dry density was applied as an extra constraint.

There are numerous potential applications of mineral transforms of geochemical logging data in addition to the immediate display of lithofacies types. These include using the minerals as surrogates for other petrophysical properties to obtain quantitative estimates of grain size, cation exchange capacity, and permeability (Chapman et al., 1987). While there may be some differences of opinion about how far these goals have been met, they certainly set forth a worthwhile agenda of research targets. Accurate clay mineral typing and geochemical clues to understand diagenesis have immediate obvious consequences as tools for improving reservoir engineering practices. Selley (1992) considers that the "third age of log analysis" has arrived with geochemical logging and that it can be a useful discriminator of a variety of diagenetic effects of cementation and solution, especially when used in conjunction with measurements from other logging tools.

REFERENCES CITED

Alger, R. P., L. L. Raymer, W. R. Hoyle, and M. P. Tixier, 1963, Formation density log applications in liquid-filled holes: SPE Transactions of the AIME, v. 228, no. 3, p. 321-332.

Anscombe, F.J., 1961, Bayesian statistics: American Statistician, v. 15, no. 1, p. 21-24.

Bacon, L. O., 1948, Formation clay minerals and electric logging: Pennsylvania State College, Mineral Industry Station Bulletin, v. 52, p. 53-75.

Bornemann E., and J. H. Doveton, 1983, Lithofacies mapping of Viola Limestone in south-central Kansas based on wireline logs: AAPG Bulletin, v. 67, no. 4, p. 609-623.

Chapman, S., J. L. Colson, C. Flaum, R. C. Hertzog, G. Pirie, H. Scott, B. Everett, M. M. Herron, J. S. Schweitzer, J. La Vigne, J. Quirein, and R. Wendlandt, 1987, The emergence of geochemical well logging: The Technical Review, v. 35, no. 2, p. 27-35.

Cross, W., J. P. Iddings, L. V. Pirsson, and H. S. Washington, 1902, A quantitative chemico-mineralogical classification and nomenclature of igneous rocks: Journal of Geology, v. 10, p. 555-690.

Doveton, J. H., 1986, Log analysis of subsurface geology—concepts and computer methods: New York, John Wiley & Sons, 273 p.

Doveton, J. H., and E. Bornemann, 1981, Log normalization by trend surface analysis: The Log Analyst, v. 22, no. 4, p. 3-8.

Doveton, J. H., and H. W. Cable, 1979, Fast matrix methods for the lithological interpretation of geophysical logs, *in* D. Gill, and D. F. Merriam, eds., Geomathematical and petrophysical studies in sedimentology: Oxford, Pergamon Press, Computers in Geology Series, p. 101-116.

Ellis, D. V., 1987, Well logging for Earth scientists: New York, Elsevier, 532 p.

Eslinger, E., and D. Pevear, 1988, Clay minerals for petroleum geologists and engineers: SEPM Short Course Notes No. 22, SEPM, Tulsa, Oklahoma, 410 p.

Griffiths, J. C., 1952, Grain-size distribution and reservoir-rock characteristics: AAPG Bulletin, v. 36, no. 2, p. 205-229.

Gysen, M., C. Mayer, and K. H. Hashmy, 1987, A new approach to log analysis involving simultaneous optimization of unknowns and zoned parameters, *in* Transactions 11th Formation Evaluation Symposium, Paper B: Canadian Well Logging Society, 20 p.

Harris, M. H., and R. B. McCammon, 1971, A computer-oriented generalized porosity-lithology interpretation of neutron, density and sonic logs: Journal of Petroleum Technology, v. 23, no. 2, p. 239-248.

Harvey, P. K., J. F. Bristow, and M. A. Lovell, 1990, Mineral transforms and downhole geochemical measurements: Scientific Drilling, v. 1, no. 4, p. 163-176.

Harvey, P. K., J. C. Lofts, and M. A. Lovell, 1992, Mineralogy logs: Elements to mineral transforms and compositional colinearity in sediments: Transactions of the SPWLA 33rd Annual Logging Symposium, Paper M, 18 p.

Herron, M. M., 1986, Mineralogy from geochemical well logging: Clays and Clay Minerals, v. 34, no. 2, p. 204-213.

Herron, M. M., 1988, Geochemical classification of terrigenous sands and shales from core or log data: Journal of Sedimentary Petrology, v. 58, no. 5, p. 820-829.

Herron, M. M., and S. L. Herron, 1990, Geological applications of geochemical well logging, *in* A. Hurst, M. A. Lovell, and A. C. Morton, eds., Geological applications of wireline logs: Geological Society of London, Special Publication 48, p. 165-175.

Hertzog, R., L. Colson, B. Seeman, M. O'Brien, H. Scott, D. McKeon, J. Grau, D. Ellis, J. Schweitzer, and M. Herron, 1989, Geochemical logging with spectrometry tools: SPE Formation Evaluation, v. 4, no. 2, p. 153-162].

Heslop, A., 1974, Gamma ray log response of shaly sandstones: Transactions of the SPWLA 15th Annual Logging Symposium, Paper M, 11 p.

Hurd, B. G., and J. L. Fitch, 1959, The effect of gypsum on core analysis results: Journal of Petroleum Technology, v. 11, no. 9, p. 221.

Imbrie, J., and A. Poldervaart, 1959, Mineral compositions calculated from chemical analyses of sedimentary rocks: Journal of Sedimentary Petrology, v. 29, no. 4, p. 588-595.

King, D. E., and J. Quirein, 1987, Systematic incorporation of geologic data in the log interpretation process with applications to lithology analysis, *in* Transactions 11th Formation Evaluation Symposium, Paper N: Canadian Well Logging Society, 14 p.

Krumbein, W. C., and F. J. Pettijohn, 1938, Manual of sedimentary petrography: New York, Appleton-Century-Crofts, 549 p.

Marett, G., and S. Kimminau, 1990, Logs, charts, and computers—the history of log interpretation modelling: The Log Analyst, v. 31, no. 6, p. 335-354.

Mayer, C., and A. Sibbit, 1980, GLOBAL: A new approach to computer-processed log interpretation: SPE Paper 9341, 55th Annual Fall Technical Conference and Exhibition, Dallas, Texas, 12 p.

McCammon, R. B., 1970, Component estimation under uncertainty, *in* D. F. Merriam, ed., Geostatistics, a colloquium: New York, Plenum, p. 45-61.

Moss, B., and R. Harrison, 1985, Statistically valid log analysis method improves reservoir description: Society of Petroleum Engineers, Offshore Europe Conference, Aberdeen, SPE 13981 preprint, 32 p.

Quirein, J. A, J. S. Gardner, and J. T. Watson, 1982, Combined natural gamma ray spectral litho-density measurements applied to complex lithologies: SPE Paper 11143, 57th Annual Fall Meeting, New Orleans, p. 1-14.

Quirein, J., S. Kimminau, J. Lavigne, J. Singer, and F. Wendel, 1986, A coherent framework for developing and applying multiple formation evaluation models: Transactions of the SPWLA 27th Annual Logging Symposium, Paper DD, 16 p.

Savre, W. C., 1963, Determination of a more accurate porosity and mineral composition in complex lithologies with the use of the sonic, neutron, and density surveys: Journal of Petroleum Technology, v. 15, no. 6, p. 945-959.

Schmoker, J. W., and C. J. Schenk, 1988, Facies composition calculated from the sonic, neutron, and density log suite, upper part of the Minnelusa Formation, Powder River Basin, Wyoming: The Mountain Geologist, v. 25, no. 3, p. 103-112.

Selley, R. C., 1992, The third age of wireline log analysis—application to reservoir diagenesis, *in* A. Hurst, C. M. Griffiths, and P. F. Worthington, eds., Geological applications of wireline logs II: The Geological Society of London, Special Publication 65, p. 377-387.

Tixier, M. P., and R. P. Alger, 1967, Log evaluation of non-metallic mineral deposits: Transactions of the SPWLA 8th Annual Logging Symposium, Paper R, 22 p. [later published in 1970 in Geophysics, v. 35, no. 1, p. 124-142].

van den Oord, R. J., 1990, Experience with geochemical logging: Transactions of the SPWLA 31st Annual Logging Symposium, Paper T, 25 p.

Wendlandt, R. F., and K. Bhuyan, 1990, Estimation of mineralogy and lithology from geochemical log measurements: AAPG Bulletin, v. 74, no. 6, p. 837-856.

Yaalon, D. H., 1962, Mineral composition of average shale: Clay Minerals Bulletin, v. 5, no. 27, p. 31-36.

Chapter 4

Multivariate Pattern Recognition and Classification Methods

The compositional solution methods of the previous chapter are "deductive" or "top-down" in their operation. In each case, a model is specified that links log responses with components. An algorithm is then followed to resolve the volumetric proportions of the components. The user dictates both the model and the solution procedure in a controlled method that deduces the composition from the data. The algorithm generally incorporates functions to generate error diagnostics. Diagnostics can be designed to alert the user to those situations where the data characteristics suggest that the model is flatly wrong. However, there will still be times when an apparently satisfactory solution is mathematically optimal but geologically incorrect. The spurious solution is not detected mathematically because it is not inconsistent with the log-response data. Typically, the problem is caused by the poor model decisions of the user, such as incorrect selection of components or entry of erroneous log parameters.

In an alternative philosophy, methods can be applied that "allow the data to speak for themselves." These techniques are "inductive" in operation; they are "bottom-up," because they originate with the observational data rather than being driven top-down by a deductive model. The two approaches represent extreme philosophic positions. In practice, investigators commonly use a hybrid strategy in which they use elements of both paradigms. The purpose of most inductive methods is to isolate distinctive patterns and develop classification procedures that will differentiate these patterns. The geological "meaning" of the patterns can either be inferred from their physical properties or defined by external information from core samples or cuttings. The patterns that are isolated can themselves be used as components of deductive models, although some care must be taken to avoid closed loops of reasoning.

Two distinct types of pattern-recognition and classification procedures should be recognized. In a "supervised" method, different categories or groups are specified before the analysis; the goal of the method is to find the best function to distinguish the categories based on the characteristics of the data. Subsequently, the function can be used to classify unknown observations on the basis of likelihood of membership in one or other of the groups. An "unsupervised" method is given no prior information concerning group membership of individual observations. Instead, the method is designed to analyze the intrinsic structure of the data, most commonly expressed as a cloud of points in the hyperdimensional space of multivariate log measurements. Trends and clusters revealed by these techniques are assigned geological meaning according to general principles or drawn from observations from data where available.

In much of this chapter we will discuss some of the most common mathematical techniques used for multivariate pattern recognition and classification. These methods did not develop within petrophysics or geology, but have been borrowed from a variety of scientific disciplines. Their mathematical antecedents have a distinguished pedigree of development predating the advent of computers. Today, these same methods are included as standard options in multivariate statistical software packages that are widely available on microcomputers as well as mainframe machines. Consequently, they can be applied easily by many readers of this book by downloading data files from conventional log-analysis packages and linking them with multivariate statistical programs. The results of analysis can then be transferred back and incorporated with the more conventional log-analysis operations. For readers unfamiliar with these methods, the widely used textbook written by Davis (1986) is recommended reading as a useful source of supplementary material on geological data analysis.

ELECTROFACIES RECOGNITION, CLASSIFICATION, AND PREDICTION

The term "electrofacies" was introduced by Serra and Abbott (1980) as a means of characterizing collective associations of log responses that are linked with geological attributes. An electrofacies is defined as "the set of log responses which characterizes a bed and permits it to be distinguished from the others."

Electrofacies are clearly determined by geology, because log responses are measurements of the physical properties of rocks. Electrofacies can usually be assigned to one or another lithofacies type, although the correspondences are often blurred. For example, some lithofacies when seen in outcrop or core may have distinctively different aspects, such as color, that do not cause appreciable differences in logging properties. Conversely, a single lithofacies may be matched with several electrofacies in cases where there are significant changes in pore volume or geometry that are not obvious in hand specimen. These types of electrofacies are generally linked more closely with "petrofacies" (e.g. see Kopaska-Merkel and Friedman, 1989), where distinctions are made at the petrographic level.

An important philosophical distinction between electrofacies and "classical" geological facies is that electrofacies are primarily observational in origin and facies are traditionally rooted in genesis. So, for example, Reading (1978) states, "A facies should ideally be a distinctive rock that forms under certain conditions of sedimentation reflecting a particular process or environment." This concept is driven by a top-down model that attempts to classify facies as process products. The aim is made possible by comparative studies of modern sediments, in which future rock types can be identified with specific depositional environments. Electrofacies are generally distinctive empirical log-response associations that must be linked with a more conventional rock terminology. Hopefully, the electrofacies classification will also reflect patterns of deposition and diagenesis, because these define the historical development and geometrical expression of reservoir units and associated rocks.

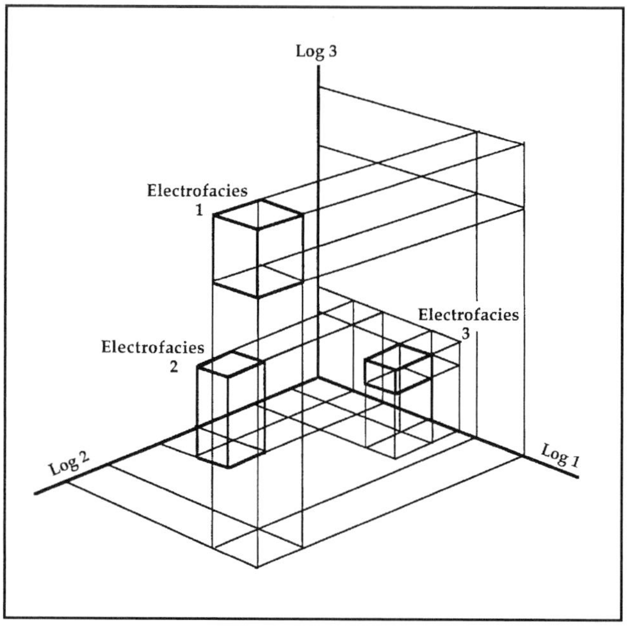

Figure 1. Representation of electrofacies in three-log space as rectangular volumes, bounded by minimum and maximum log values. Adapted from Serra and Abbott (1980). Copyright SPE.

MULTIVARIATE REPRESENTATIONS OF ELECTROFACIES

In Chapter 2, glyphs of "spider webs" and "ladders" were shown as an alternative means to condense the multilog signature of an electrofacies into two-dimensional graphic forms. However, glyphs are only crude cartoon representations and their use is restricted to simple, visual comparisons. In more systematic work, the electrofacies should be mapped out in the dimensional framework set by the log measurements. If three logs are used, then electrofacies can be located in a Cartesian coordinate system whose orthogonal axes are the logs (Figure 1). The specification of maximum and minimum log values for each electrofacies results in a rectangular volume for each electrofacies.

In practice most electrofacies are more closely approximated by a cloud of points which are relatively concentrated in the center and diffuse in density towards the margin. A multivariate normal distribution provides a convenient means to model such a distribution efficiently. It also provides the basis for a probability model that allows statistical classifications to be made from the electrofacies database. In some cases, the normal distribution could be arguably be the theoretical expectation. For example, beds of pure anhydrite should ideally be represented by a single point. In practice, the combination of tool errors and mineral impurities will cause a constricted cloud to be focused on the ideal anhydrite point. The compounded effect of these deviations will generally approximate a normal distribution.

For many electrofacies, there may be no intrinsic reason for the data points to be normally distributed about their mean. However, they commonly appear to be normal, probably as a result of compounded random measurement errors and independent systematic deviations. In other cases, obviously asymmetrical shapes can be normalized effectively through scale transformations such as a logarithmic conversion; extended clouds may be partitioned into smaller clusters separated by relatively diffuse regions. If approximately normal, then the expected density of points at any coordinate location can be specified completely by the statistics of the multivariate mean and the matrix of variances and covariances between the logs. The vector of mean values gives the location of the center of the cloud; the variance–covariance matrix gives its relative degree of dispersion and orientation within the multivariate log space.

The multivariate normal cloud is hyperellipsoidal in shape, but has no discrete surface because the normal distribution is continuous in all directions. However, the shape of the distribution ensures that the majority of points are confined to within a few standard deviations of the cloud centroid. Conventional representations of the ellipsoid equate the 95% probability contour with the outer boundary. Beyond this surface, any point has a rapidly decreasing likelihood of being drawn from the ellipsoid population. The situation can be visualized fairly easily in two dimensions (see Figure 2) and the concepts are equally applicable to higher dimensions. Many electrofacies are not normally distributed, as can be seen on crossplots of logs taken two at a time. Nonnormal data-point clouds are recognized by marked asymmetries or shapes that clearly are not elliptical. However, the primary purpose of the normal ellipsoids is to capture the essential spatial distribution character of the data-point cloud. Judicious subdivisions of electrofacies types and scale transformations of selected logs can be applied to the clusters that make up the totality of a logged succession. By these means, the clusters can generally be represented adequately by a set of multivariate normal parameters for each electrofacies.

There are three different methods to define electrofacies for entry into a reference-set database: the theoretical, empirical, and interpretive approaches (Delfiner et al., 1987). In practice, the database is built using all three procedures as a mix of predictions from tool-response equations and interpretations from crossplots integrated with core observations. The general database of Delfiner et al. (1987) was composed of 30 sandstones, 25 shales, 30 limestones, 25 dolomites, 25 evaporites, 3 coals, 10 igneous rocks, and 4 miscellaneous rocks. Local databases can be designed to include unusual lithologies and to fine-tune electrofacies parameters to specific rock types. So, for example, Stowe and Hock (1988) developed a Zechstein database for applying classification procedures to the Permian gas-bearing formations in northern Germany. Their Zechstein reference set consisted of electrofacies for 48 carbonates and 24 evaporites.

The theoretical approach to electrofacies definition is that of forward modeling, described briefly in the last chapter and reviewed in detail by Quirein et al. (1986). The determined-system solution of compositional proportions was obtained by inversion of the log-response equations. The forward model is the reverse of this process. The log responses of an electrofacies are predicted by multiplying its component log properties by the proportions with which they occur in an equivalent lithofacies:

$$\mathbf{CV} = \mathbf{L}$$

where **C** is the matrix of component properties, **V** is a vector of component proportions within the lithofacies, and **L** is the vector of electrofacies log responses. The lithofacies compositions can come from a variety of sources, such as generalized estimates from sedimentary geology textbooks (e.g. Pettijohn et al., 1972) and be convolved with mineral logging parameters (such as those listed in Edmundson and Raymer, 1979). Alternatively, electrofacies can be created from local lithofacies determined by core analyses or based on other sources of geological information. These customized electrofacies accommodate local variability,

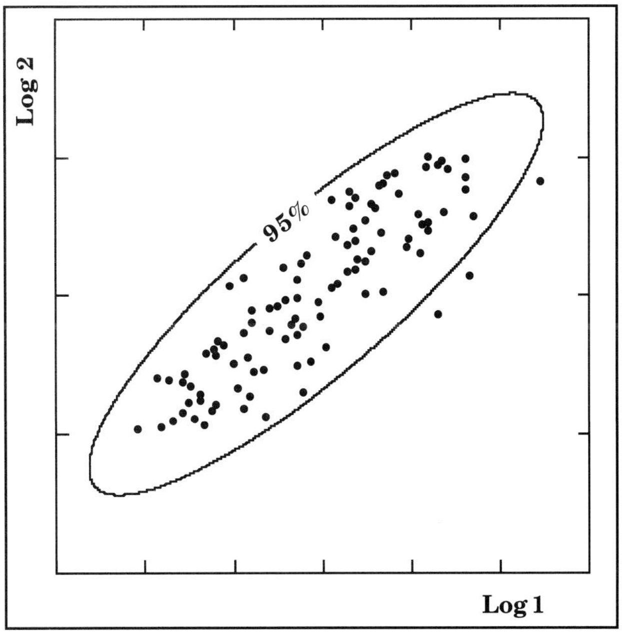

Figure 2. Representation of hypothetical electrofacies by 95% density contour of a bivariate normal ellipse in terms of two logs set as orthogonal axes.

and the added precision inspires more confidence in the results of subsequent classifications. In all cases, the analysis ranges, porosity distributions, accessory-mineral influence, and data-acquisition errors are incorporated in the computation of descriptive normal ellipsoids. Using the terminology developed earlier, this theoretical type of electrofacies can be seen to be deductive and model driven.

The empirical approach relates lithofacies observed in core to the set of log responses measured over the cored interval. The log-response statistics of means, variances, and covariances then describe hyperdimensional ellipsoids for each electrofacies and are tagged to a specific lithofacies. This process of creation can be viewed as inductive, but supervised. The major stumbling block to this approach is economic. Sections that have been both cored and logged by a full suite of tools usually form a minor component of the total information available. More typical are sequences that have been logged with geological data restricted to descriptions of cuttings. Electrofacies definition is then a pattern-recognition process based largely on interpretation of log character, supplemented by geological knowledge.

The interpretive approach to electrofacies generation is based on the examination of log crossplots for clusters that can be identified geologically or validated by cores or cuttings. On each cluster an ellipse is circumscribed that is set by the range on each log and the log-pair correlation coefficient for the cluster. Histograms and Z-plots are also used for the location of ellipse boundaries (Delfiner et al., 1987). However, in practice, the edge of an ellipse is usually chosen to be an informal estimate of the 95% probability density contour. Crossplots for all possible pairs of logs are examined to determine the parameters of the multivariate ellipsoid.

Automatic zonation of logging data is a particularly helpful preparatory step for the discrimination of electrofacies. When strings of logging data are sampled at a standard frequency of two readings per foot the transitional curve features are captured, along with the peak and trough values that are more representative of the zones. The continuous trace that is sampled by the digital data can be reduced substantially by replacing the original curve with a stepped function. This process is known variously as "blocking," "squaring," "segmenting," or "lumping" the logs. An automatic computer operation emulates the manual procedure of the traditional log analyst who subdivides the blue-line log into "zones" matched with distinctive peaks, troughs, and shoulder features. Zone boundaries are marked at depth points coinciding with curve inflections between features. Systematic elements must be discriminated from random fluctuations caused by logging-tool measurement error. To ensure compatibility when zoning a log suite, the scale of subdivision should be tied to the tool with the coarsest resolution.

A number of different mathematical procedures have been proposed for zoning. Serra and Abbott (1980) used a simple error range to differentiate log shifts that could be attributed to actual formation changes. Kerzner and Frost (1984) developed a blocking algorithm that allowed the stipulation of the length of a filter window and a noise-level cutoff (Figure 3). Lanning and Johnson (1983) described the use of square-wave Walsh functions (see Chapter 5) as a means to ramp logging data. Elek (1988) zoned a multiple log suite by first condensing the traces to a single log through the computation of the first principal component. He then applied a median filter to a log of the first principal component scores. Moline et al. (1992) also used principal component analysis, but segmented the scores log by a maximum likelihood estimation algorithm introduced by Mehta et al. (1990). In yet another approach to the process of blocking logs, Runge and Runge (1991) have utilized neural network concepts of simulated annealing.

Clearly there is a variety of different ways to block logs that should give substantially similar results, although each differs in its ability to handle log noise and the characteristics of the tool-response function. Most of the methods require some decisions by the user concerning appropriate filter-length and log-noise characteristics. So, the success of a blocking operation is often decided by the user's skill in the selection of parameters. The sequencing of zones in blocked logs can lead to significant reduction in database size. By restricting crossplots to zoned logging data, extraneous transitional changes are discarded, while local clusters are accentuated (Figure 4).

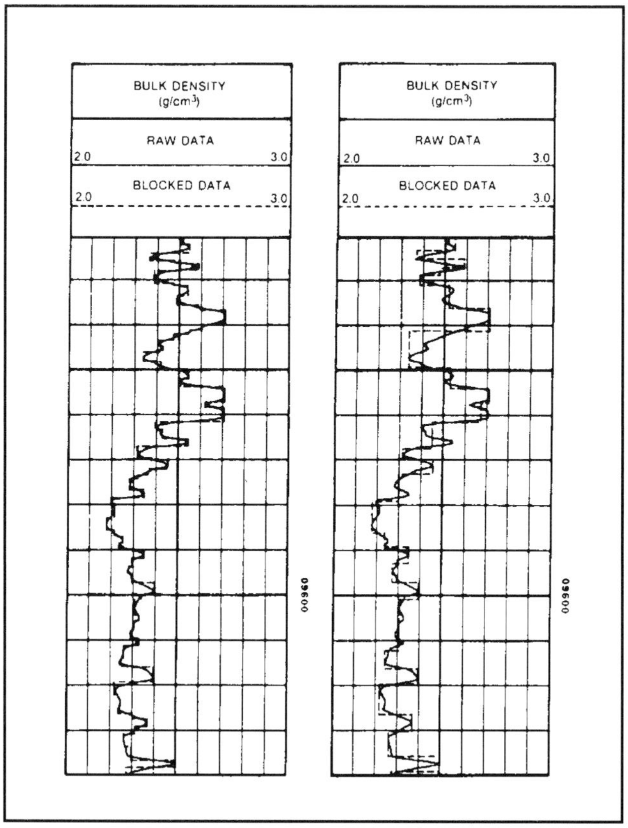

Figure 3. Example of a density log blocked by a 2-ft filter window with a noise-level cutoff of 10% (left) and a 5-ft filter window with a noise-level cutoff of 30% (right). Depth lines marked at 10-ft spacing. From Kerzner and Frost (1984). Copyright SPE.

The interpretive approach can be highly labor-intensive if electrofacies are built by examining all possible crossplots. For example, using five logs, Stowe and Hock (1988) analyzed some 4500 crossplots in the construction of a Zechstein carbonates database of 72 electrofacies. The number of crossplots is partly contingent on the number of logs and can be calculated as the number of combinations of n logs, taken a pair at a time:

$$N = \frac{n!}{(n-2)!\,2!}$$

For five logs there are ten possible crossplots which represent ten alternative and orthogonal views of the hyperdimensional data clouds. If the number of logs is expanded to 11, then the total set of crossplots expands dramatically to 55. This problem is commented on briefly by Serra et al. (1985), who bluntly call it "the curse of dimensionality." On one hand, a sufficient number of logs is required to distinguish between electrofacies with minimal ambiguity. On the other hand, the multidimensional space created by a large number of logs becomes difficult to handle by traditional methods.

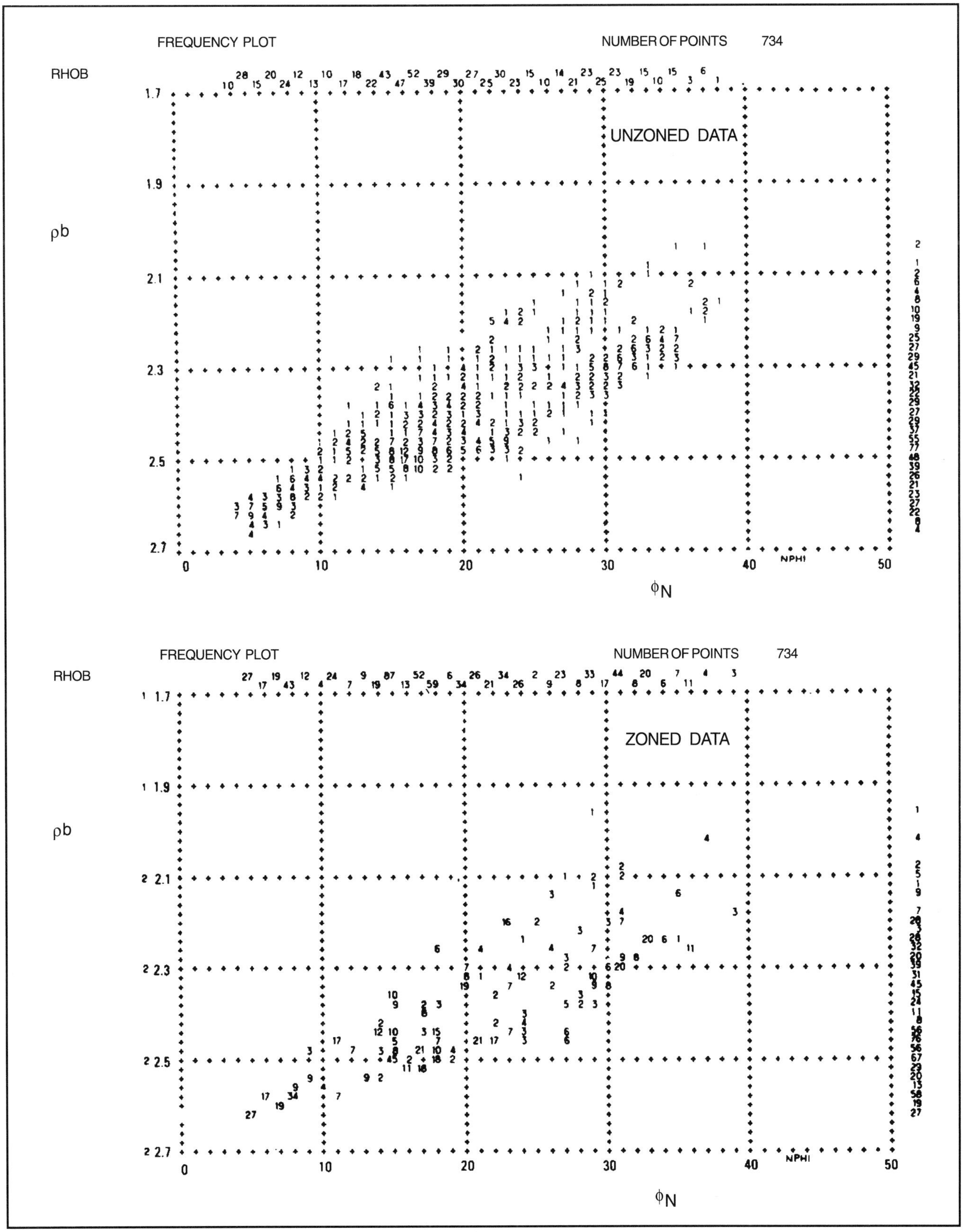

Figure 4. Resolution of electrofacies clusters through use of zoned logs demonstrated by the contrast between unzoned (above) and zoned (below) data on a neutron-density crossplot. From Serra and Abbott (1980). Copyright SPE.

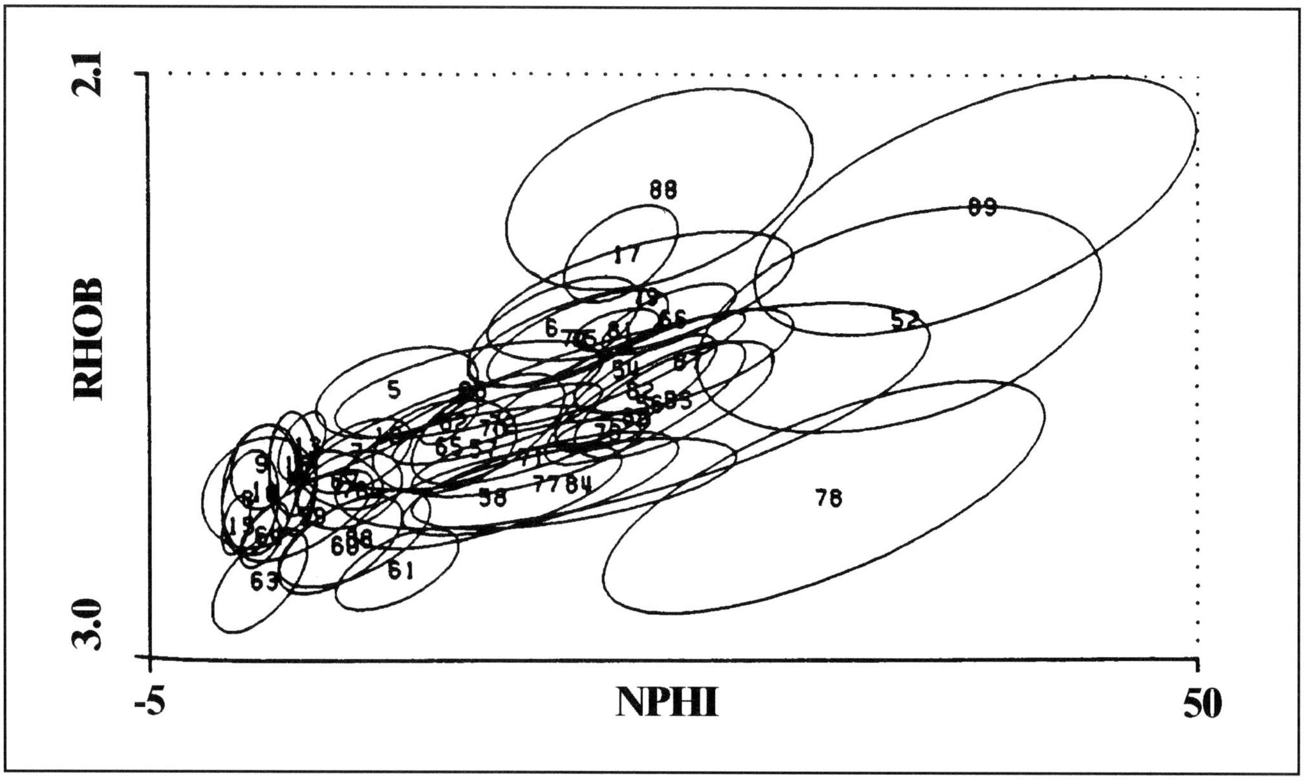

Figure 5. Density-neutron crossplot of Zechstein carbonate electrofacies. From Stowe and Hock (1988).

Wolff and Pelissier-Combescure (1982) described a strategy to circumvent this dilemma through the use of principal component analysis (PCA). This method is described in some detail later in this chapter, but the basic ideas are touched on here. Although the locations of data points in multilog space collectively delineate clouds with the same number of dimensions as logs, they can often be mapped effectively in a much reduced dimensionality. This is because intercorrelations between the log variables cause the clouds to be extended along certain trends. Principal component analysis computes an ordered set of orthogonal axes that absorb the variation in a systematic manner. The first two principal components describe a plane that contains the maximum variability of all possible planes. Typically, this will be a plane that cuts through the cloud obliquely, and will be an improvement on the traditional crossplot views that are fixed to be parallel or perpendicular to the original axes.

A crossplot of the first two principal components often accounts for most of the variability, so that only a minor amount is lost by neglecting the remaining principal components. Systematic local concentrations of data can then be located in this reduced space through the use of cluster analysis methods. As implemented in the method described by Wolff and Pelissier-Combescure (1982), the clustering is run as two passes. In the first phase, the clustering is used to isolate local modes as zonal representatives of the digital data. In the second phase, the local modes are agglomerated into clusters that are equated with electrofacies. Decisions concerning potential cluster subdivision or fusion are monitored by referral to geological information from core or from geological experience.

CLASSIFICATION OF UNKNOWN ZONES USING AN ELECTROFACIES DATABASE

Once the electrofacies database is established, it can be used as a means to classify zones, based solely on their log responses. The problem becomes that of allocating a multivariate coordinate location to one or another of a number of electrofacies ellipsoids. In many cases there will be overlap between the ellipsoids, although the potential conflicts in multivariate space will be fewer than those suggested in a two-dimensional crossplot projection (Figure 5). It should also be recalled that the boundaries of the ellipses are 95% density contours—a simple convention to represent what are really diffuse clouds of points. Consequently, the allocation of a zone to any electrofacies is a matter of probability.

The multivariate normal distribution associated with each electrofacies is infinite in all directions, although realistic probability contours are more localized around the electrofacies centroid. Consequently all zone log-response coordinates have a finite probability of belonging to any of the electrofacies. The probability of observing a set of log responses, *L*,

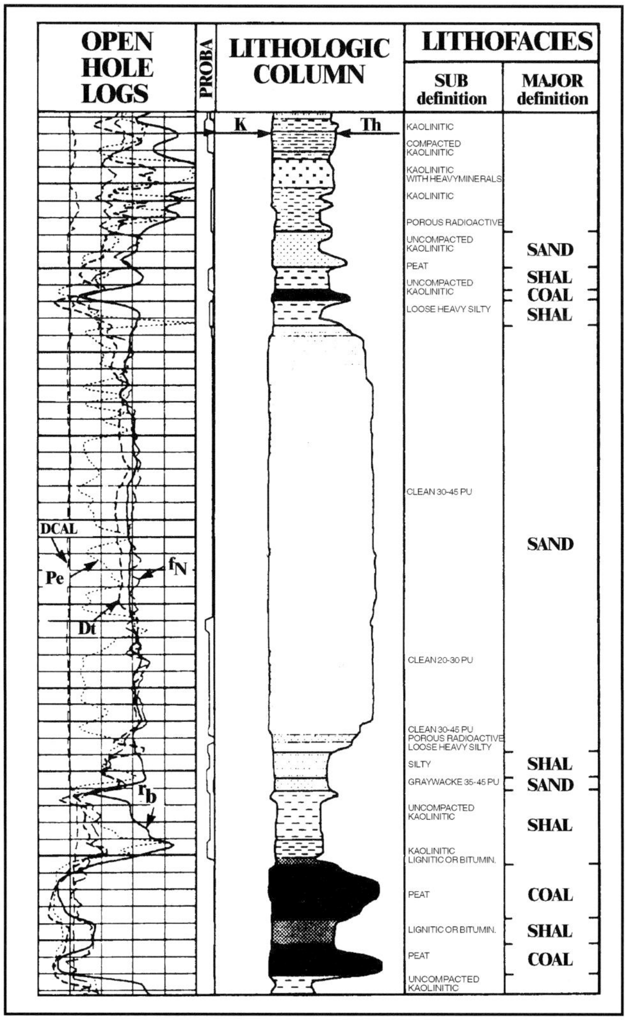

Figure 6. Example of lithofacies sequence in a sandstone–shale–coal predicted from logs using classifications drawn from an electrofacies database. From Delfiner et al. (1987). Copyright SPE.

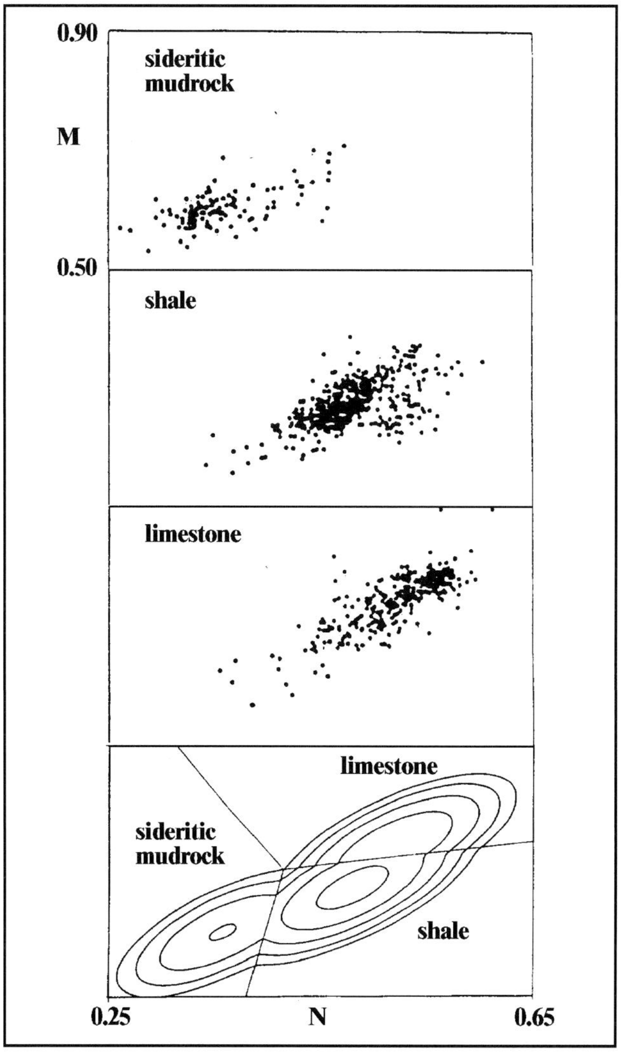

Figure 7. *M-N* crossplots of sideritic mudrock, shale, and limestone lithofacies from the Shublik Formation, together with *M-N* plot showing the bivariate normal densities and the classification partitions of the three lithofacies. From Busch et al. (1987). Copyright SPE.

given an electrofacies *i*, is symbolized as:

$$P(L|F_i)$$

and is given by the normal probability distribution structure of the *i*th electrofacies hyperellipsoid. But the actual problem is the reverse: What is the probability of being a product of electrofacies *i*, *given* a set of log responses, *L*? This posterior probability is given by Bayes' theorem as:

$$P(L|F_i) = \frac{p_i \cdot P(L|F_i)}{\Sigma p_j \cdot P(L|F_j)}$$

where p_i is the prior probability of electrofacies *i*. The prior probabilities are determined by external geological experience and are an important means for excluding lithologies that do not occur in the analytical sequence or for weighting electrofacies that occur particularly commonly. Posterior probabilities of zero value are an immediate means to extract a subset of relevant electrofacies from a large database and so speed computations. Systematic prior probabilities can be estimated from counts of electrofacies types within comparable control wells or from other sources. In the event of profound ignorance, equal prior probabilities can be assigned to all reasonable electrofacies.

The classification step follows from the computation of the posterior probabilities, with electrofacies assignment dictated by the maximum probability value. The probability figure gives the degree of confidence associated with the classification decision. In

Table 1. Cross-classification Best Wireline Log Model Applied to Validation Data Set for Prediction of Shublik Formation Lithologies (from Busch et al., 1987).

Core Lithology	Sandstone	Shale	Limestone	Siltstone	Phosphatic limestone	Phosphatic mudrock	Sideritic mudrock
Sandstone							
Shale	3	233	45	18	2	4	5
Limestone	3	61	374	30	13	1	4
Siltstone		12	45	64			
Phosphatic limestone			16		115	6	
Phosphatic mudrock					25	88	5
Sideritic mudrock		3	2			13	74

Percent agreement = 75.00%

cases where the zone falls outside the 95% limits of all electrofacies ellipsoids, the zone is normally considered to be "unidentified." After each zone is allocated to one or another of the electrofacies, the result can be graphed in a format that emulates a conventional geological column (see Figure 6). It is common practice to use either the gamma-ray log or the thorium and potassium traces as boundary curves on the strip-log presentation. This additional option provides a generalized shaliness curve as a supplementary and continuous record of information. The format also tends to mimic a weathering profile as would be seen in outcrop because of the broad relationship between shaliness and rock competence. This coincidence is a visually appealing aid to traditional geologists in their attempt to comprehend results of complex computer analyses.

The probabilistic classification of unknown observations to one or another class represented by a multivariate normal distribution is a common statistical technique used in discriminant analysis. Clearly, a cut-and-dried decision on electrofacies type would be preferable to one phrased in terms of probability. However, there frequently is some degree of overlap between electrofacies clouds; it is in these regions that some uncertainty regarding the correct classification of unknown zones exists. This situation is especially true when electrofacies are defined empirically from lithofacies observed in core. Readily distinctive lithofacies may be poorly differentiated by common log properties. If they are indistinguishable, then they fail to be "electrofacies" by the original definition of Serra and Abbott (1980). In reality, the electrofacies signatures of most lithofacies can be used for classification, provided that several logs are used and that some latitude is allowed for a minor degree of uncertainty.

It is common practice to develop a model for reservoir description and engineering purposes based on available core descriptions and measurements. The costs of core recovery and analysis are generally higher than those of wireline logging and analysis. Consequently, a common objective of log interpretation is to predict the rock properties that would be seen in the cores, but must be inferred from the logging curves. The logging data from the cored wells constitute a "training set," from which a multivariate statistical method can "learn" the relationships between logs and core descriptors. A satisfactory classification function can then be applied to the analysis of other wells that have been logged but not cored.

A useful case study of this approach is described by Busch et al. (1987) who applied it to the prediction of rock types within the Shublik Formation of the Prudhoe Bay field in Alaska. They illustrated their method by a simplified example in which they differentiated Shublik limestones, shales, and sideritic

$$
\begin{array}{c}
 \\ E_1 \\ E_2 \\ E_3 \\ \\ \\ \\ E_m
\end{array}
\begin{array}{c}
\begin{array}{ccccc} E_1 & E_2 & E_3 & E_4 & \quad\quad E_m \end{array} \\
\left[\begin{array}{cccccc}
p_{11} & p_{12} & p_{13} & p_{14} & \cdots\cdots & P_{1m} \\
p_{21} & p_{22} & p_{23} & p_{24} & \cdots\cdots & P_{2m} \\
p_{31} & p_{32} & p_{33} & p_{34} & \cdots\cdots & P_{3m} \\
\cdots & \cdots & \cdots & \cdots & \cdots & \cdots \\
\cdots & \cdots & \cdots & \cdots & \cdots & \cdots \\
\cdots & \cdots & \cdots & \cdots & \cdots & \cdots \\
p_{m1} & p_{m2} & p_{m3} & p_{m4} & \cdots\cdots & P_{mm}
\end{array}\right]
\end{array}
$$

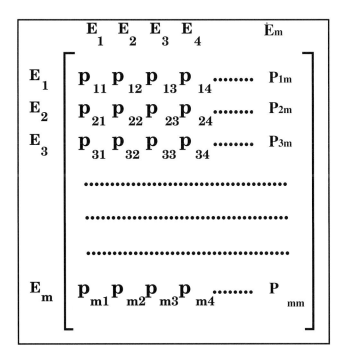

Figure 8. Diagrammatic representation of a transition probability matrix for a set of *m* electrofacies. The transition probability, p_{ij} is the probability that electrofacies *j* overlies electrofacies *i*.

mudrocks based on the lithoporosity parameters, *M* and *N* (see Chapter 2). The separate crossplots of the *M* and *N* values for these three lithofacies are shown in Figure 7. By computing the bivariate means, variances, and covariances, a normal probability density function can be mapped for each lithofacies. The *M-N* crossplot space can then be subdivided into exclusive regions assigned to each lithofacies (Figure 7). The probability contours and lithofacies decision boundaries developed from logs in cored wells can then be applied to lithofacies prediction in uncored wells.

Up to this point, the case-study methodology follows essentially the same logic as the automated electrofacies methods of Delfiner et al. (1987) and others described previously. However, rather than being developed from the pattern recognition of separable clusters on log crossplots, the database was tailored for the Prudhoe Bay Shublik and was tied to lithofacies. In addition, Busch et al. (1987) divided the training data set into a calibration subset, used for prediction, and a validation data set, used to test the prediction performance independently and to make comparisons between different models. In the fully developed model they attempted to differentiate seven different lithofacies, based on a number of log-derived variables. The data in Table 1 show the classification results from the best model applied to the validation set. These results were used to investigate the causes of erroneous predictions which accounted for 25% of the total predictions. In the majority of cases, it was found that misclassifications were caused by depth discrepancies between core records and logging data. Otherwise, ambiguities were genuine, reflecting similarities of lithofacies or gradational trends between them or interbedding at a relatively fine scale.

Accurate classification using wireline logs to distinguish the wide variety of potential lithofacies and their possible textural subdivisions is a challenging task for any computer program. Major improvements in performance are possible through the incorporation of additional geological information to resolve ambiguities or to censor geologically unreasonable choices. One important source of information is the sequence position of the electrofacies with respect to other electrofacies. Thus Busch et al. (1987) included a variable they called "proportional height in zone," which captured and incorporated the tendency for certain lithofacies to be near the top of a zone, while others tended to be close to the bottom. Other work has been done in this area by Berteig et al. (1985), who incorporated contextual measures based both on lithofacies occurrence and log autocorrelation.

Vertically adjacent lithologies have a strong interdependence that is ultimately a consequence both of Walther's Law and the repetitive nature of sedimentary units. Walther (1894) was one of the first to realize that facies that succeed one another conformably must have been deposited in adjacent environments. This concept, known as Walther's Law, was summarized by Selley (1976): "a conformable vertical sequence of facies was generated by a lateral sequence of environments." Because empirical electrofacies are the petrophysical expression of lithofacies, they are subject to the same constraints. It follows that, in a conformable sequence, certain electrofacies will never be expected to succeed others. Up to this point, we have considered the classification of any unknown zone in isolation. However, Walther's Law suggests that the electrofacies nature of the immediately preceding zone is an important piece of information. This could be incorporated easily as the prior probability term of the Bayesian equation described earlier. The prior probabilities would then utilize the probability of electrofacies assignment of a zone, given the electrofacies classification of the preceding zone.

Posterior probabilities based on sequence position are easily generated by Markov chain analysis (see Chapter 5 for a more extensive treatment). Based on data from control wells, a tally matrix of observed vertical transitions between electrofacies types can be converted to a matrix of transition probabilities (Figure 8). Each transition probability is then a prior probability as to the electrofacies identity of a zone, given the identity of the immediately preceding zone. The transition matrix automatically quantifies Walther's Law for insertion into the Bayesian classifier. It should be noted that Walther's Law holds exactly only for conformable sequences. However, the exceptions that are caused by unconformities and other discontinuities are automatically accommodated within the transition statistics of the observational sequences. Finally, the common tendency for loosely repetitive motifs of lithofacies (or cycles) is expressed

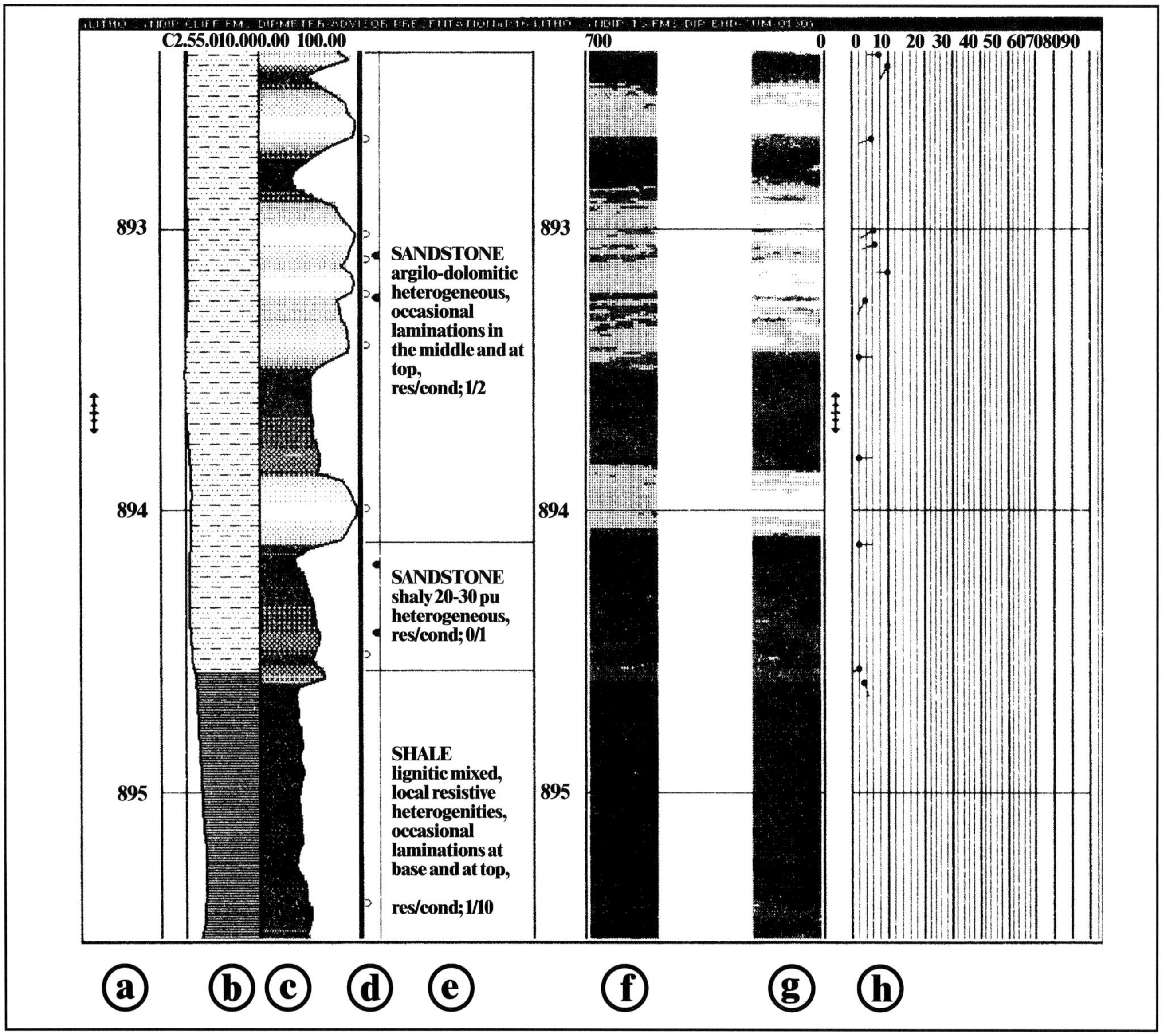

Figure 9. Computer-terminal display of lithofacies predictions from an electrofacies database coupled with electrical borehole images and dipmeter vectors. Track (a): depth in meters; (b) predicted lithology bounded by gamma-ray log to left; (c) stratification generated from dipmeter resistivity curve processing bounded by median conductivity log to right; (d) conductive (black) and resistive (white) heterogeneities; (e) corelike description based on composite log characteristics; (f) and (g) borehole wall microresistivity images; (h) dipmeter vector (tadpole) plot. From Anxionnaz et al. (1990).

by the transition probability structure. In the event of a randomly ordered succession of electrofacies, the transition probabilities are equal to the relative proportion of the occurrence of each electrofacies. These estimates would be appropriate prior probabilities for a Bayesian model that disregards vertical contiguities and is weighted by overall electrofacies abundances.

In a more typical situation the transitions between electrofacies would not be deterministic; they would be the product of some intermediate stochastic model described by the transition probabilities. It is probable that the results from a discriminant analysis incorporating Markovian measures would show a marked improvement, perhaps comparable with those reported from remote sensing applications. Haslett (1985) showed how substantial improvements could be made in maximum likelihood classifications of pixel elements on LANDSAT imagery by incorporating Markovian transition probabilities of the states of adjacent pixels. By this means the vector of responses at each pixel location is not processed in isolation, but is considered in the context of broader geographic elements and lateral interrelationships.

The integration of high-resolution dipmeters and borehole-wall imaging logs has added a major new

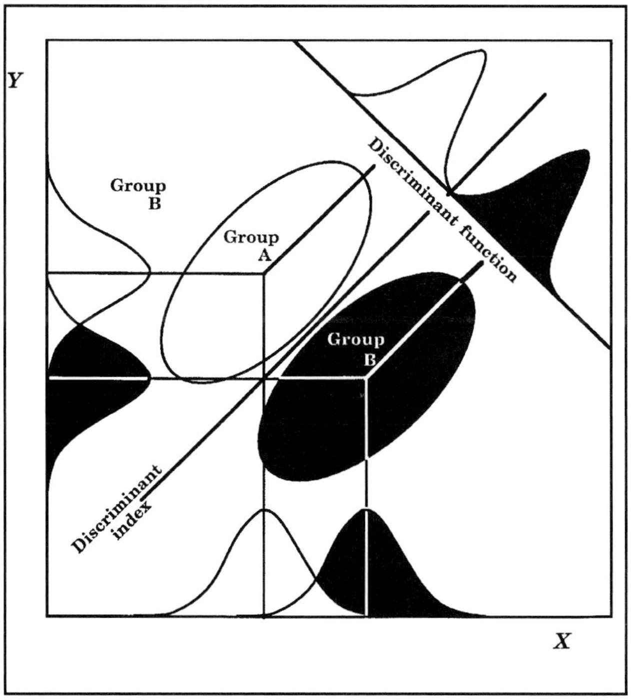

Figure 10. Graphic representation of discriminant function analysis applied to an idealized bivariate example. Modified from Davis (1986).

dimension to geologic characterizations from logs (Anxionnaz et al., 1990). The fine scale of vertical resolution of the multiple microresistivity traces allows the recognition of both open and healed fractures, bedding structures, and coarser clastic grains. Collectively they show textural and structural sedimentary rock features that complement the compositional information from the electrofacies predictions (Figure 9). These applications are being developed actively because of the increasing emphasis on improved methods for reservoir description and modeling.

DISCRIMINANT FUNCTION ANALYSIS

Discriminant analysis covers a wide range of techniques aimed at the classification of unknown samples to one of several possible groups or classes. Classical discriminant function analysis has the tighter focus of attempting to develop a linear equation that best differentiates between two different classes. Fisher (1936) first derived the linear discriminant function as a statistical method to separate two populations by a linear weighted function of their measurement variables. The method is a supervised technique that requires a training data set for which assignations to the two populations are already known. The data consist of multivariate values for every individual in both population samples. If the two groups were plotted in multidimensional space, they would appear as two clouds of data points with either a distinctive separation or some degree of overlap. An axis is located on which the distance between each cloud is maximized while the dispersion within each cloud is simultaneously minimized. This axis defines the linear discriminant function and is calculated from the multivariate means, variances, and covariances of the two groups. The data points of the two groups may be projected onto this axis as locations on a single line. This operation results in the collapse of the many variable dimensions of the recorded data into a single, composite variable that best discriminates between the two groups.

The process is perhaps best understood by reference to a cartoon representation of a two-dimensional case (Figure 10). When two groups are plotted with respect to a single variable there is frequently a range of overlap, as shown by the frequency curves on each axis. When an observation of unknown affinity occurs in this range there is a degree of uncertainty in its classification. It can be assigned to one of the groups by a decision based on probability, and the assignment will be incorrect in (hopefully) a minority of occasions. When plotted in the Cartesian space of two variables, the degree of differentiation will improve or, at worst, stay the same. In Figure 10 an ideal situation is shown with a perfect separation between the two groups. In practice, there will commonly be an area of intermingling of the two data-point clouds. The discriminant function is then the equation of the axis that cuts obliquely across the crossplot plane at an angle determined by the relative contribution of the two variables to discrimination. In the computation of the function, the two clouds are modeled by bivariate normal distributions whose probability density contours map out elliptical shapes. This is the same representation used in the electrofacies data banks described previously. When a higher number of variables is used for discrimination the two clouds are matched with hyperellipsoids in multivariate space.

The equation of the discriminant function, Z, is:

$$Z = \lambda_1 X_1 + \lambda_2 X_2 + \ldots + \lambda_m X_m$$

where X_1 to X_m are the m variables used for discrimination and the lambdas are the weighting coefficients to be applied to each of the m variables. The optimum discriminant function is set at that location which maximizes the value of the distance between the cloud centroids divided by the dispersion of the two clouds. This condition occurs when the following equation is satisfied:

$$S\Lambda = D$$

where **S** is a matrix of pooled variances and covariances from the two groups, Λ is a vector of the unknown lambda coefficients to be solved, and **D** is a vector of the differences between the means of the two groups with respect to the m variables.

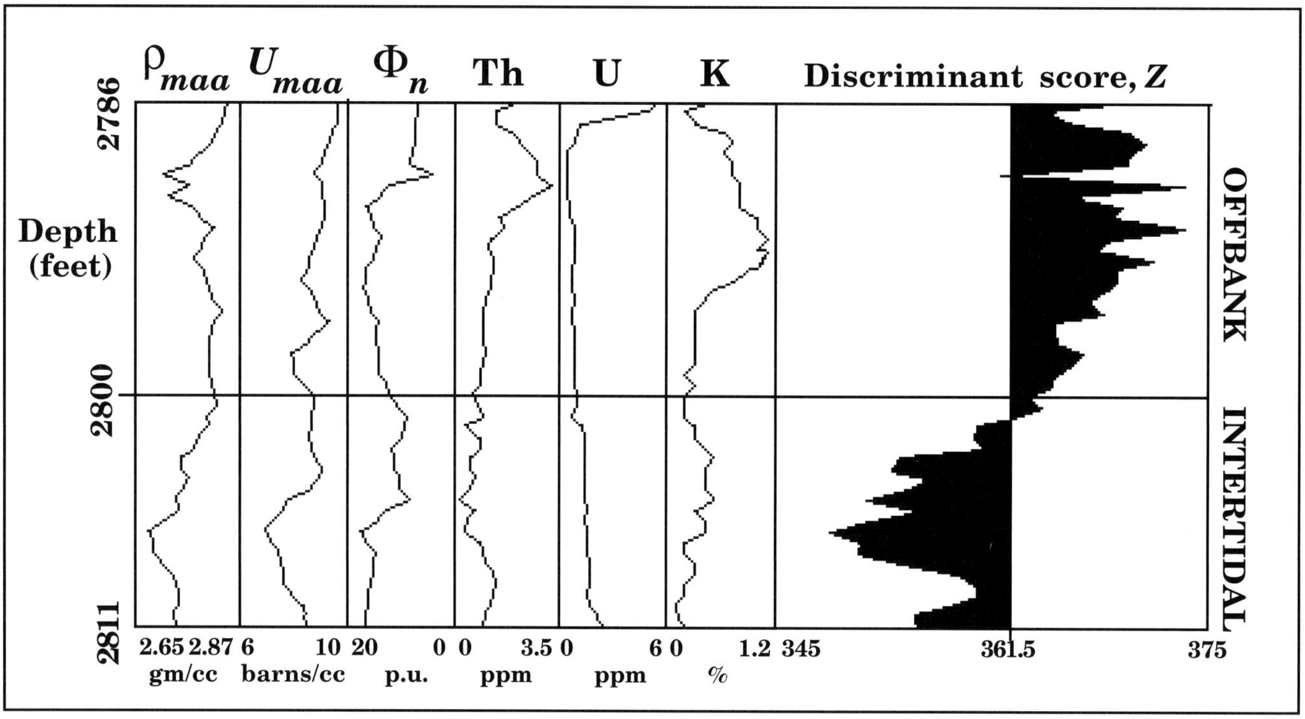

Figure 11. Input logs and discriminant function score log computed to differentiate offbank from intertidal facies in a Lower Permian Winfield Limestone section.

There is a simplifying assumption that the variance–covariance matrix is the same in the two groups. So, the pooled variance–covariance matrix is an average for the two groups weighted by their number of observations. This assumption implies that the two group data clouds have similar inflations and orientations. Whether the stipulation is reasonable can be checked by a simple visual inspection of crossplots. The method is sufficiently robust that moderate differences do not generally degrade the power of the function as an effective classifier. Indeed, as Reyment (1974) pointed out, the original example used by Fisher (1936) clearly violates this assumed equality of the dispersion matrices. Heterogeneous discriminant functions can be applied for particularly troublesome classification problems. However, the simple linear discriminant is usually quite adequate for most applications.

The solution of the matrix equation for the vector of discriminant coefficients then follows as:

$$\Lambda = S^{-1}D$$

The function describes a linear axis that crosses the multivariate space. The coordinates of any individual zone can be converted into a discriminant score, Z_i, which is a projection onto the discriminant axis and gives its relative position:

$$Z_i = \lambda_1 X1_i + \lambda_2 X2_i + \ldots + \lambda_m Xm_i$$

where $X1_i$ to $X2_i$ are the values of the m variables in the ith zone. The multivariate data clouds are now condensed into two ranges of points on a single axis. A discriminant index is calculated as the midpoint between the Z-score projections of the two group multivariate means. The discriminant index serves as a boundary between the domains of the two groups and can be used as a decision criterion to identify an unknown observation with one or the other of the groups. The choice of midpoint reflects the assumption of equal dispersion matrices for the two groups. However, the discriminant boundary can be adjusted to a better value, based on the rate of misclassifications observed in validation tests of the function.

The mathematics of the discriminant function will generate some form of solution, even when there are no systematic differences between the two groups. This contingency can be checked statistically by an F-test, based on a generalized measure known as Mahalanobis's distance, D^2. Details of the procedure are given by Davis (1986, p. 485-487). Mahalanobis's distance represents the relative degree of separation between the centroids of the two groups, given by:

$$D^2 = \lambda_1 d_1 + \lambda_2 d_2 + \ldots + \lambda_m d_m$$

where d_1 is the difference between the means of the two groups for the ith variable.

Mahalanobis's distance can also be used to assess the relative contribution of each measurement variable to discrimination. This is computed through a comparison of the weighted difference of the group means:

$$E_i = \frac{\lambda_i d_i}{D^2}$$

where E_i is the contribution of the ith variable. In addition, the distance function can be used to provide a means to judge whether an unknown observation belongs to one of the two groups or not. In this case, the distance is computed between the multivariate location of the zone and the centroid of each group. Because the distance is computed in multivariate space, it is useful as an error diagnostic to alert the user to situations where a zone is likely to belong to a third group.

Finally, a table of correct and incorrect classifications for the calibration data set and in addition, preferably, a validation set, will also give a direct measure of classification performance. The comparison can be formatted as shown by the Shublik Formation example in Table 1. The results of such tables can be used to evaluate overall success in predicting classifications, compare different models of log measurements, and monitor possible improvements in prediction made through corrective adjustments of the discriminant index.

The capability to compute Z-scores is particularly useful in log-analysis applications because a continuous Z-score log can be generated which is keyed to depth. In fact, decisions regarding whether any particular zone should be identified with one of the groups can be deferred to a time following the generation of the log. The Z-score trace is a continuous record of the relative degree of affinity of a zone with the two groups. The curve within the intervals used for calibration should be examined carefully, both as a quality check and as a source of potentially useful feed-back information about the samples used for classification. Outside the calibration interval the Z-score log may be used cautiously for group identification. The apparent group identifications will obviously fail in zones that belong to neither group. Typically, the multivariate coordinates of such zones will be relatively far removed from the centroids of both groups. These potential problem zones can be monitored through the display of Mahalanobis' distance as an ancillary quality curve.

APPLICATION OF DISCRIMINANT FUNCTION ANALYSIS (DFA) TO LITHOFACIES DISTINCTION

The linear discriminant function is a widely popular statistical method because of its relative simplicity and robustness. The theoretical aspects described so far are illustrated by the following simple lithofacies identification example. Core studies of a Winfield Limestone (Lower Permian, Chase Group) section in a well in southwestern Kansas contrasted an upper part, interpreted to be offbank marine in origin, from a lower part, considered to have been formed in intertidal conditions. A discriminant function analysis was used to distinguish the two units in terms of their wireline responses. There were several purposes for the analysis: to assess whether the distinction was even possible, based on the logs; to see which logs appeared to contribute to any distinction; to determine the geological reasons that would account for petrophysical distinctions. If the discriminant function analysis was found to be useful, then the function could be used for classification purposes in other geologically similar parts of the uncored section.

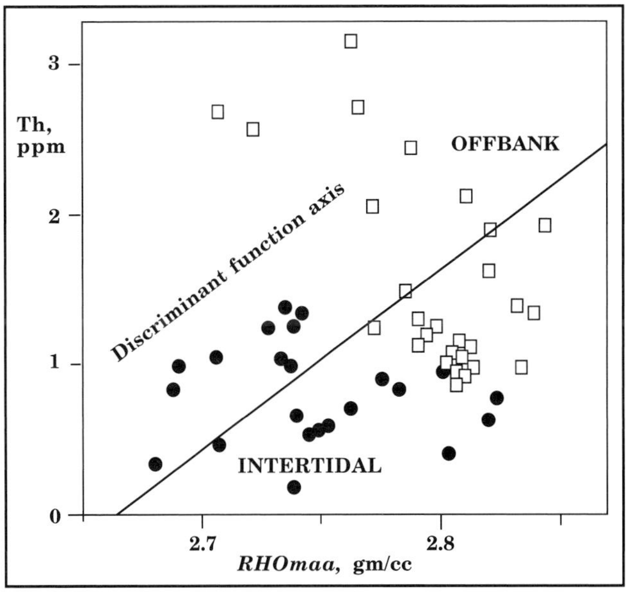

Figure 12. Axis of the discriminant function located on a crossplot of apparent grain density and thorium computed to differentiate offbank from intertidal facies in a Lower Permian Winfield Limestone section.

The log measurements used for discrimination were apparent grain density (*RHOmaa*), apparent volumetric photoelectric cross-section (*Umaa*), compensated neutron porosity, and spectral gamma-ray estimates of thorium, uranium, and potassium. The computed discriminant function was:

$$Z = 131.9\ RHOmaa - 1.987 Umaa + 0.3\Phi_n + 5.9Th - 0.6U + 5.8K$$

Higher scores were associated with the offbank facies, lower scores with the intertidal facies. The computed F-test value, based on Mahalanobis' distance, proved to be highly significant, rejecting the null hypothesis that there was no appreciable difference between the two groups. This result was not surprising in view of the Z-score log computed for the calibration interval (Figure 11), which showed an almost perfect separation between the two lithofacies.

Calculation of the relative contribution of each log measurement to the total discrimination revealed marked contrasts between them. The strongest discrimination was made by apparent grain density, thorium, and potassium, with negligible contributions by the other logs. A crossplot of the thorium and apparent grain density values of the zones gives visual confirmation of the importance of these two measurements (Figure 12). Note the improvement in discrimination when the logs are used together compared to their separate use. Clearly, this is a textbook example of near-perfect discrimination, but many field examples also can be expected to show strong separation, although a certain degree of overlap is seen more commonly on the discriminant axis.

The petrophysical distinctions between the groups have been diagnosed automatically by the discriminant function. The geological reasons for this differentiation are a matter for interpretation that takes into account the petrographic data from available core. In the case of this Winfield Limestone section, it appears that the offbank carbonates are mostly dolomites with some illitic shale, in sharp contrast to the intertidal zones which are shale-free cherty dolomites.

Davis (1986) gives a useful explanation of discriminant function analysis with geological examples. DFA is one of the most widely used multivariate methods in the earth sciences, but is little used in log analysis. Doveton (1986) described an application of a discriminant function to the log distinction of oolitic limestones in a Mississippian carbonate succession from western Kansas. Davis (1986) also describes the extension of discriminant function analysis to the differentiation of more than two groups. Multigroup DFA requires the computation of several discriminant functions to separate the different groups in a plane or a higher dimensional space.

NONPARAMETRIC DISCRIMINANT ANALYSIS

The discriminant methods described up to this point have all been parametric. In other words, the discrimination has been made in terms of parameters that summarize the distributions, rather than the original data clouds themselves. The parameters of each electrofacies are the multivariate means and the variance–covariance matrix specifying the location and dispersion of a multivariate normal hyperellipsoid. As discussed before, the normal distribution is a useful descriptive model, because it is specified by a few parameters and has a well-known probability density structure.

It is not uncommon for electrofacies to take a variety of cluster shapes that are represented poorly by ellipsoids. In an alternative strategy, their discrimination can be based on the actual distribution of their component data points, rather their summary parameters. This can be achieved by building a multivariate histogram into a computer database. The range of each log variable is partitioned into divisions. In multidimensional space, with logs as orthogonal axes, the divisions subdivide the space into a lattice of cells (Figure 13). The log responses for each zone from a training set are coordinates of a single data point that can be allocated to one of these cells. The frequencies of points for each electrofacies are totaled for all cells

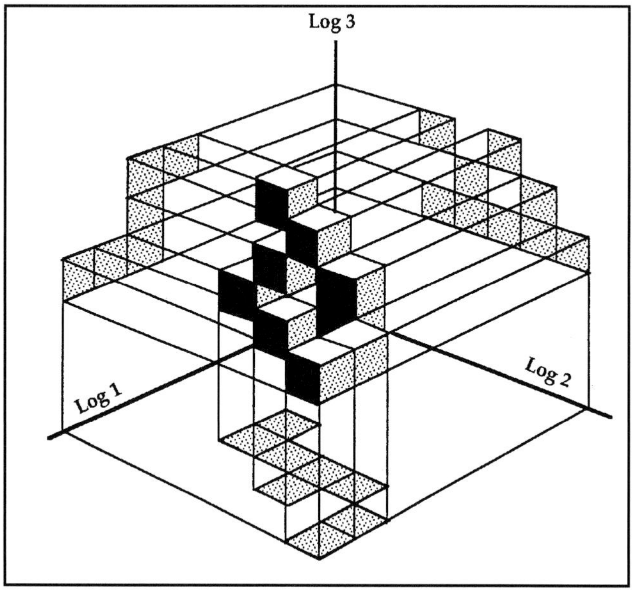

Figure 13. Representation of electrofacies in three-dimensional log space as a multivariate histogram of counts in a lattice work of cells. Adapted from Serra and Abbott (1980). Copyright SPE.

to create the database histogram.

The database should be designed efficiently to avoid a huge and unnecessary allocation of computer memory. For example, the relatively coarse partition of each of six logs into ten subdivisions results in a framework of a million cells. If the training set has a thousand zones, then at least 999,000 cells will be empty. Many of these empty cells can also be discounted because they represent unreasonable combinations of log responses. In practice, efficient histogram databases store only information for those cells that are not empty. Tetzlaff et al. (1989) reported that histogram cell coordinates and frequencies from several wells can be recorded on a disk file that can be handled by a personal computer.

The database can then be used for classification of an unknown zone by inspection of the cell that matches the zone's log coordinates. If the cell is not empty, then the relative frequencies of each of the possible electrofacies can be used in the Bayesian formula described earlier to compute their respective posterior probabilities. The *a priori* probabilities can be taken from the training set and comprise the number of zones that belong to a given electrofacies divided by the total number of zones. If the training set is not considered to be representative of the entire section, then the user may decide to fall back on the assignment of equal prior probabilities to all electrofacies. This situation might occur, for example, if the training set were created from cored intervals which may have a lower proportion of shales. The electrofacies with the highest posterior probability is selected as the prediction for the category of the input zone.

If the unknown zone responses coincide with an empty cell, then a decision must be made by examining the neighboring cells. Tetzlaff et al. (1989) interpolated (or extrapolated) frequencies from neighboring cells that were weighted by their inverse squared distance. The sum of these interpolated values is a projected estimate of the electrofacies densities for the empty cell. The approach is essentially the same as the potential method used in pattern recognition (Duda and Hart, 1973). If all the neighboring cells are empty, then the category of the zone remains unknown.

As with the other discriminant methods, the best measure of quality for the nonparametric method is the discrimination matrix, where frequencies of predicted electrofacies are tabulated against actual lithofacies (cf. Table 1). Tetzlaff et al. (1989) found that the use of zoned logs substantially improved discrimination between electrofacies. The zoned blocking of logs tends to tighten the concentrations of electrofacies clouds by eliminating points that match transitions between zones (Figure 4). The training or calibration set will give the most optimistic measure; more realistic numbers will result from the application of the multivariate histogram to a validation data set.

The advantages of the nonparametric approach tend to be the disadvantages of the parametric methods and vice versa. The nonparametric histogram captures all the shape features of an electrofacies data cloud at resolutions coarser than the basic cell dimension. This can lead to large demands on computer memory and so must be encoded in an efficient manner. By contrast, a multivariate normal distribution can be described with relatively few parameters. This simplicity is achieved by fitting an ellipsoid to the data cloud. Zones that are misclassified by the two approaches often result for different reasons. The parametric method generalizes by basing the probability density at any multivariate location on the normal statistics of the entire cloud. Misclassifications can occur at local flexures in cloud shape. By contrast, the nonparametric method is highly specific, with its prediction based on the contents of an individual cell. When using typical sizes of training set, the cell frequencies may be small and poor estimates of their theoretical population values. Ideally, a perfect discrimination method would incorporate the strengths of both methods by sufficiently generalizing the datapoint distributions to make predictions robust, but to retain localized features that reflect systematic, but smaller scale effects.

PRINCIPAL COMPONENT ANALYSIS (PCA)

The ability to both visualize and manipulate the multidimensional representation of logging data in a satisfactory manner has been discussed frequently in this text. The scatter of points plotted with reference to four logs is just about at the outer limit of our perception (three axes of space plus a fourth dimension provided by color). Graphical representations are therefore helped by methods that allow high-dimensional representations to be compressed into a lower dimensionality. Principal component analysis is the most commonly used technique for this purpose, and has many additional useful properties that can be applied in systematic log interpretation.

If m log responses from a sequence of zones are plotted as points in a space with mutually orthogonal axes, they form a cloud in m-dimensional space. Principal components are the eigenvectors of this cloud, computed to locate the major axes in order of importance. These axes provide a new framework of reference which is aligned with the natural axes of the cloud ("eigen" is German for "intrinsic"), rather than the original log measurement axes. The orientation of the principal components are computed from either the covariance or correlation matrix of the zone log data. The correlation matrix is the more common choice, because most logs are recorded in radically different units. In order to avoid artificial and undue weighting by any of the logs, the original data should be standardized to dimensionless units by subtracting the mean and dividing by the standard deviation. The covariance matrix of standardized data is the correlation matrix.

The raw data cloud is modeled by a single hyperellipsoid. The ellipsoid orientation is at some angle to the measurement axes and the relative directions of axis elongation reflect systematic relationships between the logs. The axes of the ellipsoid are the eigenvectors, whose relative magnitudes are given by their associated eigenvalues. A cartoon depiction of the procedure to this point is shown in Figure 14. A compression of dimensionality is made possible because the components are based on the intercorrelations between the variables. As a simple and extreme example, if there is a perfect correlation between three variables, then data points will be strung out on a single oblique axis. So, although the points are found in a three-dimensional space of the original variables, the intrinsic dimensionality of the data variability is only one. In a more realistic case of high correlations between logging variables there will still be a strong tendency for data to be aligned along an axis, with the residual scatter of points absorbed by subsidiary and shorter axes. The first principal component therefore accounts for the maximum amount of variability of any single possible axis. The remaining principal components pick up the rest of the variability in an ordered allocation.

The total variance of the original set of m variables is the sum of their separate variances. This quantity is absorbed by m possible principal components. In practice, many measurement variables show a significant degree of intercorrelation, so that the last few principal components may account for trivial amounts of the total variability. Put another way, this property highlights the amount of information redundancy within the logs. If the majority of the variability is picked up by p principal components, then the dimensionality has been shrunk from

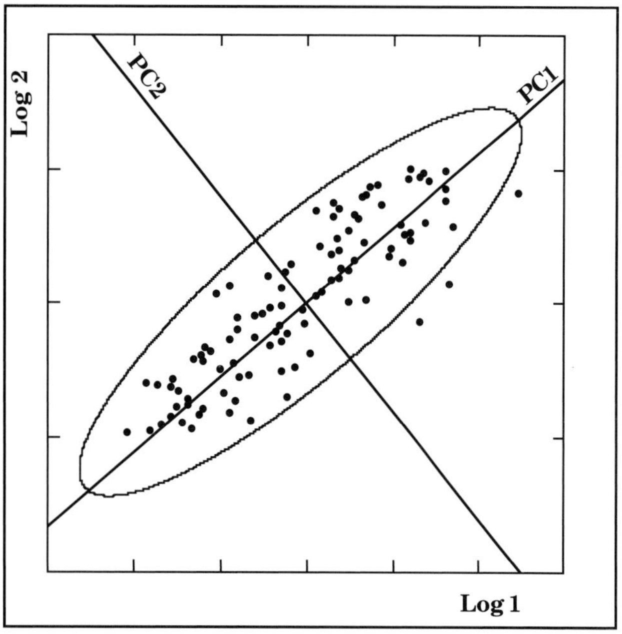

Figure 14. Principal component axes (PC1 and PC2) of a hypothetical data cloud of points measured with respect to two logs shown as dimensionless axes.

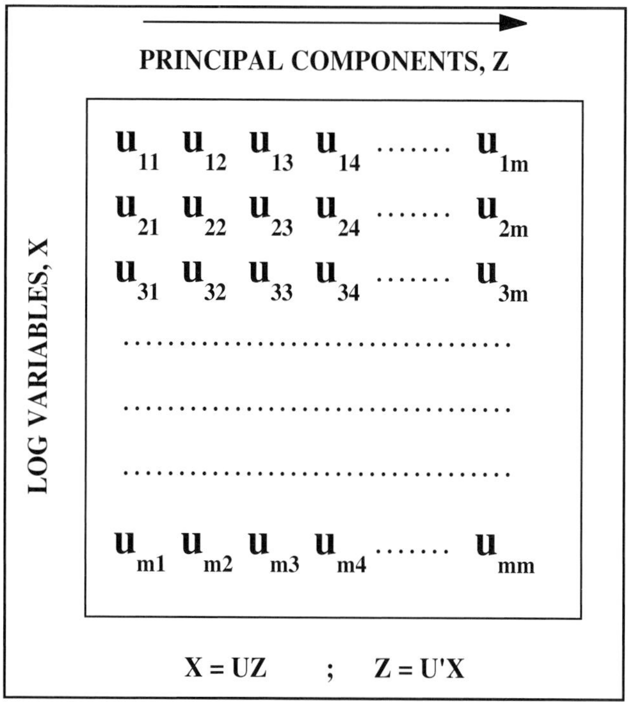

Figure 15. Diagrammatic representation of the transformation matrix, U, whose elements can be used both to convert variables (X) to principal component scores (Z) and PCA scores to the original variables.

m to *p*. The collapse reflects the dimensionality of the information content of the variables as a replacement for the original reference framework. It is not uncommon for the first two principal components to account for most of the variability, thus a multilog data set can be mapped on a crossplot with little loss of information.

The derivation of the principal components follows from a property of matrix algebra that a symmetric, nonsingular matrix, **S**, can be converted into a diagonal matrix, **L**, by multiplying by an orthonormal matrix, **U**, through the following equation:

$$U^{T}SU = L$$

where T signifies the transpose of a matrix. If **S** is the covariance matrix, then the conversion to a diagonal matrix is the geometrical equivalent of a rotation of the original axes to new descriptive axes. The diagonal matrix has zeroes in the off-diagonal elements, which means that the new axes are independent of one another. The values of the diagonal elements register the eigenvalues of these principal components which express their variances. The sum of these eigenvalues is then the same as the sum of the variances of the original variables. The relationship gives an immediate measure as to how much variability is assigned to each principal component. The numbers are particularly easy to follow when the correlation matrix is selected. The variance of each variable is then unity, and the total variability equals *m* (the number of variables). Each eigenvalue divided by *m* is the proportion of a principal component's share of the total variability.

The fact that **U** is an orthonormal matrix leads to the useful result that the inverse of **U** is the same as the transpose of **U**. This means that both the transformation from the measurement space to principal component space and the reverse mapping are variations of the same operation. The matrix **U** contains the loadings that relate the eigenvectors to the original variables (Figure 15). The location of any point within the data cloud can be related to the principal component axes by the transformation:

$$Z = U^{T}X$$

where **X** is a vector of the zone log responses and **Z** is a vector of principal component scores. This means that the score of the *i*th zone on the *p*th principal component is given more simply by:

$$z_{pi} = u_{1p}x_{1i} + u_{2p}x_{2i} + \ldots + u_{mp}x_{mi}$$

where the *u* coefficients are loadings from the *p*th principal component. These are the values of the *p*th *column* in the table matrix, shown diagrammatically in Figure 15. The original variables can be recovered from the principal component scores through the

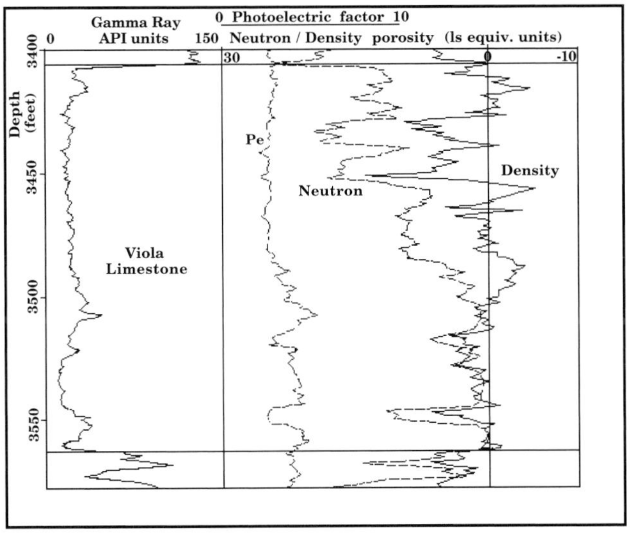

Figure 16. Gamma-ray, neutron, density, and photoelectric-factor logs from a section of Middle Ordovician Viola Limestone in northern Kansas.

TABLE 2. Principal Components of the Correlation Matrix of the Gamma-ray, Neutron Porosity, Density, and Photoelectric Factor Logs Run in a Middle Ordovician Viola Limestone Section in Northern Kansas (the table contains the loadings of the PCs with respect to the original logs, as well as the percentage of the total variability accounted for by each principal component).

LOGS	PC1	PC2	PC3	PC4
Total gamma-ray	−0.23	0.80	−0.54	−0.13
Compensated neutron	0.63	0.28	−0.02	0.72
Bulk density	−0.44	0.42	0.76	0.24
Photoelectric factor	−0.59	−0.32	−0.37	0.64
% Total variability	52.4	26.6	17.6	3.4

inverse of this procedure:

$$X = UZ$$

So, the gth log response of the ith zone can be computed by the equation:

$$x_{gi} = u_{g1}z_{1i} + \ldots + u_{mp}z_{mi}$$

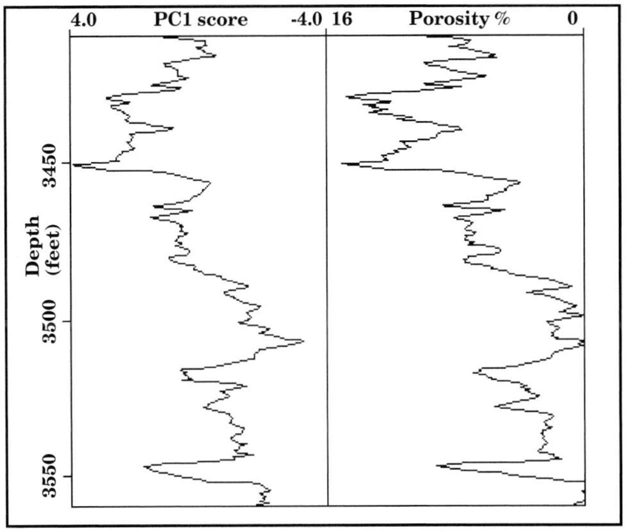

Figure 17. Log of first principal component scores (left) and log of effective porosity calculated by conventional analysis (right) in a section of Middle Ordovician Viola Limestone from northern Kansas.

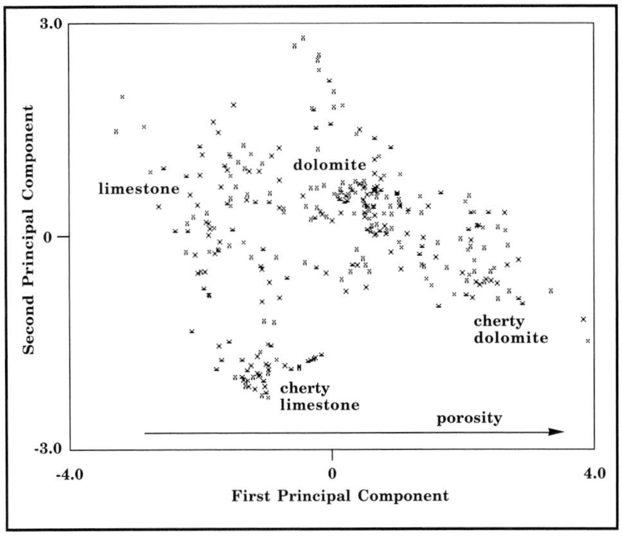

Figure 18. Crossplot of first and second principal component scores computed from gamma-ray, neutron, density, and photoelectric-factor logs from a section of Middle Ordovician Viola Limestone in northern Kansas.

which on Figure 15 would correspond to the multiplication of the scores by the loadings of the gth *row*.

The ability to move easily between the two systems of coordinates is more than a mathematical convenience. As will be seen in the following examples,

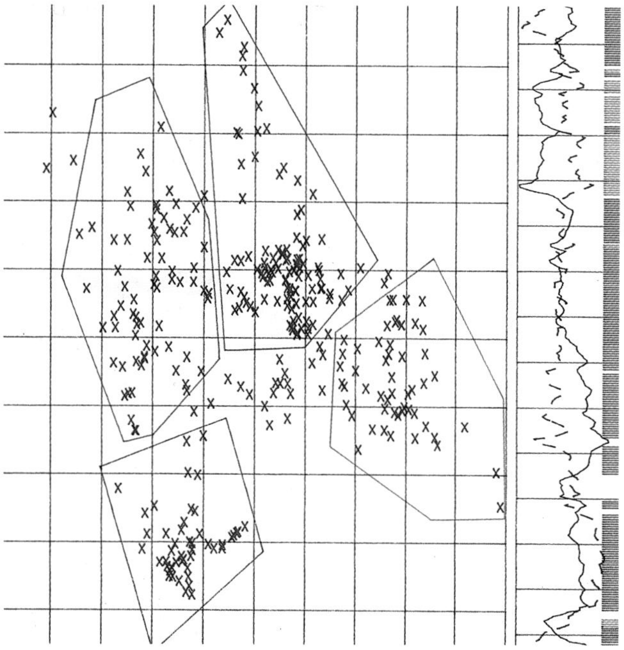

Figure 19. Manual cluster analysis of data points on the principal components crossplot of Figure 18 displayed as depth classification (right) shows good discrimination between lithology types and partition of the Viola section between dolomites and cherty dolomites overlying limestones and cherty limestones.

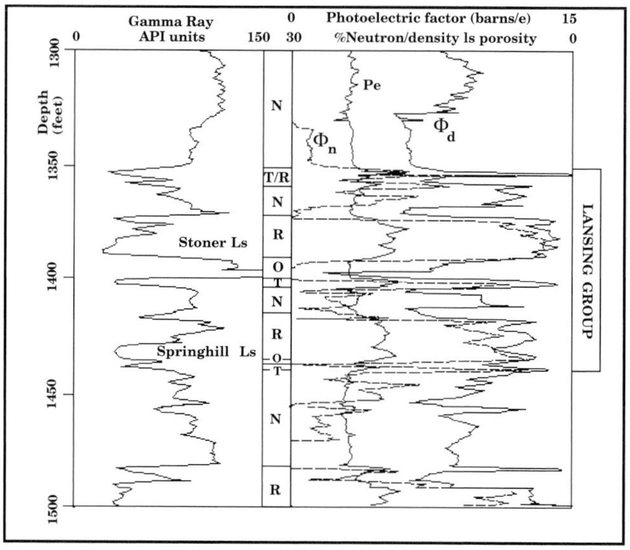

Figure 20. Gamma-ray, neutron, density, and photoelectric factor logs of an Upper Pennsylvanian sequence of limestones and shales from northern Kansas. Genetic unit types are marked in the central column, where T= transgressive limestone, O = offshore shale, R = regressive limestone, and N = nearshore shale.

subsets of principal components can be mapped and related directly to the log variables. Also, logs can be partially reconstructed from a few principal components and the differences from the actual logs can be used for log interpretation. The loadings of the **U** matrix summarize the relationships between the log variables and the principal components. As such, they can often be "read" for their geological or petrophysical meaning. The various log responses are indirect measurements of fundamental, but unseen, rock properties. In the PCA of the correlation structure of log relationships, the principal components will often reveal these properties implicitly. Common sense must be used in such interpretations. The computation of principal components is simply a geometrical operation which relocates the reference axes to the apparent axes of elongation of the data cloud. The preceding explanation only covers the bare bones of the mathematics of principal component analysis. The ideas and further ramifications are best understood by consideration of actual examples.

PRINCIPAL COMPONENT ANALYSIS OF LITHOFACIES IN A COMPLEX CARBONATE

The Middle Ordovician Viola Limestone of the U.S. Midcontinent typically consists of cherty, dolomitic limestones whose compositions are highly variable with depth. A gamma-ray–neutron–density–photoelectric factor log suite is shown in Figure 16 for a Viola section from northern Kansas. Principal components were computed from the correlation matrix of the four log measurements; the principal component loadings are shown in Table 2, together with percentage of total variability accounted for by each component. A log of the scores computed from the first principal component is shown in Figure 17 compared with the trace of effective porosity calculated by conventional log analysis. The strong visual similarity between the two traces is confirmed by their correlation coefficient of 0.96. The lock on porosity of the first principal component is to be expected in this example. The rock consists entirely of calcite, dolomite, and chert with minimal shale content. Consequently the major differentiation is between the physical properties of the fluid in the pore network in contrast with those of the minerals, which do differ markedly from one another. Uses for the automatic prediction of porosity from principal components has been described by Elek (1990) and will be reviewed later in this chapter.

The first two principal components collectively account for 79% of the total variability. A plot of the scores from these two components (Figure 18) therefore represents an oblique plane that cuts the original four-dimensional data cloud along its two principal axes. The scatter of points about this plane is only 21% of the variability and is a measure of the information that is lost when the data are projected onto the crossplot. This optimum view of the multi-

TABLE 3. Principal Components of the Correlation Matrix of the Gamma-ray, Photoelectric Factor, Density, and Neutron Porosity Logs Run in an Upper Pennsylvanian Limestone–Shale Section in Northern Kansas (the table contains the loadings of the PCs with respect to the original logs, as well as the percentage of the total variability accounted for by each principal component).

LOGS	PC1	PC2	PC3	PC4
Total gamma-ray	–0.48	0.76	0.43	–0.09
Photoelectric factor	0.51	0.14	0.48	0.70
Bulk density	0.49	0.64	–0.59	–0.09
Compensated neutron	–0.52	0.04	–0.49	0.70
% Total variability	77.3	9.7	7.7	5.3

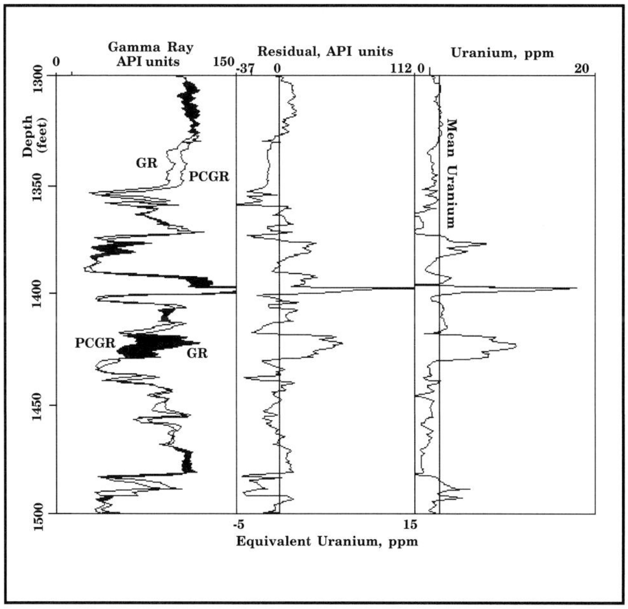

Figure 21. Gamma-ray log reconstructed from the first principal component (PCGR) overlaid on actual gamma-ray log (GR) at left; log of residuals from the first principal component scaled both in API units and equivalent ppm uranium in center; log of uranium recorded by a spectral gamma-ray tool at right.

dimensional cloud is useful for recognizing distinctive clusters that can be assigned to separate electrofacies. Principal component analysis was first used for this purpose by Wolff and Pelissier-Combescure (1982) as a means of dimensional compression prior to the isolation of electrofacies by cluster analysis. The nature of the clusters can be determined through an examination of the component loadings on the log variables and an inspection of the depths and log responses of the clustered zones. There is good distinction of Viola lithology types by the plot shown on Figure 18. The strong association of the first principal component with porosity brings out a pattern of dolomites which are mostly more porous than the limestone zones, with cherty dolomites and limestones tending to be more porous than their chert-free equivalents.

Some interactive log-analysis packages now allow the user to outline clusters on a terminal screen. The captured zones are then highlighted on a parallel depth display of the crossplot logs (Figure 19). This computer strategy capitalizes on the strengths of both numerical methods and human skills. The principal component analysis attempts to locate the most effective viewing planes in multilog space. These become the targets for the pattern-recognition abilities and geological expertise of the user. The interaction between user and machine allows experimentation and learning in the facies subdivision of stratigraphic sections that can be quite complex.

PRINCIPAL COMPONENT ANALYSIS IN SHALE VOLUME ESTIMATION

Various logs have been used in the estimation of shale, including the spontaneous potential, the gamma-ray, and a combination of the density and neutron responses. There are conceptual problems in defining a meaningful shale end member. In addition, individual logs are sensitive to different shale properties. However, these same properties all differentiate shales markedly from other lithologies. Rather than being considered individually, the logs are best combined in a composite measure based on their collective response to shales. Principal component analysis provides a way to extract a common shale response based on observed log properties, rather than on theoretical models.

A case study of shale estimation using principal component analysis is drawn from the Upper Pennsylvanian limestone–shale sequence of northern Kansas. Cyclothems or repetitive sequences of lithologies have been recognized in these successions for many years (Heckel, 1977). Depositional models developed from outcrop studies have been extended to the subsurface through detailed stratigraphic correlation and interpretation of log-trace character. The basic sequence consists of a thin transgressive limestone (T), followed by a black shale (O) that is usually interpreted to represent deeper water conditions, succeeded by a thick regressive limestone (R), and terminated by a nearshore shale (N). An example of gamma-ray, neutron, density, and photoelectric factor logs from a Lansing Group section and marked with these cyclothem unit types is shown in Figure 20. Gamma-ray logs are frequently used as a generalized lithology log for the distinction of limestones and shales, but they must be used carefully to avoid classifying "hot" limestones as shaly intervals. Local increases in natural radioactivity are caused by concentrations of uranium and are one reason that spectral gamma-ray logs are run—to differentiate shale effects (picked up by thorium and potassium) from the contribution of uranium. In the Lansing Group example, two uranium anomalies appear to occur, in the Stoner Limestone and Springhill Limestone members. Marked increases in gamma radiation occur at intervals that are suggested to be limestone by the neutron, density, and photoelectric factor logs.

Four principal components were calculated from a correlation matrix of the four logs. The loadings of each component and their percentage accounts of the total variability are shown in Table 3. The first principal component alone absorbs 77% of the variability, while the signs of the loadings suggest that it represents a composite shale component. The effective collapse of four log dimensions to a single axis of variation is not surprising. The sequence consists of two basic end members: a limestone that is generally low in porosity and nonradioactive and a radioactive shale. Breaks from this pattern are provided by radioactivity that is not linked with the shales, variability within the end members, and minor, ancillary lithologies. These other sources of variation are contained in the 23% of residual variability picked up by the other three principal components.

As explained earlier, the principal component model may be inverted through a reconstruction of the original variables from the principal components. A full reconstruction would have no useful purpose, but a partial reconstruction based on a subset of principal components demonstrates how much of the variable is linked with the component. In this case study, the first principal component has been suggested to be a composite measure of shale. A gamma-ray log can be computed from this principal component by the equation:

$$G_i = u_{11}z_{1i}$$

where G_i is the first principal component estimate of the gamma-ray reading of the ith zone, the u_{11} coefficient is the loading of the first principal component with respect to the gamma-ray log, and the z_{1i} value is the first principal component score for the ith zone. For this example (see Table 3), this becomes:

$$G_i = -0.48 \cdot z_{1i}$$

Because the principal components were computed from the correlation matrix, the reconstructed logs are in units of standard deviation with an origin at the mean. Rescaled in API units, the estimated log is calculated by:

$$G_i = \left(-0.48 \cdot z_{1i} \cdot s\right) + \overline{G}$$

where $\overline{G}$ is the mean and s is the standard deviation of the original gamma-ray log.

The PCA-reconstructed gamma-ray log is shown superimposed on the actual trace in Figure 21. For the most part, there is a close match between them. Zones in which they coincide mark horizons where the gamma-ray value matches the expectation based on all four logs together. Most of the major divergences can be attributed to zones which have either higher or lower uranium content than would be associated with the implied volume of shale. Note that the anomalies in the Stoner and Springhill limestones are now clearly delineated as potential uranium anomalies. Deviations within the shales are less marked, but show some distinctive patterns that may reflect changes in clay mineralogy as well as changes in uranium.

Subtraction of the gamma-ray log reconstruction based on the first principal component from the measured log gives a residual value that represents the

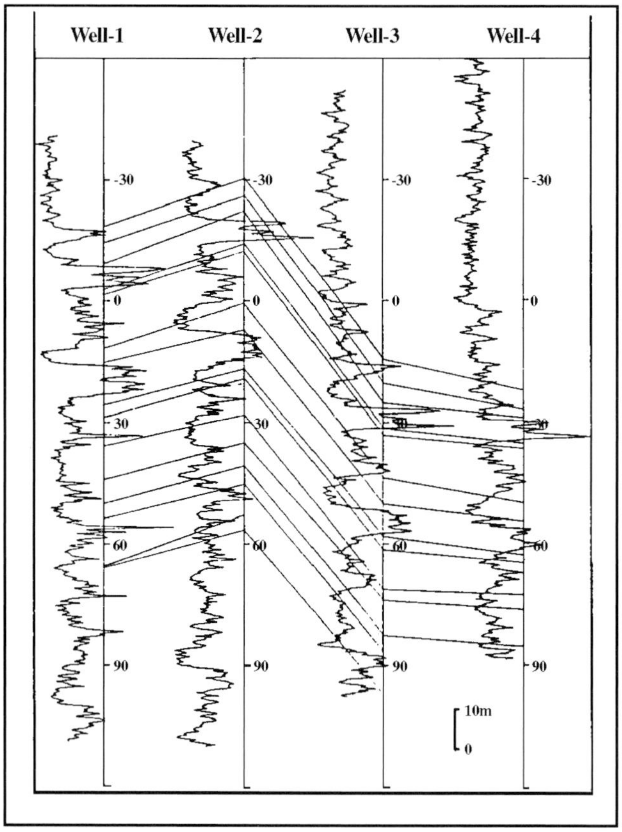

Figure 22. Example of stratigraphic correlation between four wells using logs of the scores from the first principal component of a suite of logs. From Elek (1988).

combined contribution of all three remaining principal components. As a check on the hypothesis that much of this residual can be attributed to uranium, the residual log is shown both in API units and in uranium equivalent in parts per million (Figure 21). The conversion to equivalent uranium is made by the formula:

$$API = 16.5 \cdot K(\%) + 7.48 \cdot U(ppm) + 1.43 \cdot Th(ppm)$$

which is a serviceable equation quoted by Eslinger and Pevear (1988). A log measurement of uranium was made in this borehole by a spectral gamma-ray tool and is shown in Figure 21 as a "blind test" of the hypothesis. There is a strong similarity between the major anomalies of the residual and uranium logs in the radioactive zones of the regressive limestones and the black shale. The increased uranium within the black shales was fixed by organic and phosphatic matter accumulated in reducing conditions. The cause of the radioactive zones in the regressive limestones is under investigation. Typically, the zones occur as a distinctive unit whose upper boundary is located about a meter from the top of the limestone. It is possible that the uranium was concentrated immediately below the interface of the phreatic and vadose zones at the time when the limestone was partially exposed in a coastal complex setting. Similar anomalies are known to occur in modern cays in the Bahamas (Chung and Swart, 1990). If the modern analog can be demonstrated as applicable to these Pennsylvanian limestones, then gamma-ray logs may provide a tool to map ancient water tables and interpret paleogeography.

The example borehole was selected as a useful test because its spectral gamma-ray log could be used for validation of interpretations from principal component analysis. The vast majority of wells in the area have been logged by a conventional gamma-ray tool that pools all sources of radioactivity. The principal component model can therefore be applied to these other wells, both to generate composite shale logs for improved stratigraphic correlation and to highlight potential uranium anomalies.

GENERALIZED APPLICATIONS OF PRINCIPAL COMPONENT ANALYSIS

The methods of discriminant analysis described previously are types of *supervised* pattern-recognition procedures. Training data sets are used to develop discrimination statistics that best partition the classes recognized before the analysis. The success of the discrimination is first checked by classification predictions from the training or calibration set. Whenever possible, a more realistic assessment is made by measuring classification performance in another known, but independent validation set.

By contrast, when principal component analysis is used to look for patterns and discriminations within data it operates in an *unsupervised* manner. The PCA solution is simply a geometric rotation in multidimensional space that locks onto the orthogonal axes of relative elongation in a cloud of data points. The cloud is represented by a covariance or correlation matrix, so that the derived principal components respond to the structure of intercorrelations between the measurement variables. Any correlations other than zero indicate that there is some degree of information redundancy in the system and that the full dimensionality can be collapsed to a representation in fewer dimensions. In many instances the majority of the variability is allocated to the first two components, in which case a crossplot will show the basic structure of a multidimensional cloud. This attribute of dimensional reduction is a major reason for the popularity of principal component analysis in data processing.

Sufficient variability may be absorbed by the first principal component so that its scores can be used as a synthetic log to summarize an entire log suite. This condensation to a single log allows methods to be applied that work well with univariate data. It was for this reason that Elek (1988) advocated the use of first principal component logs as a useful framework for stratigraphic correlation. Interpreters who routinely correlate with a single log could then apply their pattern-recognition skills to a trace containing significantly more information (Figure 22). Both Elek (1988) and Moline et al. (1992) also described the PCA log as a good medium for zonation, which is an easier objective to achieve using a single log than a multiple log suite.

In a more ambitious analysis it is reasonable to suppose that the loadings of the principal components reflect some underlying causative variables. In some cases the principal components will also prove to be effective surrogates for hidden variables of interest and can be used to create synthetic logs. However, users should be vigilant at all times against simplistic interpretations. In the two examples just described, the relatively simple geology meant that the behavior of the principal components could be anticipated and used to further pattern recognition and analysis. The data were allowed to "speak for themselves," but under controlled circumstances. This control was applied in the selection of the interval and the choice of the logs, and employed some knowledge of the geology and test expectations for the link between the logs and the rock properties of interest.

In the first example shale contents were negligible, so the first principal component was keyed to porosity, while the other components picked up compositional changes in calcite, dolomite, and chert. In the second example, the first principal component primarily reflected shale content because of its relative dominance compared to the smaller variability in effective porosity. Elek (1990) described the results of PCA for the more generalized case of a shaly sandstone succession in a Hungarian field where both porosity and shale content had significant roles. Prin-

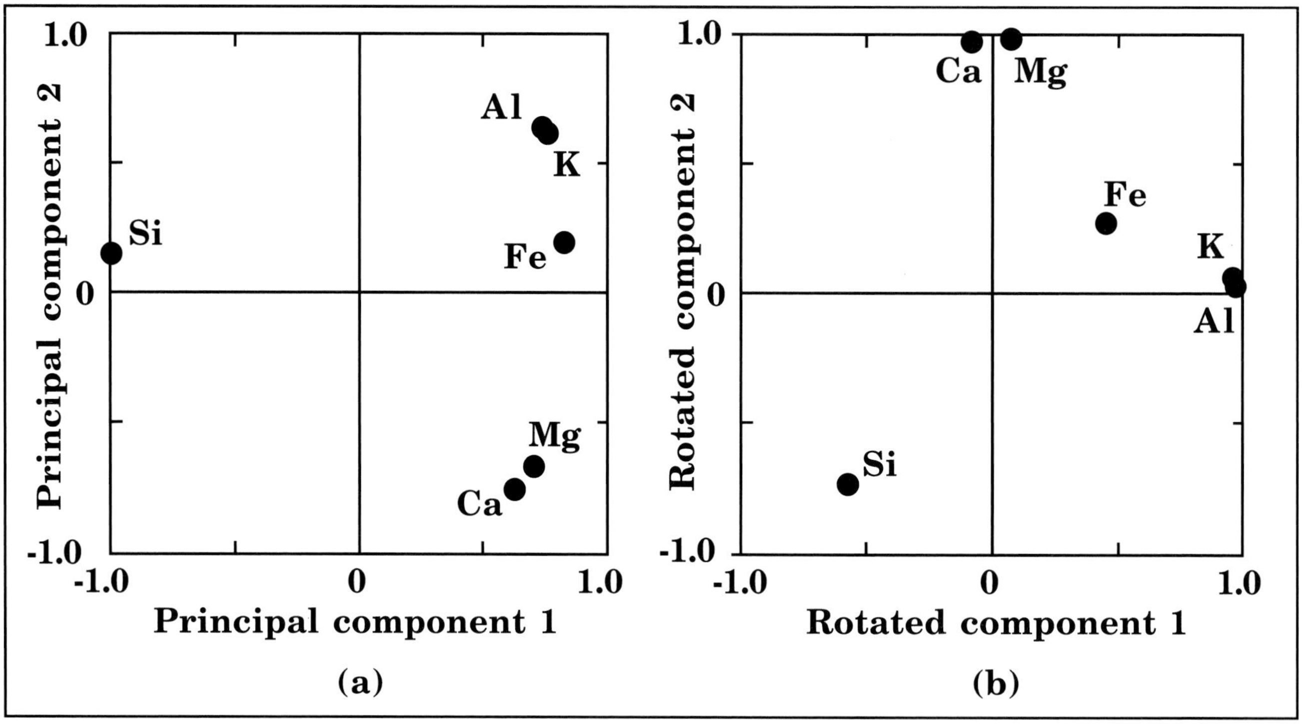

Figure 23. (a) Loadings of elemental analyses on first two principal components contrasted with (b) loadings on axes rotated to a position determined by the varimax criterion. Analytical data from core analyses of a Cretaceous sequence in Utah (Wendlandt and Bhuyan, 1990).

cipal components were computed from a combination of density, sonic, neutron, gamma-ray, and spontaneous potential logs. The strongest loadings on the first principal component were linked with density, gamma-ray, and spontaneous-potential measurements. This result matched the expectations of experienced log analysts that almost everything of importance could be determined from these logs. Elek (1990) found strong associations between the first principal component scores and effective porosities (a correlation of 0.93) and shale volumes (a correlation of –0.91) calculated by traditional log-analysis methods. The principal axis of the data cloud is located along a major trend between clean sandstones and a shale end member where increases in shale content are matched by decreases in effective porosity. Elek (1990) concluded from this study that principal component analysis could provide a means for fast porosity estimation. The dimensionless units of the first principal component could be rescaled to units of effective porosity by calibration with two or more measurements from core.

In a more exploratory study of principal components, Moss and Seheult (1987) came to similar conclusions in experiments with logs from successions of complex clastic mineralogies. Typically, the first component proved to be a surrogate for porosity and a second component accounted for most of "everything else." While generally optimistic, the authors advised caution in the interpretation of the potential "meaning" of individual components. The warning is a word to the wise and is confirmed by common sense. Principal component analysis is a data-driven method. Consequently its results are not dictated by some theoretical model, but by the structure of the data as they are actually observed. These measurements will tend to reflect fundamental causative variables which may be revealed by PCA in its synthesis of intercorrelations between the variables. It is tempting to run the principal component analysis first and ask questions afterwards. Detractors of this approach often call it a "fishing expedition." The underlying linear combinations that are revealed may not be variables that are of interest. The principal components are mutually orthogonal (and uncorrelated) and may not lock onto the underlying variables that may be correlated, and therefore are represented by oblique axes. The likelihood of success of the fishing expedition can therefore be improved by a thoughtful selection of tackle and bait.

First, the fundamental geology and petrophysics of the section should be understood in broad terms. Secondly, log types should be selected that are likely to be related to variables or components of interest. Finally, whenever possible, external validation checks from other information sources should be applied to any PCA interpretations. All three of these rules may be disregarded if the user wishes to embark on a high-risk exploration of logging data. Such a choice may be appropriate in cases where the geology is poorly known and the user wishes to interpret trends, clusters, and statistics that can be used to create theoretical models for further investigation. The risk

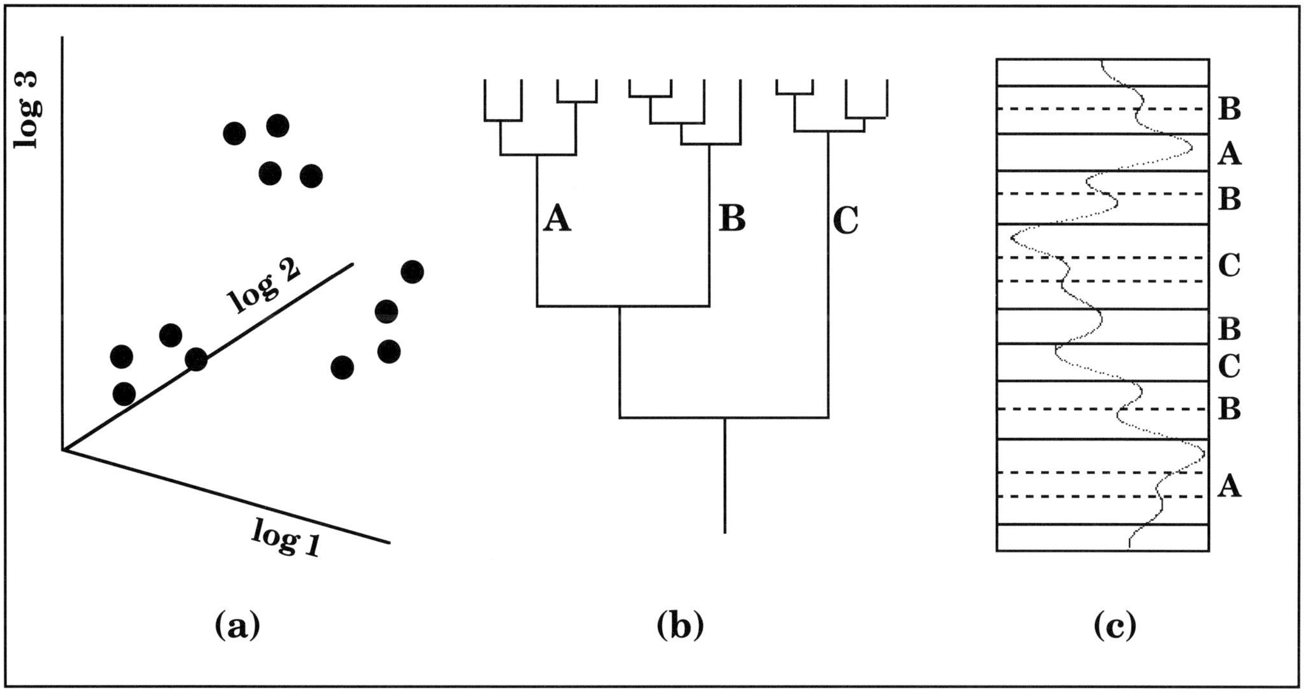

Figure 24. Stages of cluster analysis of log data: (a) multivariate database of zones; (b) dendrogram of zones according to hierarchical clustering of the zones based on their similarities; (c) classification of zones related to input logs and plotted in order of depth.

comes from the possibility of a complete misinterpretation of the data.

There are many other extensions of PCA that can provide useful analytical aids; these are described well by Jackson (1991). A major variant is the facility to rotate the principal components to new positions that show better discriminations with respect to the raw variables. In this rotated orientation, the axes are no longer the eigenvector principal components. However, what is lost from the eigenvectors' property of soaking up the maximum amount of *composite* variability is often made up by a closer link between the rotated components and the original measurement variables. The following example shows the reasoning behind rotation and its possible benefits.

Detailed core analyses were reported by Wendlandt and Bhuyan (1990) from a well which penetrated a Cretaceous sequence of siliciclastic rocks in Utah. The rock types were highly variable and included subarkose, sublitharenite, litharenite, feldspathic litharenite, quartz wacke, lithic wacke, siltstone, mudstone, and coal. Dolomite cement was common, while ankerite and siderite were important locally. For the purposes of this case study, a subset of elemental analyses were selected to explore the relationships that could be revealed by PCA. A principal component analysis of silicon, aluminum, calcium, magnesium, potassium, and iron oxides resulted in an absorption of 93% of the total variability by the first two principal components. The loadings of the elements on these components (Figure 23a) can be crossplotted as vectors that show their interrelationship. The first principal component reveals a bipolar split between silicon and the other elements, which is certainly a reasonable first-order descriptor of a siliciclastic association. However, the bifurcation that contrasts the calcium and magnesium variables with the aluminum, potassium, and iron is obscured by the averaging of the first component.

The most commonly used method for axis rotation applies a "varimax" criterion and was devised by Kaiser (1958). The eigenvectors are rotated rigidly so that they remain orthogonal. The axes are rotated to a position in which the sum of the variances of the loadings is the maximum possible. This criterion is best met when the loadings tend to be close either to a value of unity or to zero. By this means individual variables can be identified more closely with the various rotated components, which (hopefully) will clarify interpretation. The results of a varimax rotation are shown in Figure 23b, and show an improved differentiation of three associations of vectors. These can be identified readily as silica, carbonates, and a composite of clays and feldspar.

Although this example used chemical analyses from core, a similar kind of analysis can be applied to equivalent measurements from wireline geochemical logging tools. The loadings on either the original eigenvectors or the rotated components may be used for the computation of component scores. These scores can be plotted as synthetic logs that combine the information of several elements and whose meaning may be deduced from the component loadings. Rotation is most commonly applied to principal com-

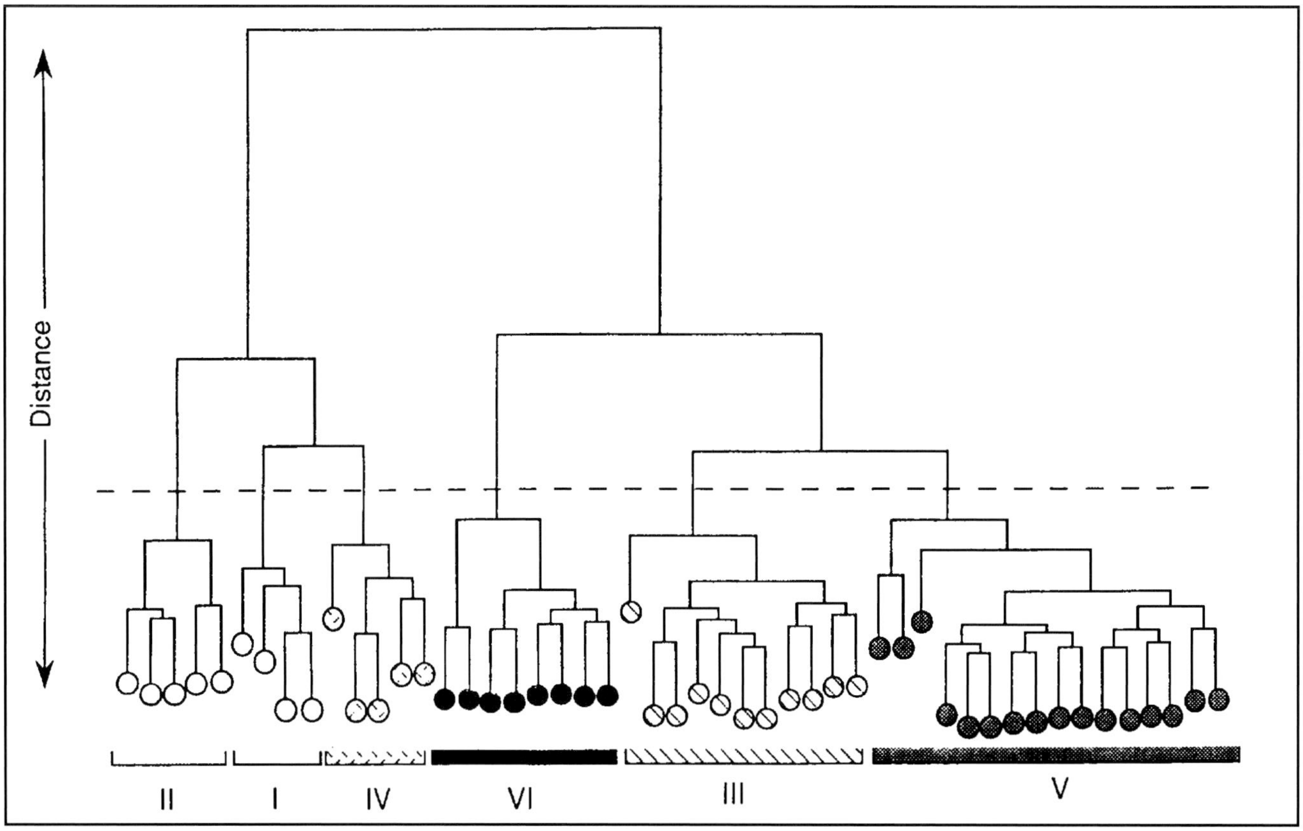

Figure 25. Discrimination of electrofacies of the St. Peter Sandstone in the Michigan basin by cluster analysis of zone similarities based on neutron porosity, bulk density, acoustic travel time, and square-root conductivity. From Moline et al. (1992).

ponent analysis as a secondary step when the investigator wishes more than a reduction in the dimensionality of the problem. The aim of the rotation is to find a simple structure to the data. A satisfactory varimax solution will tend to partition the variables as clusters between the components. The orthogonality of the rotated components means that they remain uncorrelated and so can be studied as independent partitions of the total variability.

CLUSTER ANALYSIS

Cluster analysis is the name given to a wide variety of mathematical techniques designed for classification. The techniques all have a common goal—to group objects that are similar and to distinguish them from other dissimilar objects on the basis of their measured characteristics. The philosophy is one of unsupervised classification, since the operation is neither dictated by an external model nor determined by reference to a training set of objects whose identity is known beforehand. However, the user is involved to a degree in that he or she selects the variables that are to be used in the clustering process. Different selections may generate radically different results. On the basis of everyday experience it should be obvious that individual objects can be grouped or distinguished in different ways according to various criteria we may choose to apply. Hopefully, there is a clearly understood purpose in any cluster analysis that intends to use the result for a specific objective. If, for example, the purpose of the clustering is to make lithological discriminations, then we should select logs that are sensitive to lithology. In some cases the link may not be so obvious, thus the choice of logs will be more intuitive. In this event, the results should be evaluated by reference to the external information that is generally supplied by core data.

The most common class of clustering methods used in geology and other sciences is that of hierarchical analysis (Romesburg, 1984). A cartoon of the basic steps is shown in Figure 24 as an illustration of the clustering of hypothetical zones based on their log responses. First, a database of attribute measurements is compiled for the objects to be clustered. Then a matrix of similarities or statistical distances between the objects is computed on the basis of the collective treatment of the attributes. The choice of similarity measure is itself a decision that needs careful consideration, as the selection will influence the result. The clustering algorithm is applied to the similarity matrix as an iterative process. The pairs of objects with the highest similarities are merged, the matrix is recomputed, and the procedure repeats. Ultimately all the objects will be linked together as a hierarchy,

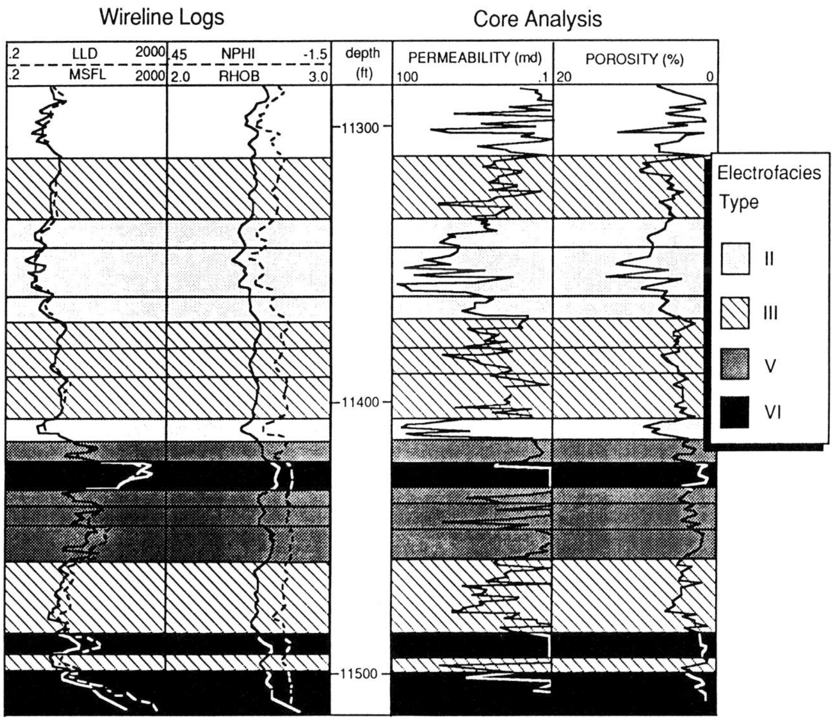

Figure 26. Electrofacies classification of logged zones in a well section of the St. Peter Sandstone in the Michigan basin compared with core measurements of porosity and permeability. The Roman numerals that identify electrofacies types are ordered with respect to average permeability; Type VI has the lowest permeability. From Moline et al. (1992).

which is most commonly shown as a dendrogram. At this point, the objects are in one giant cluster. Some decision must now be made concerning where to cut the tree diagram into branches that coincide with distinctive groupings. The choice may be based either on visual inspection, a mathematical criterion that appears to reveal a natural breaking point, or (preferably) some measure that can be used to check potential clusters against some external standard.

Moline et al. (1992) provides a good practical example of the application of cluster analysis to logs as a means of identifying pressure seals. Recently low-permeability seals have attracted interest, particularly because they may be produced by diagenesis linked with depth and provide traps for hydrocarbons (Powley, 1990). These seals may account for the pressure compartments known to occur throughout the Ordovician St. Peter Sandstone in the Michigan basin. Both the St. Peter and the overlying Glenwood Formation have been the targets for active gas exploration in the last ten years. The location and mapping of potential pressure seals of low-permeability zones is difficult because of the limited availability of core samples and pressure measurements. By contrast, wireline logs are widely available and provide extensive coverage, both vertically and areally.

By a detailed study of logs and cores in four wells, Moline et al. (1992) attempted to define distinctive electrofacies seemingly tied to permeability and porosity. They applied cluster analysis to dissimilarities calculated from zone log responses of neutron porosity, bulk density, acoustic transit time, and the square root of the conductivity. Association with porosity was clear, and it was hoped that permeability discriminations would be apparent from both the porosity and lithological information contained in these logs. The resulting clustered hierarchy is shown in the dendrogram on Figure 25; each terminal branch represents an individual log segment. The hierarchy is a stepped continuum that can be cut at any point to subdivide the total sample into greater or fewer numbers of clusters. Moline et al. (1992) used the core measurements of porosity and permeability as the criteria for the appropriate selection of groups. The best separation was judged by analysis of variance comparisons at different levels of clustering. A subdivision into six electrofacies clusters corresponded to a maximization of the ratio of variances between clusters to variances within clusters. The groups were identified by Roman numerals that reflected an ordering from highest to lowest permeability. Core samples from the lowest permeability zones of electrofacies Group VI showed extensive diagenetic features of cementation, stylolitization, and quartz over-

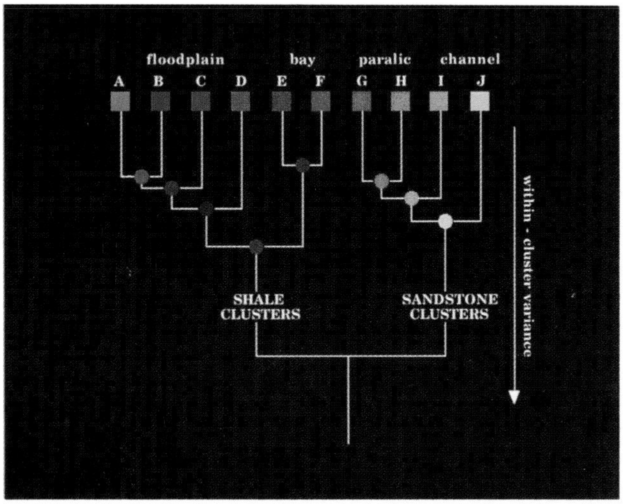

Figure 27. Dendrogram of hierarchical clustering of 124 zones from a Lower Cretaceous sequence of sandstones and shales, based on similarities computed from apparent matrix density, apparent matrix photoelectric cross-section, neutron porosity, potassium, uranium, and thorium logs. In this illustration, the dendrogram is shown truncated at a partition level of ten groups.

growth. The cluster classification was used to construct electrofacies sequences for each of the four wells. An example is shown in Figure 26. By their very nature as derivations from wireline logs, zones from electrofacies Group VI could not be positively identified as pressure seals. However, their depths in two of the study wells located in an area of large hydraulic anomalies gives strong circumstantial evidence for this interpretation. Although this cluster analysis was a pilot-project study in the four wells, its success encouraged enlargement of the application and the assignment of zones to these electrofacies in the many uncored wells of the Michigan basin so they could be used for areal mapping of electrofacies types linked to permeability and the occurrence of potential pressure seals.

In many applications of cluster analysis there are no relationships between objects other than the similarities implied by their attributes. Zones from logs have an additional property: they are ordered along the dimension of depth. Because depth position is not used as part of the clustering procedure, spatial contiguity can be used as an external criterion to select a meaningful number of groups. The evaluation can be made visually by the comparison of logs of clustered zones at different levels of group agglomeration. Alternatively, the best group partition level can be equated with the optimum value of some appropriate measure calculated from spatial contiguity. These ideas are clarified in the following example.

The Lower Cretaceous formations in central Kansas consist of sandstones and shales deposited in a mosaic of deltaic and paralic environments. The implications of log measurements from this succession have been discussed in Chapter 2, where *RHOmaa–Umaa* and spectral gamma-ray ratio plots were applied to the interpretation of the highly variable clay mineralogy and sandstone facies. Collectively, the complex log characteristics are a reflection of a wide variety of sedimentary environments ranging from distributary channels to flood plains, bays, and marginal marine deposits. Cluster analysis can provide an alternative treatment of logging data in which the logged zones are allowed to group themselves and interpretation is made after the fact.

If the original logs had been used as input to the clustering process, then both common sense and the results from the PCA examples would suggest a strong tendency toward formation of groups keyed primarily to shale content and pore volume. Therefore, the input log variables were selected to accentuate potential differences in matrix compositional changes. Similarities between 124 contiguous logged zones of a Lower Cretaceous section were calculated from their values of apparent grain density (*RHOmaa*), apparent matrix photoelectric cross-section (*Umaa*), neutron porosity, potassium, uranium, and thorium. The zones were then clustered as a hierarchy of groups, using Ward's (1963) method of minimum variance. A dendrogram of the result is shown in Figure 27 for levels of partition ranging up to ten groups.

At this point, there is no compelling reason to select a specific subset of groups other than an intuitive feeling for what appears to be a natural break in the branching structure. However, in an alternative display, zones from the group selections that are contingent on all possible choices can be plotted as strip logs (Figure 28). Ideally, distinctive levels in a hierarchical structure should become apparent in the vertical arrangement and relative cohesion of the cluster subdivisions. If there is an upper limit of effective subdivision, then finer clusters should tend to be scattered randomly because their distinction is artificial and local, rather than systematic and regional. This concept has been applied successfully to classification of data by regionalization (Harff and Davis, 1990). The hierarchical classification of locations in terms of their geologic attributes is followed by their geographic mapping. The procedure is a search for the group partition level at which geographic subdivision results in contiguous regions that are relatively homogeneous. The extension of this idea to zones within a borehole succession is easier, because of the spatial arrangement on the single dimension of depth, rather than the two dimensions of geographic location. However, the problem is complicated by the probability that certain associations of zones are likely to repeat at different depths, rather than being constrained to a single level.

An examination of the clustered strip logs in Figure 28 shows distinctive patterns that appear to coincide with certain levels of the group hierarchy. The most obvious example is the division of two groups that clearly distinguish between sandstones and shales. The stratigraphic layers assigned to these

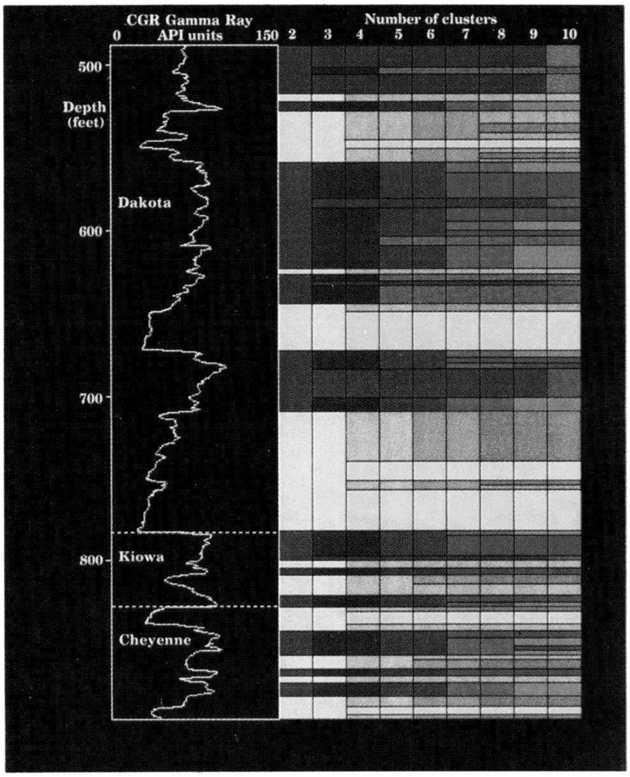

Figure 28. Cluster analysis classification of zones from a Lower Cretaceous sequence of sandstones and shales at different partitions of the clustering hierarchy. See Figure 27 for identification key of color patterns related to clusters.

groups consist of relatively compact aggregations of zones. As a general principle, we would tend to favor a group subdivision that summarizes the succession into a coherent sequence of zones. This goal can be aided by comparing the statistics of the clustered log results with those that would be expected for an incoherent, or random process.

For each alternative clustering agglomeration, the number of input zones, n, is the same. However, the number of layers, l, created by assigning zones to one or other of the groups changes, increasing with the number of cluster groups. A result for each hierarchical level of clustering can be calculated as the quantity, j:

$$j = \frac{n}{l}$$

which is a summary statistic of the average number of zones per layer. The theoretical expectation for j if the layers are composed of zones arranged randomly in the sequence is given by:

$$k = \sum \frac{p_i}{(1 - p_i)}$$

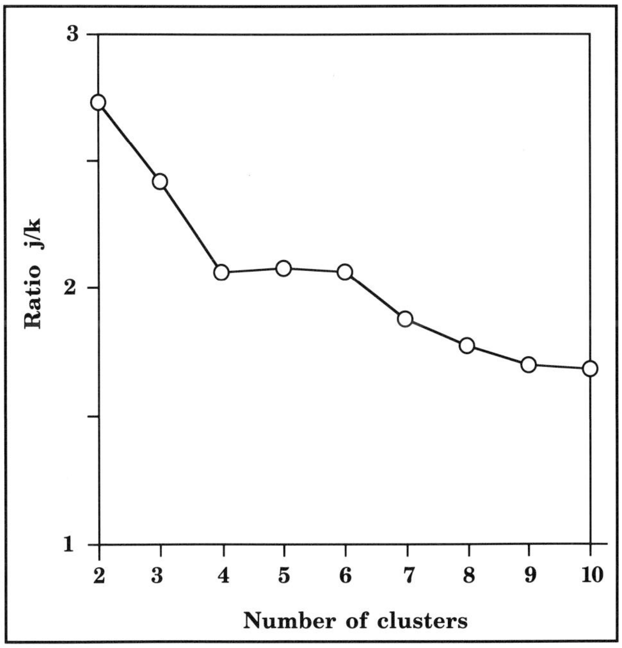

Figure 29. Ratio of observed number of zones per layer (j) to expected number (k) from an independent-events expectation versus number of clusters at successive hierarchical levels of clustering of zones in a Lower Cretaceous section.

where p_i is the proportion of zones assigned to the ith group. The equation follows from Markov chain theory concerning the average residence time that a process remains in a given state (Kemeny and Snell, 1960, p. 61). The value of k is a simple and composite estimate that is averaged across cluster types. A more detailed analysis could be made through the comparison of j and k values for each individual cluster. A plot of j/k ratio values versus number of cluster partitions is shown in Figure 29.

Conclusions concerning appropriate partition levels of the clustered hierarchy for the Lower Cretaceous section can be made by collectively considering the ratio plot, dendrogram, and cluster strip logs (Figures 27-29). Division into two groups picks up the succession's fundamental dichotomy into sandstones and shales. The next natural level of partition appears to occur in the range between four and six clusters. Sedimentological interpretations based on drill cuttings and regional geology suggest that the fourfold division corresponds to facies matched to flood-plain, bay, paralic, and channel environments. The increase to six clusters is a modification of this framework to include distinctions between flood-plain shales and types of paralic sandstones. Another break in the clustering hierarchy possibly occurs at about ten clusters and marks further refinement in the subdivision of the larger groups.

It is important to recognize that the groups generated by this cluster analysis are electrofacies, not lithofacies or genetic facies. Discriminations have

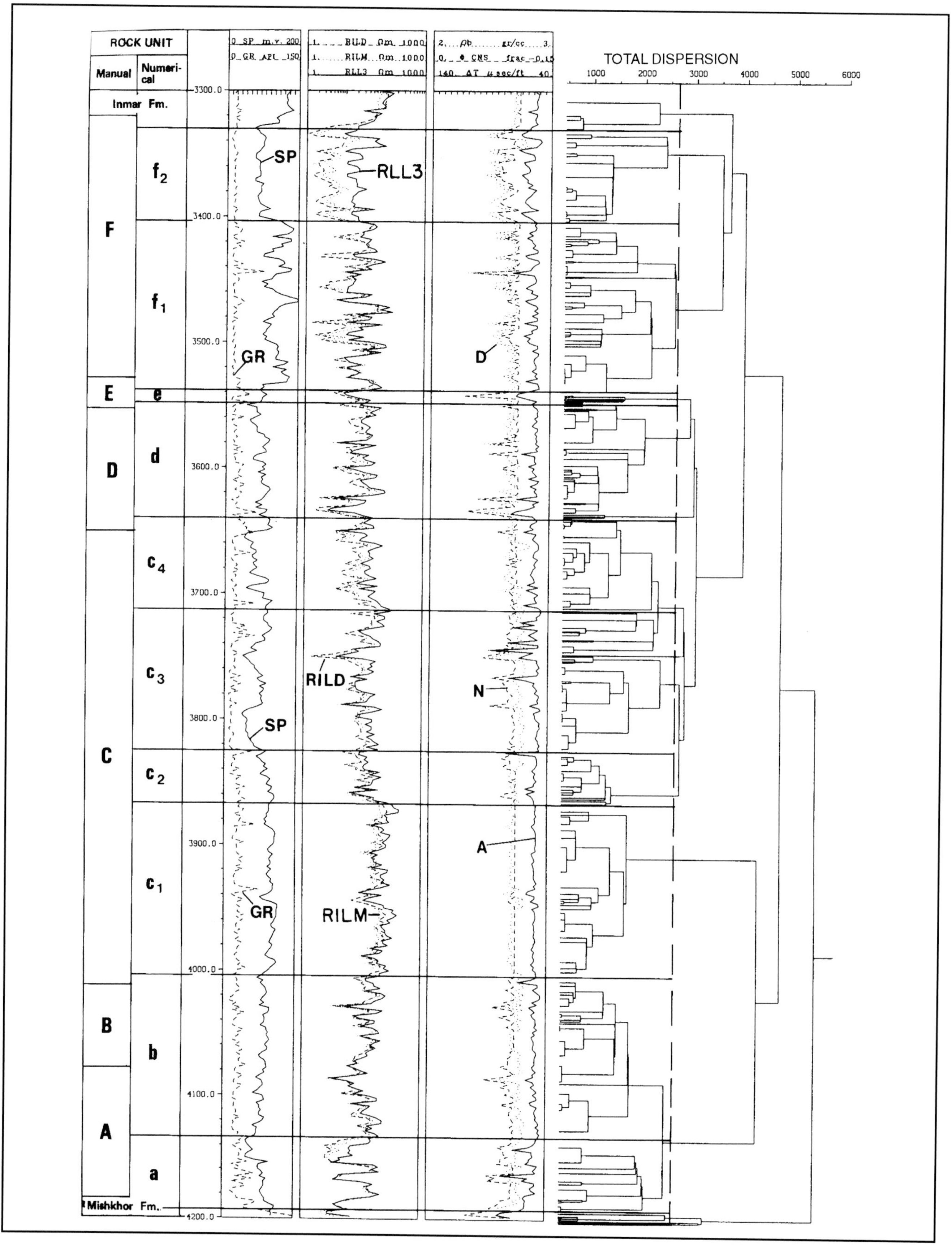

Figure 30. Input logs and zonation dendrogram of a Lower Jurassic succession of limestones and dolomites. From Gill et al. (1993).

been made purely by reference to the log properties of the zones in the sequence. In some cases, a one-for-one correspondence may be argued between an electrofacies and a given lithofacies based on validation by core or drill cuttings. In others, the distinction may not be so clear. The subdivision of the sandstone clusters, for example, appears to be linked most closely with overall grain size. If so, then it may be possible to establish broad equivalencies with sandstone lithofacies. However, fine-grained sandstones occur both within the paralic sandstones at the top of the section and within the upper parts of the channels in the center. These rocks are assigned to common clusters, even though the depositional environments are distinctly different and might be differentiated by properties that could be seen in the cores or cuttings, but are not measured by these logs.

As long as these distinctions are kept in mind, the interpretation of clustering is a matter of common sense and cluster analysis can give valuable insights into the discrimination of geological units. The fundamental difference in philosophy is that it is an *unsupervised* method of pattern recognition. As such, the assignment of meaning to its results must be validated by other sources of information. The situation is further complicated by the fact that clustering is controlled by the variables selected by the user, and even by the type of similarity or dissimilarity measure that is chosen. The text by Sneath and Sokal (1973) is a classic guide to both the theory and the application of the many types of cluster analysis. A more recent general review is given by Romesburg (1984).

A major drawback to hierarchical techniques is the inordinate size of the similarity matrix that must be manipulated if the number of objects is large. If there are n objects, then there will be $n(n-1)/2$ similarities and the cluster analysis will take $(n-1)$ steps. All possible similarities between the current clusters must be considered at each step. Consequently, the computation time increases exponentially with the number of objects. A logged section of even moderate length can result in a cluster analysis that easily exceeds the bounds of reasonable computation time on a conventional computer. For this reason, the use of zoned logs is definitely preferable to the use of continuously sampled digital data as a means of saving substantial computer resources.

An alternative strategy is provided by the k-means clustering method (McQueen, 1967), which is nonhierarchical in its operation. Prior to the cluster analysis, the user specifies k as the number of groups to be clustered. As a starting point, k seeds are defined as cluster centroids in the attribute space and are either selected by the program or specified by the user. A matrix of similarities is calculated between the k centroids and the n objects. The closest objects are clustered with their neighboring centroids. A new centroid is calculated and the process iterates in a way similar to a hierarchical procedure. In general the centroids should move towards the true centers of the local groups. The method usually runs considerably faster than a traditional cluster analysis because it operates on a smaller $k \times n$ matrix of similarities. A major potential disadvantage is that the initial seeds may be poorly chosen, so that outlying clusters fail to be detected.

Some k-means clustering programs allow the user to select not only the number, but the attributes of the seeds. This has great potential application in the clustering of logs, because it allows the use of core data as an additional guide. If limited core is available, then logged zones in the cored interval can be tagged with groupings recognized on the core. These zones can then be used as seeds to cluster zones in other uncored sections. Used in this manner, the technique is a supervised pattern-recognition method, where clustering is a classification device tied to predefined groups. In contrast to conventional clustering, the reduced demands of the smaller similarity matrix also make it a more practical procedure for routine log processing.

Gill et al. (1993) describe an interesting new application of cluster analysis to stratigraphic zonation of logging data. The method is extremely efficient in computer usage. It is an adaptation of a conventional agglomerative and hierarchical cluster analysis, but has an additional adjacency constraint. This constraint limits the analysis to the consideration of stratigraphically neighboring units, thus only vertically adjacent zones and clusters may be merged into larger clusters. This variation on clustering was introduced by Grimm (1987), who wrote a computer program for stratigraphic zonation of palynological data. The procedure also incorporates the algorithm of incremental sums of squares introduced by Ward (1963). Clusters are defined so that the sum of the variances within the clusters is the minimum possible. By using Ward's method and the adjacency constraint, the succession of zones is replaced by a stratified sequence of partitions that merge into coarser units at higher ranks.

Gill et al. (1993) compared the results of stratigraphic subdivision with those of conventional lithostratigraphy drawn from drill cuttings. An example for a Lower Jurassic succession of fairly homogeneous limestones and dolomites is shown in Figure 30. The zoned data consist of readings from the spontaneous potential, deep-induction resistivity, gamma-ray, acoustic travel time, neutron porosity, and bulk density logs. The relative monotony of the sequence made this a realistic test for the method and a challenge for interpretation. In particular, a decision must be made concerning the most appropriate hierarchical level at which to cut across the partition. The potential choices range across the entire hierarchy, from the finest division given by the individual zones to the coarsest agglomeration of all the zones in the section. Gill et al. (1993) compared the dendrogram shown in Figure 30 with a subdivision based on petrography. They found that a reasonable match with the six petrographic subdivisions could be achieved, but this required a total of ten clusters. Some of the finer divisions were attributed to distinctive changes in degrees of dolomitization and porosity, but others were more difficult to explain. As in previous examples, the discriminations are rooted in electrofacies, so correspondences and differences between lithofacies and petrofacies must receive careful thought.

REFERENCES CITED

Anxionnaz, H., P. Delfiner, and J. P. Delhomme, 1990, Computer-generated corelike descriptions from open-hole logs: AAPG Bulletin, v. 74, no. 4, p. 375-393.

Berteig, V., J. Helgeland, E. Mohn, T. Langeland, and D. van der Wel, 1985, Lithofacies prediction from well data: Transactions of the SPWLA 26th Annual Logging Symposium, Paper TT, 25 p.

Busch, J. M., W. G. Fortney, and L. N. Berry, 1987, Determination of lithology from well logs by statistical analysis: SPE Formation Evaluation, v. 2, no. 4, p. 412- 418.

Chung, G. S., and P. K. Swart, 1990, The concentration of uranium in freshwater vadose and phreatic cements in a Holocene ooid cay: A method for identifying ancient water tables: Journal of Sedimentary Petrology, v. 60, no. 5, p. 735-746.

Davis, J. C., 1986, Statistics and data analysis in geology: New York, John Wiley & Sons, 646 p.

Delfiner, P. C., O. Peyret, and O. Serra, 1987, Automatic determination of lithology from well logs: SPE Formation Evaluation, v. 2, no. 3, p. 303-310.

Doveton, J. H., 1986, Log analysis of subsurface geology—concepts and computer methods: New York, John Wiley & Sons, 273 p.

Duda, R. O., and P. E. Hart, 1973, Pattern classification and scene analysis: New York, John Wiley & Sons, 482 p.

Edmundson, H., and L. L. Raymer, 1979, Radioactive parameters for common minerals: Transactions of the SPWLA 20th Annual Logging Symposium, Paper O, 25 p.

Elek, I., 1988, Some applications of principle component analysis; well-to-well correlation, zonation: Geobyte, v. 3, no. 20, p. 46-55.

Elek, I., 1990, Fast porosity estimation by principal component analysis: Geobyte, v. 5, no. 3, p. 25-34.

Eslinger, E., and D. Pevear, 1988, Clay minerals for petroleum geologists and engineers: SEPM Short Course Notes No. 22, SEPM, Tulsa, Oklahoma, 410 p.

Fisher, R. A., 1936, The use of multiple measurements in taxonomic problems: Annals of Eugenics, v. 7, no. 2, p. 179-188.

Gill, D., A. Shomrony, and H. Fligelman, 1993, Numerical zonation of log suites by adjacency-constrained multivariate clustering: AAPG Bulletin, in press.

Grimm, E. C., 1987, CONISS: A FORTRAN 77 program for stratigraphically constrained cluster analysis by the method of incremental sum of squares: Computers & Geosciences, v. 13, no. 1, p. 13-35.

Harff, J., and J. C. Davis, 1990, Regionalization in geology by multivariate classification: Mathematical Geology, v. 22, no. 5, p. 573-588.

Haslett, J., 1985, Maximum likelihood discriminant analysis on the plane using a Markovian model of spatial context: Pattern Recognition, v. 18, no. 3-4, p. 287-296.

Heckel, P. H., 1977, Origin of phosphatic black shale facies in Pennsylvanian cyclothems of Mid-Continent North America: AAPG Bulletin, v. 61, no. 7, p. 1045-1068.

Jackson, J. E., 1991, A user's guide to principal components: New York, John Wiley & Sons, 561 p.

Kaiser, H. F., 1958, The varimax criterion for analytic rotation in factor analysis: Psychometrika, v. 23, no. 2, p. 187-200.

Kemeny, J. G., and J. L. Snell, 1960, Finite Markov chains: Princeton, Van Nostrand, 210 p.

Kerzner, M. G., and E. Frost, Jr., 1984, Blocking—a new technique for well log interpretation: Journal of Petroleum Technology, v. 36, no. 2, p. 267-275.

Kopaska-Merkel, D. C., and G. M. Friedman, 1989, Petrofacies analysis of carbonate rocks: Example from Lower Paleozoic Hunton Group of Oklahoma and Texas: AAPG Bulletin, v. 73, no. 11, p. 1289-1306.

Lanning, E. N., and D. M. Johnson, 1983, Automated identification of rock boundaries—an application of the Walsh transform to geophysical well-log analysis: Geophysics, v. 48, no. 2, p. 197-205.

McQueen, J., 1967, Some methods for classification and analysis of multivariate observations: Proceedings of the 5th Berkeley Symposium on Mathematics, Statistics and Probability, v. 1, p. 281-297.

Mehta, C. H., S. Radhakrishnan, and G. Srikanth, 1990, Segmentation of well logs by maximum-likelihood estimation: Mathematical Geology, v. 22, no. 7, p. 853-869.

Moline, G. R., J. M. Bahr, P. A. Drzewiecki, and L. D. Shepherd, 1992, Identification and characterization of pressure seals through the use of wireline logs—a multivariate statistical approach: The Log Analyst, v. 33, no. 4, p. 362-372.

Moss, B., and A. Seheult, 1987, Does principal component analysis have a role to play in the interpretation of petrophysical data?: Transactions of the SPWLA 28th Annual Logging Symposium, Paper TT, 25 p.

Pettijohn, F. J., P. E. Potter, and R. Siever, 1972, Sand and sandstone: New York, Springer-Verlag, 618 p.

Powley, D. E., 1990, Pressures and hydrology in petroleum basins: Earth Science Reviews, v. 29, no. 2, p. 215-226.

Quirein, J., S. Kimminau, J. Lavigne, J. Singer, and F. Wendel, 1986, A coherent framework for developing and applying multiple formation evaluation models: Transactions SPWLA 27th Annual Logging Symposium, Paper DD, 16 p.

Reading, H. G., 1978, Facies, *in* H. G. Reading, ed., Sedimentary environments and facies: New York, Elsevier, p. 4-14.

Reyment, R. A., 1974, The age of zap, *in* D. F. Merriam, ed., The impact of quantification on geology: Syracuse University Geology Contribution 2, p. 19-26.

Romesburg, H. C., 1984, Cluster analysis for researchers: Belmont, CA, Lifetime Learning Publications, 334 p.

Runge, R. J., and K. J. Runge, 1991, Obtaining lumped (blocked) well logs by simulated annealing: The Log Analyst, v. 32, no. 4, p. 371-378.

Selley, R. C., 1976, An introduction to sedimentology: London, Academic Press, 408 p.

Serra, O., and H. T. Abbott, 1980, The contribution of logging data to sedimentology and stratigraphy: SPE 9270, 55th Annual Fall Technical Conference and Exhibition, Dallas, Texas, 19 p.

Serra, O., P. Delfiner, and J. C. Levert, 1985, Lithology determination from well logs: Case studies: Transactions of the SPWLA 26th Annual Logging Symposium, Paper WW, 19 p.

Sneath, P. H. A., and R. R. Sokal, 1973, Numerical taxonomy: San Francisco, W. H. Freeman & Co., 573 p.

Stowe, I., and M. Hock, 1988, Facies analysis and diagenesis from well logs in the Zechstein carbonates of northern Germany: Transactions of the SPWLA 29th Annual Logging Symposium, Paper HH, 24 p.

Tetzlaff, D. M., E. Rodriguez, and R. L. Anderson, 1989, Estimating facies and petrophysical parameters from integrated well data, *in* Transactions, Log Analysis Software Evaluation and Review (LASER) Symposium, Paper 8: SPWLA, London Chapter, 22 p.

Walther, J., 1894, Einleitung in die Geologie als historische Wissenschaft (3rd Edition), Lithogenesis der Gegenwart: Fischer, Jena, p. 535-1055.

Ward, J. H., Jr., 1963, Hierarchical grouping to optimize an objective function: Journal of the American Statistical Association, v. 38, no. 301, p. 236-244.

Wendlandt, R. F., and K. Bhuyan, 1990, Estimation of mineralogy and lithology from geochemical log measurements: AAPG Bulletin, v. 74, no. 6, p. 837-856.

Wolff, M., and J. Pelissier-Combescure, 1982, FACIOLOG—automatic electrofacies determination: Transactions of the SPWLA 23rd Annual Logging Symposium, Paper FF, 22 p.

Chapter 5

Theory and Applications of Time Series Analysis to Wireline Logs

Each wireline log records the properties of rocks that are ordered with respect to age. Unless the succession is inverted or faulted, depth values are a monotonic function of time, and so, log measurements are examples of time series. However, there are some important distinctions between logs and conventional time series. The results of standard mathematical methods should be interpreted thoughtfully, in a way that takes account of the true nature of the rock record of time. Patterns and trends may be revealed as systematic functions of depth. However, the depth scale should be recognized as both a nonlinear and highly discontinuous representation of time. Variable sedimentation rates and the differing compactional characteristics of lithologies are nonlinear in their effects on time–thickness relationships. Possibly more important is the fact that typical successions are riddled with discontinuities, caused by breaks in sedimentation and erosional events. These discontinuities result in the generation of multiple gaps in the time record to the extent that Ager (1973) suggested that the gaps were the dominant feature. He likened sedimentation to the life of the soldier: long periods of boredom, punctuated by short intervals of terror.

The processing of logs by techniques of conventional time series analysis must therefore be interpreted in the light of appropriate concepts of sedimentation and erosion. The task has been made easier by recent research which has greatly improved our understanding of relationships between sedimentation and time. These newer concepts have their origin in results from seismic stratigraphy and sequence stratigraphy. In addition, significant progress has been made in the computer modelling of sedimentary processes that require numerical estimates of sedimentation parameters and their constraints. Consequently, there is a growing body of theory to link depth-related patterns on log traces revealed by time series analysis to geological mechanisms on a variety of time scales. The results have many practical applications in the resolution of the vertical elements of reservoir architecture, as well as spatial interpolation between wells of systematic units, and the deduction of the geological histories of entire basins.

TREND ANALYSIS OF LOG VARIABILITY WITH DEPTH

Long-term trends in time series data commonly are extracted by the fit of a low-order polynomial curve. The first-order polynomial describes a straight line and is given by the linear equation:

$$\hat{x} = a_0 + a_1 d$$

where $\hat{x}$ is the trend value of the log at depth d, and a_0 and a_1 are the coefficients of intercept and slope, solved by a least-squares regression method. The second-order polynomial is a quadratic function:

$$\hat{x} = a_0 + a_1 d + a_2 d^2$$

which is a parabola and has either a single maximum or minimum value. These two functions are the first in a hierarchy of progressively more complex polynomials. The mth-order polynomial has the equation:

$$x = a_0 + a_1 d + a_2 d^2 + \ldots + a_m d^m$$

and is a curve with (m-1) maxima and minima. Any of these functions can be solved by regression analysis, in which the trend is fitted in a manner that minimizes the sums of the squared deviations of the log values from the trend.

The useful properties of polynomials in their extraction of trends from logs can be illustrated by the analysis of a spectral gamma-ray log from a Lower Permian Chase Group section in southwestern Kansas (Figure 1). The Chase Group consists of carbonates, formed in a variety of shallow marine environments, interbedded with shales that are mostly supratidal in origin. The log shown on the left of Figure 1 is a depth profile of the logarithm of the thorium/uranium ratio computed from the spectral gamma-ray curves. As discussed in Chapter 2, the ratio is an estimate of relative uranium concentration and so can be interpreted as a generalized indicator of ancient redox potential. Darker areas of the ratio log highlight zones of apparent uranium enrichment attributable to reducing conditions at times of deposi-

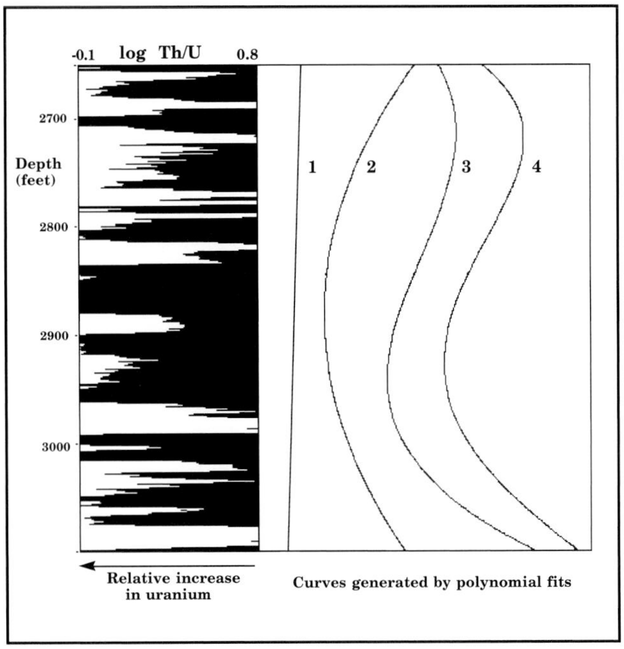

Figure 1. Spectral gamma-ray ratio log of thorium/uranium for a Lower Permian Chase Group section, southwestern Kansas, fitted by polynomial regression functions for first through fourth orders.

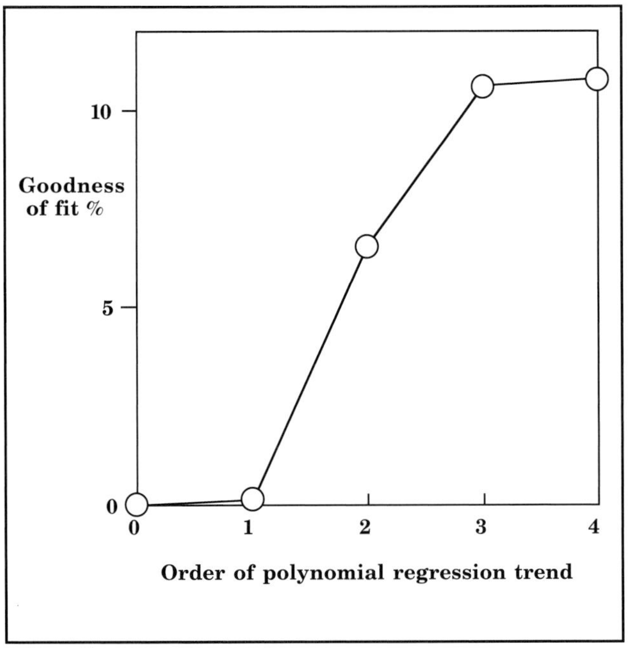

Figure 2. Graph of percentage of total variability accounted for by polynomial regression of thorium/uranium on depth for a Lower Permian Chase Group section from southwestern Kansas.

tion and/or diagenesis. Visual inspection shows a highly fluctuating trace with some indications of broader trends.

A hierarchy of successively higher order polynomials was fitted to the ratio log by standard regression analysis. The polynomial curves are shown at the right of Figure 1; an underlying trend appears to be reached by the third order. This impression is supported by the graph in Figure 2 which shows a rapid increase in the fit of the polynomials to the data up to third order and little increase at the fourth. Notice that even at third order, the total fit to the log is only marginally greater than 10%. This is not surprising when the wild fluctuations of the log are contrasted with the relatively lazy curve of the cubic polynomial (Figure 1). However, an analysis of variance of the fit of the polynomials to the data is useful to verify whether the fitted trends account for a systematic and statistically significant trend in the data. An analysis of variance (ANOVA) can also be made to test the fit of each polynomial against its predecessor to check whether it has absorbed a significant proportion of additional variability (Table 1). The ANOVA results show that the polynomial trends are systematic and suggest that the cubic polynomial is the best simple approximation of the long-term trend in the data.

In reality there is no reason why the polynomial functions should not be generated to even higher orders. A graph of their fits should show additional plateaus at levels where the terms accommodate other distinctive features. However, at higher orders the increasing sensitivity to progressively shorter term components causes the polynomial to become more a smoothed representation of the log than a long-term trend descriptor. Although "trend" is more of a notion than a formal term, Granger (1966) made the useful suggestion that the "trend in the mean" should be equated with all frequency components whose wavelength exceeds the time series record. In general, this will be matched by a relatively low-order polynomial.

We should also reflect that if there is, indeed, a systematic underlying trend, it is highly unlikely that the physical process that caused it would generate a polynomial record. Instead, the polynomial function is a useful descriptor that is calculated easily and can generally provide a good approximation to a variety of trends, whose real form is often unknown. In addition, polynomial equations have simple properties that can be applied usefully in log studies.

In the Chase Group analysis, a cubic polynomial was shown to be a good representation of the trend in the Th/U log. Its equation is given by:

$$\hat{x} = a_0 + a_1 d + a_2 d^2 + a_3 d^3$$

From basic calculus concepts, the first derivative is:

$$\frac{dx}{dd} = a_1 + 2a_2 d + 3a_3 d^2$$

which is the slope of the cubic trend at depth d. The slope is zero at locations of either a maximum or a

TABLE 1. Analysis-of-Variance Table of Polynomial Regressions of Thorium/Uranium Ratio Variation with Depth in a Section of the Lower Permian Chase Group.

Source of Variation	Sum of Squares	DF	Mean Squares	F-Ratio
Linear regression	0.2	1	0.2	0.8
Linear deviation	226.4	894	0.25	
Quadratic–Linear	14.6	1	14.6	61.6*
Quadratic deviation	211.9	893	0.24	
Cubic–Quadratic	9.2	1	9.2	40.5*
Cubic deviation	202.6	892	0.23	
Quartic–Cubic	0.6	1	0.6	2.6
Quartic deviation	202.0	891	0.23	
Total variation	226.7	895		

*significant at 5% level
Critical F-test value at 5% significance, 1 and 895 df = 3.85

minimum. Because the derivative equation is a quadratic, there are two possible solutions for these extremes. The equation is solved following the insertion of the coefficients generated by the regression analysis of the Chase Group data. The quadratic roots locate the depths of the two extremes at depths of 2712 and 2941 ft (Figure 3).

If the derivative equation is analyzed for the second derivative, the result is:

$$\frac{d^2x}{dd^2} = 2a_2 + 6a_3d$$

and is an equation of the rate of change of slope at depth d. When the second derivative is zero, the value of d gives the depth location of the inflection point and is the boundary between the two limbs of the cubic trend. For the Chase Group data, the location of the trend inflection is at a depth of 2827 ft (Figure 3).

The calculation of these polynomial roots is more involved for higher order polynomials. However, a quick method to locate these characteristic depths is to generate derivative logs directly from the log of the polynomial trend. A log of the first derivative is computed by convolving the trend with a simple filter, whose elements are:

$$\begin{bmatrix} 1 & -1 \end{bmatrix}$$

The filter subtracts successive log values to generate a first difference log, which is the same as the first derivative. The extremes are located at depths where the derivative log makes a zero crossing (Figure 3). The second derivative (or second difference) log is computed from the polynomial trend by convolution with the filter:

$$\begin{bmatrix} 1 & -2 & 1 \end{bmatrix}$$

The zero crossings on the second derivative log mark the inflection points and are trend zone boundaries.

These derivative properties have great potential for the interpolation of trends between wells. Traditional correlations are generally keyed to relatively thin and perhaps local key marker beds or horizons. By contrast, the computed extremes and inflection boundaries are global parameters that summarize the entire variability of the section. Because the descriptors are expressed in terms of depth, they can be mapped laterally as regional surfaces with distinctive meanings. In the case of the Chase Group data, the log variable is an expression of relative uranium concentration. If ultimately the log is a response to relative redox potential at time of deposition, then the depth location of the cubic trend minimum is the approximate locus of maximum transgression. On the other hand, if it is diagenetic in origin, then the surface described by this depth when correlated between wells outlines the zone of maximum diagenetic enrichment in uranium. Similarly, the correlation of inflection points between

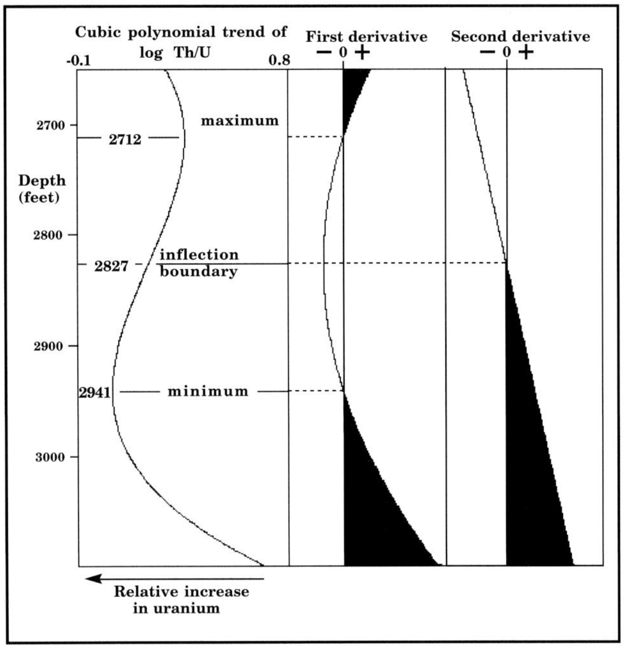

Figure 3. Cubic trend fitted to thorium/uranium ratio log for a Lower Permian Chase Group section in southwestern Kansas, together with logs of its first and second derivatives. The depth of extremes and inflection point are marked by zero-crossings on the derivative logs.

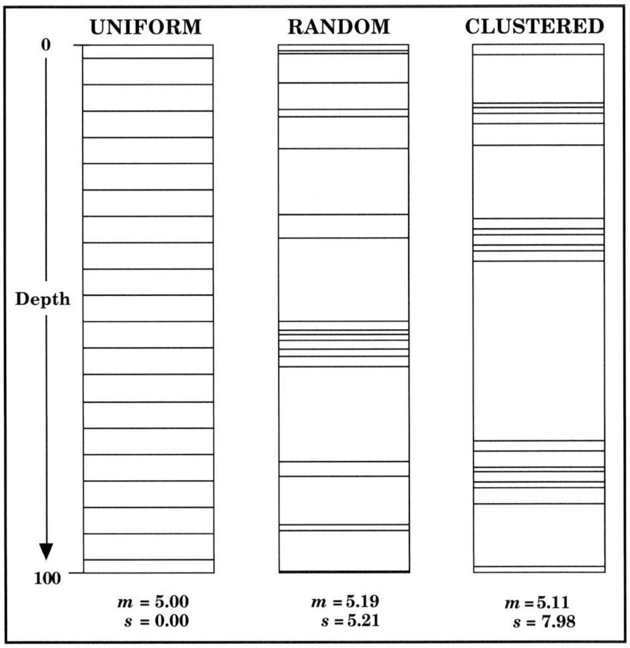

Figure 4. Synthetic sequences of events in an interval of 100 units generated by uniform, random, and clustered processes. In each case, the mean (m) and the standard deviation (s) are tabulated.

wells results in the mapping of major zone boundaries that reflect total log variability.

Doveton (1986) coined the phrase "polynomial stratigraphy" as a description for the process of extracting systematic trend extremes and major zonal boundaries in logs from polynomial regression functions. Clearly, the result is controlled by the type of log that is selected for trend analysis. So, for example, a porosity log would be a good choice in the analysis of a thick reservoir unit. Subdivisions and extremes would then be based on long-term patterns of porosity and (indirectly) permeability, with a stratigraphy that reflected major flow units. Alternatively, if the stratigraphic succession consisted of rapidly alternating sandstones and shales, trend analysis of a gamma-ray log ensures that the stratigraphy would be driven primarily by shale content and, by implication, grain size. The polynomial trend approach would be particularly appropriate for sequences in which lateral correlation of individual sandstones is often difficult, such as complex channel systems or turbidite successions.

THE ANALYSIS OF SEQUENCES OF DISCRETE EVENTS

Wireline tools make analog measurements of the physical properties of rock successions. The essential analog character is preserved by the fine digital sampling of logs recorded on magnetic tape. The logs are well suited for a variety of time series applications because they can be considered to be continuous functions of depth. However, there are occasions when interest is focussed on distinctive log characteristics that signal the occurrence of thin beds of certain lithologies, discontinuities, or other features. These can be thought of as "events" that form a sequence of discrete occurrences separated by thicknesses of "non-event."

The descriptive information concerning such events is simply the depths at which they occur. Are they arranged randomly or is there a tendency toward fixed spacing between these events? Are they grouped in distinctive clusters? Does the rate of their occurrence show systematic changes over the long term? Inferences made from statistical analysis are particularly helpful in resolving these questions, because of the innate human tendency to see patterns even when they are not there. These ideas can be clarified by a review of the set of synthetic examples shown in Figure 4. Each sequence on the figure is an arrangement of 20 events on a hypothetical scale. The left sequence shows the events at equal spacings; in the center they are randomly arranged; at the right, the events are clustered to some degree. Casual inspection of the central sequence might well suggest a nonrandom distribution. However, they were created by a random-number generator and have distribution parameters that are close to independent event expectations, even though the sample is small. The appearance of a pattern is caused partly by the special interest that humans devote to "freak" runs of events, while ignoring non-clustered sequences. This quirk has resulted in various superstitions such as notions that circus accidents and hospital patient deaths are clustered. Solterer (1941) provided an interesting statistical analysis of mortality tables in which he laid to rest the legend that Jesuits tend to die in threes.

If anything, scientists are even more prone than others to seeing patterns in random data. This assertion would seem to contradict the belief that scientists are trained to be objective. However, as discussed by Zeller (1964), the practice of science puts a high emphasis on an ability to find order, even when order may not be there. Sequences of events are particularly difficult to assess by eye in any systematic fashion. Miall (1992) described an experiment in which he was able to make surprisingly reasonable correlations between sequences generated by random numbers with the Exxon global-cycle chart (Haq et al., 1988). While these considerations do not necessarily invalidate the reality of current geological models of time, sedimentation, and stratigraphy, they do point to the need for systematic numerical tests for patterns.

The limiting model for the distribution of events over time is that there is no pattern to their occurrence at all: they are random. The relative frequencies of random events are described by the Poisson distribution. In a Poisson process, the mean and the variance of the number of events per unit time are equal. If the variance is significantly smaller then the mean, then there must be a tendency towards regularity in spacing. The variance is zero when the events are equally spaced. If the variance is significantly greater than the mean, then the events are clustered to some degree. Extremely large variances will be generated by tight clusters of events with long periods between them. These general relationships match simple intuition or, failing this, can be understood after a consideration of the three alternatives shown in Figure 4. In practice, the mean and variance of a typical random time series will only be approximately equal, because they are estimates of true parameter values, based on a limited sample.

The application of Poisson models is best explained in terms of real examples. Beds of anhydrite occur commonly throughout the Chase Group in southwestern Kansas and mark evaporite deposits thought to be linked with sabkha-like environments. Their relative depth positions may be useful zones marking exposure events in the depositional history of the group. The locations of anhydrite beds can be found through inspection of neutron–density log overlays or from *RHOmaa–Umaa* crossplots. A record of anhydrite occurrences in a Chase Group sequence is shown as the strip log to the left of Figure 5. A total of 34 anhydrite zones occur over a range of 450 ft. When the succession was subdivided into 18 equal intervals, it was found that the mean number of anhydrite zones per interval was 1.89 and the variance was 2.81. Based on the preceding discussion, it appears that the distribution is more clustered than would be expected if they were randomly distributed. However, some allowance must be made for the relatively small sample of anhydrite beds.

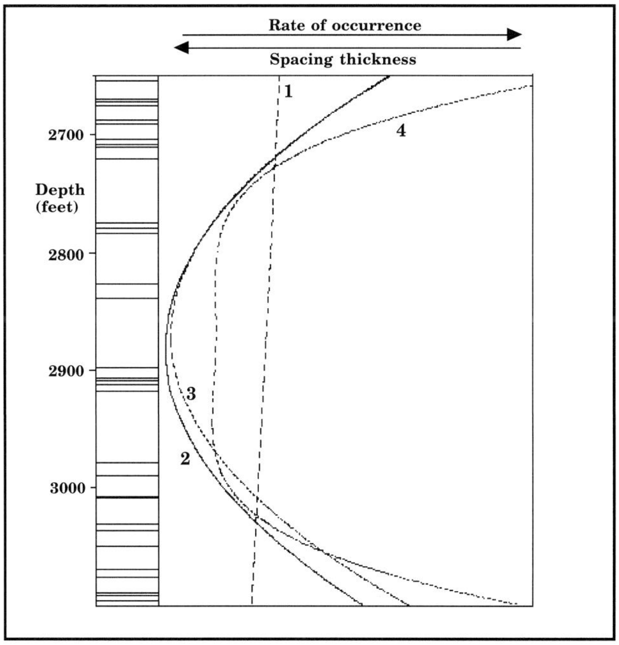

Figure 5. A Poisson model with trend fitted to distances between successive anhydrite beds in a Lower Permian Chase Group succession in southwestern Kansas. First- through fourth-order polynomial regressions are shown on the right; anhydrite occurrences are shown as a strip log to the left.

The predictions from a Poisson model of random events presume that the mean rate of occurrence is a constant. Put in another way, the depth record is presumed to be "stationary." If there is a systematic drift in the mean rate, then this is a trend that underlies the event sequence and it may have long-term geological significance. The nonstationary alternative can be modelled as a Poisson process with trend. Rather than work with frequencies of events in each (arbitrary) interval, it is often more convenient to consider the distance between successive events. Then the average distance, h, is given by the reciprocal of the rate of occurrence, r, that is:

$$h = \frac{1}{r}$$

Now, if there is no trend, the rate r is a constant, and is conventionally written as:

$$r = e^{a}$$

where a is a constant. If there is a linear trend with time, then:

$$r = e^{a_0 + a_1 t}$$

where t represents time, or, as in this example, depth.

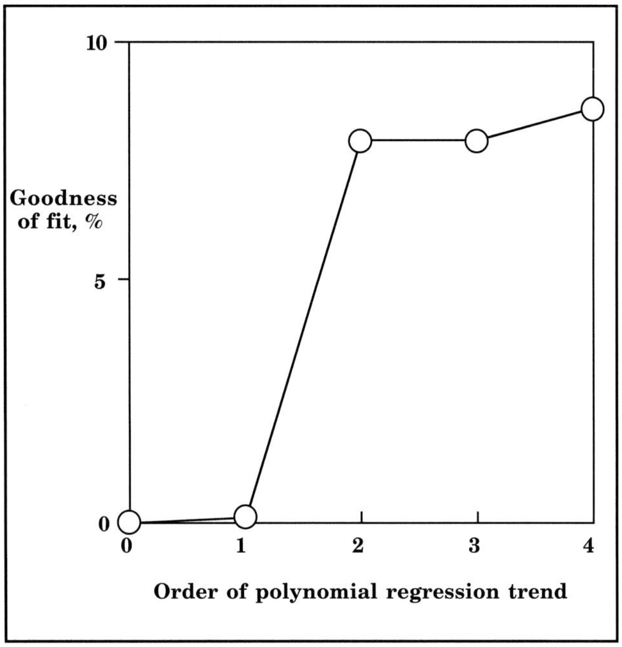

Figure 6. Graph of percentage of total variability accounted by for Poisson-with-trend polynomial regressions of anhydrite occurrences in a Lower Permian Chase Group section in southwestern Kansas.

Taking logarithms of both sides:

$$\log r = a_0 + a_1 t$$

from which it follows:

$$-\log h = a_0 + a_1 t$$

because the length between events is the reciprocal of the occurrence rate. In fact, the trend can be expanded to a polynomial series in a very similar fashion to that used for trend analysis of continuous log data earlier in this chapter. For the *m*th polynomial the trend equation is:

$$-\log h = a_0 + a_1 t + \ldots + a_m t^m$$

The polynomial equations are the basis for regression analysis of the thickness between each pair of successive events versus the depth of the midpoint between the events. Consequently, the analysis does not require any special programming, but can be run using a conventional multiple regression procedure.

The results of fitting first- through fourth-order trend polynomials to the Chase Group anhydrite data are shown by the curves on Figure 5. A quadratic

TABLE 2. Analysis-of-Variance Table of Polynomial Regressions of Interval Thickness between Successive Anhydrite Beds with Depth in a Section of the Lower Permian Chase Group.

Source of Variation	Sum of Squares	DF	Sum of Squares	F-Ratio
Linear regression	0.5	1	0.5	0.1
Linear deviation	333.0	32	10.04	
Quadratic–Linear	26.0	1	26.0	2.9
Quadratic deviation	307.1	31		9.9
Cubic–Quadratic	0.0	1	0.0	0.0
Cubic deviation	307.1	30	10.2	
Quartic–Cubic	2.5	1	2.5	0.2
Quartic deviation	304.6	29	10.5	
Total variation	333.5	33		

Critical F-test value at 5% significance, 1 and 31 df = 4.16

trend appears to be the best overall descriptor. This conclusion is based both on the curve shapes and the abrupt break in fit that is graphed on Figure 6. The degrees of fit are comparatively low and fail to pass muster as significant trends, as shown by the analysis of variance in Table 2. This reflects the fact that the sample size is fairly small (33 occurrences of anhydrite), so that the high degree of fluctuation in successive spacings between anhydrite beds may overwhelm any systematic trend. However, notice that the variance ratios of the ANOVA table also indicate that the quadratic trend is distinctive. Strictly speaking, the aggregate spacing of events taken at a minimum of four at a time is preferable in order to average out some of the extremes (Cox and Lewis, 1966). However, this step becomes impractical for short sets of events and may mask some changes in occurrence rate.

Because the functions used to describe the trends are polynomials, the depth locations of maxima, minima, and inflection points can be located using the methods described earlier. These depths provide global estimates of the overall distribution of events that can be interpolated between wells in the mapping of lateral changes in event rates. The log record of anhydrite beds in the Chase Group used for this example comes from the same well as the spectral gamma-ray ratio log used to illustrate trends located by polynomial regression. The results from the two analyses appear to show common global features. The overall maximum relative concentration in uranium occurs in the middle of the group, which is also the position of the largest average interval between successive anhydrite beds. The concordance can be explained in terms of alternative models keyed to either depositional or diagenetic processes.

Uranium may have been fixed in reducing conditions that reflect deposition in deeper marine environments. Intervals deposited at greater depths would be expected to have a lower frequency of anhydrites produced by exposure and evaporation. Alternatively, the uranium may have been concentrated in carbonates below the anhydrite beds as a product of evaporitic pumping in a sabkha environment. This groundwater mechanism has been suggested by Rawson (1980) as the cause for commercial occurrences of uranium in the Jurassic Todilto Limestone of New Mexico. Further research is needed to evaluate these and alternative models. While the analysis of logs may reveal systematic patterns, the potential meaning of the patterns is a matter for geological hypotheses and tests.

ANALYSIS OF CYCLIC PATTERNS ON LOGS

A common objective in the interpretation of sedimentary records is the location of distinctive repetitive sequences that are commonly called "cycles" (Duff et al., 1967). It seems reasonable that long sequences would be built from simple lithological associations. The structure of these basic units would reflect vertical transitions between a distinctive association of related depositional environments. Finally, the driving mechanism for the process could be attributed to lateral facies migrations caused either by autocyclic processes or larger scale phenomena such as structural movement or eustatic change in sea level. Care must always be taken not to recognize cycles that do not exist through overzealous interpretation, as discussed by Zeller (1964).

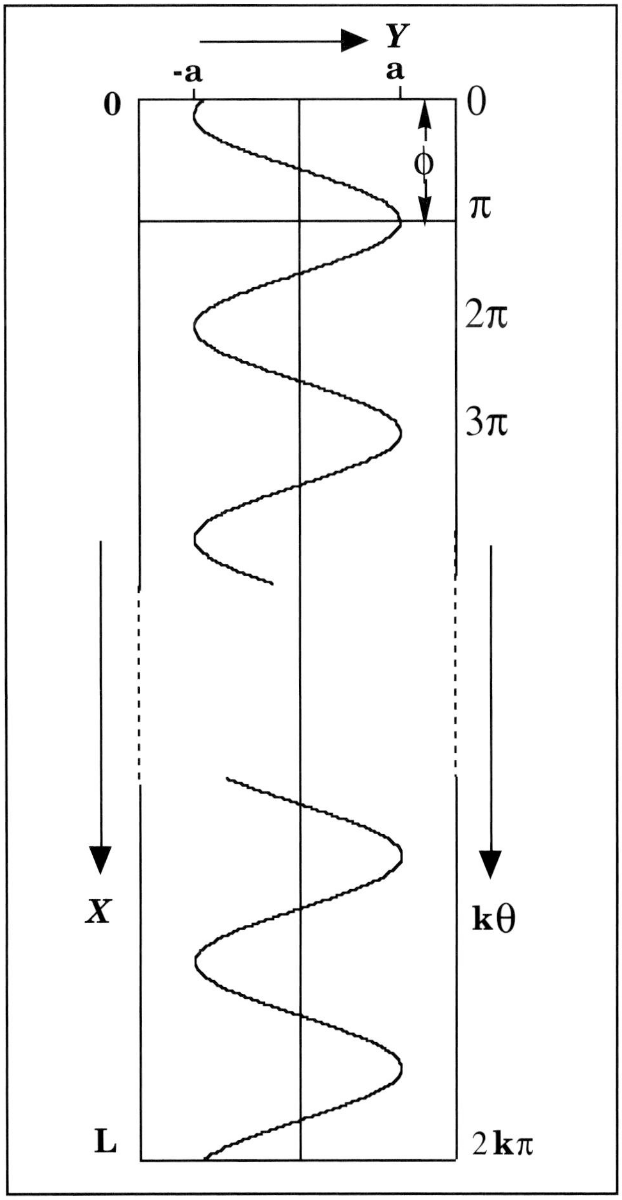

Figure 7. Representation of a depth interval, *L*, by a cosine wave with *k* harmonics. The "starting point" of the sine wave is given by the phase angle, ϕ, and the depth units, *X*, are converted to angular scale, θ, by equation 2π radians with one wavelength.

Wireline logs are particularly well-suited for the analysis of possible cycles, because such logs are numerical, lengthy, and continuous. Consequently

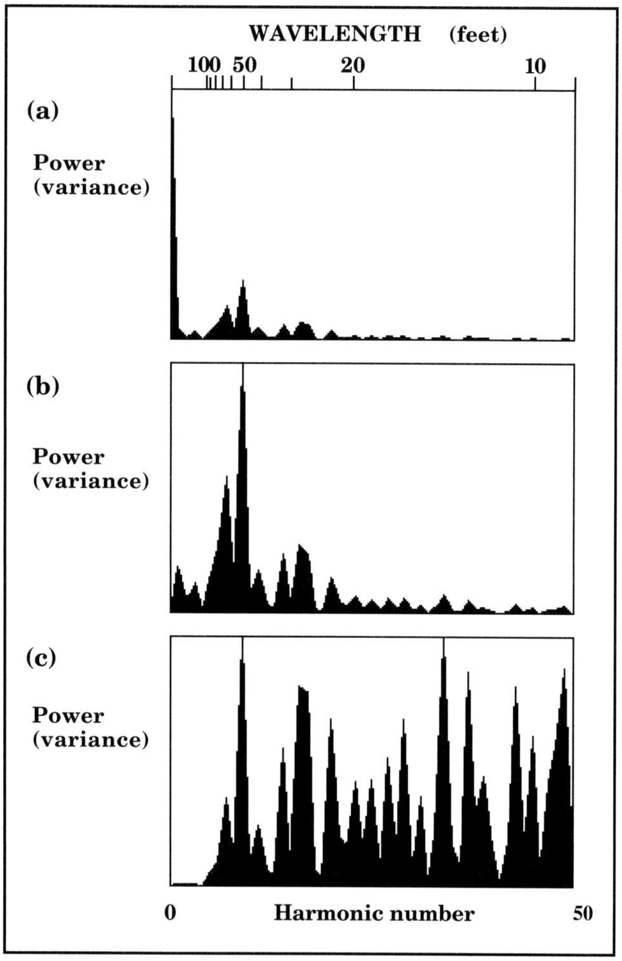

Figure 8. Periodograms of the computed gamma-ray log from a Lower Permian Chase Group section in southwestern Kansas: (a) original log; (b) log with mean subtracted; (c) first difference (derivative) of log.

they contrast with records from outcrop and core successions, which are generally descriptive, short, and discrete. The fundamental model for cyclicity is that of the sine wave which sketches out the operation of a circular process as it develops through time. A repeating sine wave may be fitted to log data by a least-squares method that is a minor modification of conventional linear regression discussed in Chapter 1. This is shown in Figure 7, and is developed using some simple trigonometry as discussed below.

Both sine and cosine functions generate sinusoidal waves. The cosine equation to generate a wave is:

$$Y = a \cos \theta$$

where a is the amplitude of the wave and θ is an angular measure that changes with time in a cyclical manner. The wave will repeat at an interval set by its frequency, which is the number of cycles per unit of time or distance. Because we are working with logs, we will take depth measurements as some monotonic function of time. The conversion from depth to angular measure can be understood from examination of Figure 7. If k cycles occur over a total thickness of h, and each cycle has a wavelength of p, then:

$$h = kp$$

If the depth axis is symbolized as X, then the conversion from depth to angles is given by:

$$\theta = \left(\frac{2\pi}{p}\right) X$$

when expressed in radians, where 2π represents one complete revolution. Because radian measures are cyclic, the value of Y will repeat at depths that are displaced by multiples of p depth units. The cosine equation as written above is a limiting case, because it equates a depth of zero with an angle of zero and a single cycle. In order to make the equation more general for k cycles that originate at some unknown depth, it can be modified to:

$$Y = a \cos (k\theta - \phi)$$

where ϕ is called the phase angle and represents the initial shift displacement between the depth origin and the beginning of the nearest wave. Now, from simple trigonometry, this can be expanded:

$$Y = a \cos \phi \cos k\theta + a \sin \phi \sin k\theta$$

Because the phase angle is a constant, the equation can be consolidated:

$$Y = A \cos k\theta + B \sin k\theta$$

where the tangent of the phase angle is equivalent to:

$$\tan \phi = \frac{B}{A} = \frac{a \sin \phi}{a \cos \phi}$$

The power or variance of the sinusoidal wave is the square of its amplitude, which is given by the sum of the squares of the coefficients:

$$A^2 + B^2$$

The expanded equation:

$$\hat{Y} = A \cos k\theta + B \sin k\theta$$

can be used to fit k cycles to a log trace, Y, by least

squares in a conventional multiple regression of the general form:

$$\hat{Y} = AX_1 + BX_2$$

because the trigonometric quantities are simply angular transforms of the depth variable.

When fitting k cycles to a log segment, the resulting wave form is known as the kth harmonic of the data. A complete set of harmonics is known as a Fourier series, which, when summed together, recreates the original log in its entirety:

$$Y = \sum A \cos k\theta + B \sin k\theta$$

The summation range extends between zero and a harmonic number of $n/2$, where n is the number of data points in the sequence. This highest harmonic corresponds to the Nyquist frequency. This is a limit beyond which higher frequency wavelets cannot be estimated and is set by the data spacing. Because the harmonic terms are orthogonal, the series represents a set of sine waves that are uncorrelated with one another. The series is conventionally shown by a periodogram that graphs the power (variance) against the harmonic number (see Figure 8). Alternatively, the harmonic numbers may be represented by a scale of their corresponding wavelengths, and the wave amplitude plotted instead of the power (the square of the amplitude).

This is an introductory treatment of the mathematics that underlie basic and discrete Fourier analysis. The concepts and their potential for geological interpretation are best understood by reference to a real example. The Lower Permian Chase Group used earlier in this chapter is an excellent subject for Fourier analysis because its component lithologies have been thought for many years to be ordered in cyclothems. The group consists of a repetitive sequence of marine carbonates and supratidal shales. In southwestern Kansas, the generalized setting was one of tidal flats and shallow marine environments. The lithological record presumably reflects the operation of oscillatory processes on the carbonate shelf, driven by tectonic, eustatic, or more localized phenomena. Fourier analysis can be used to isolate any systematic cyclic signal and to estimate its wavelength and stratigraphic position. The computed gamma-ray (thorium and potassium) log is primarily a measure of shale content and should be a useful record of the alternations between prograding clastic sources to the northwest and marine carbonate transgressions from the southeast.

Three alternative periodograms of the gamma-ray log (Figure 9) are shown in Figure 8. The periodogram at the top gives the power or variance associated with each harmonic of the original log. The spectrum is dominated by the zero harmonic. This harmonic represents a wave with an infinite period,

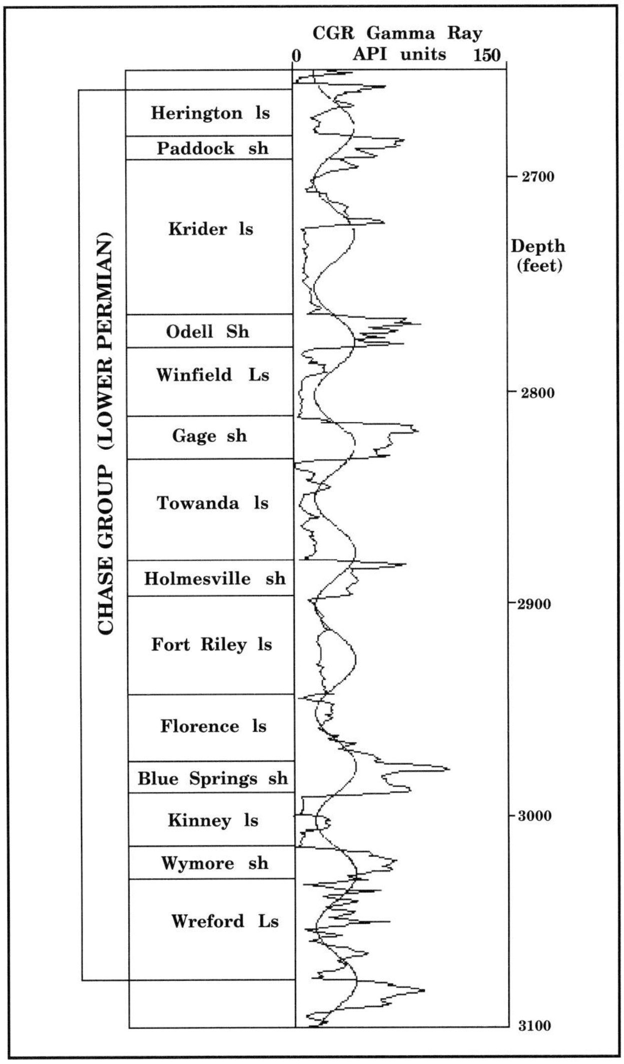

Figure 9. Computed gamma-ray log from a Lower Permian Chase Group section from southwestern Kansas fitted with ninth harmonic sine wave.

and so is a constant value positioned at the mean gamma-ray value. The power of the zero harmonic thus simply reflects the mean value. If a periodogram is computed for the log after the mean has been subtracted, then the zero harmonic is eliminated from the resulting periodogram (middle, Figure 8).

The spectrum shows a pronounced peak at the ninth harmonic, which corresponds to a wavelength of 50 ft. In computing a periodogram it is customary to remove any long-term trend from the data in order to suppress broad changes and emphasize distinctive cyclic phenomena with higher frequencies. There are a variety of methods to do this that depend to some degree on the understanding of what constitutes a long-term trend. A polynomial regression will fit the raw variation with a lazy curve, acting as a high-pass filter that concentrates higher frequencies within the residuals. Both long- and intermediate-term fluctuations can be absorbed through the computation of the first derivative of the log by subtracting adjacent log

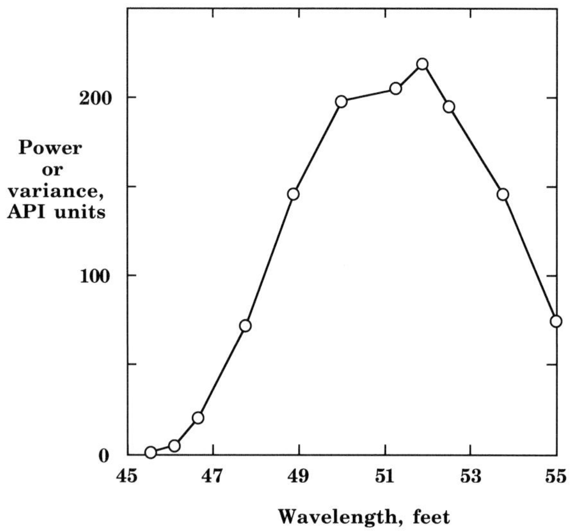

Figure 10. Location of fundamental wavelength found from multiple periodograms computed for slightly different lengths of section in the computed gamma-ray log from a Lower Permian Chase Group section.

values. The third periodogram (bottom, Figure 8) graphs the power of harmonics computed from the log of the first derivative and strongly accentuates the higher frequency harmonics. The ragged and complex character in the high-frequency range is largely a reflection of the stochastic nature of gamma-ray measurements. Much of the character can be explained in terms of error noise, which could be dampened by some local smoothing of the log prior to the computation of this periodogram.

The conversion of the gamma-ray log into a periodogram is a transformation from the time domain into the frequency domain. Conversely, each harmonic represents a repeating and simple sine wave whose wavelength is given by the total length of the record divided by the harmonic number. The high proportion of the total variance picked up by the ninth harmonic means that the corresponding sine wave should be a good representation of the major pattern of shale–carbonate alternation. The representation by a single harmonic suggests the operation of a systematic mechanism to produce such a regular spacing. If the period was more irregular, then the spectrum would tend to be smeared over a range of frequencies.

A single kth harmonic may be converted to its equivalent wave form in the time domain through the use of the Fourier equation:

$$Y = A \cos k\theta + B \sin k\theta$$

where θ is the angular conversion of the depth into k cycles shown diagrammatically on Figure 7 and calculated by:

$$\theta = \frac{(D - D_0)\, 2\pi}{k}$$

where D_0 is the depth of the top of the sequence marked as a zero angular reference point. The coefficients of A and B are taken from the Fourier series computation used to generate the periodogram. The sine wave of the ninth harmonic is shown superimposed on the gamma-ray log of the Chase Group section at the right in Figure 9. As anticipated, the extremes of the sine wave show a good match with the shale and carbonate subdivisions.

The periodogram is a discrete or line spectrum because it is composed of a hierarchy of integer harmonics. There is often no reason to believe that a sequence is described by an integer number of cycles. This is inconvenient if one wishes to estimate the period of the sine wave that best characterizes the cyclic character of a sequence. The discrete steps between harmonics causes matching jumps in the associated periods. So, in the Chase Group example, the ninth harmonic gives a period of 50 ft, but the adjacent eighth and tenth harmonics are matched by periods of 56.25 and 45 ft respectively. A pragmatic solution to this problem is to rerun the computation of the periodogram over intervals that are slightly extended or truncated in length compared to the original sequence. The changes in total length cause the periods associated with the integer harmonics to be shifted slightly. This strategy was used for the Chase Group section to refine the estimate of the period of the fundamental wave. Power versus period was plotted from the results of these modified runs and is shown in Figure 10. It appears that an optimum wavelength is between 51 and 52 ft.

The discrete Fourier series is widely used in time series analysis because so many records are measurements that are taken at intervals of a day or some other time unit. However, the wireline log is a continuous trace that can be considered to be an analog signal. The geological information content is lost at the high frequency end, through both the smoothing by the tool response function and the stochastic noise recorded by nuclear tools. Kerford and Georgi (1990) described some useful logging applications of this effective partition between the low-frequency formation signal and high-frequency tool effects. In a comparison of the Fourier transforms of main and repeat logging runs, they showed that they could discriminate measurement noise and make estimates of effective vertical resolution at different logging speeds.

The conventional digital sampling rate of two readings per foot is a reasonable sampling rate that takes account of this lower limit of resolution. A continuous spectrum of component frequencies can

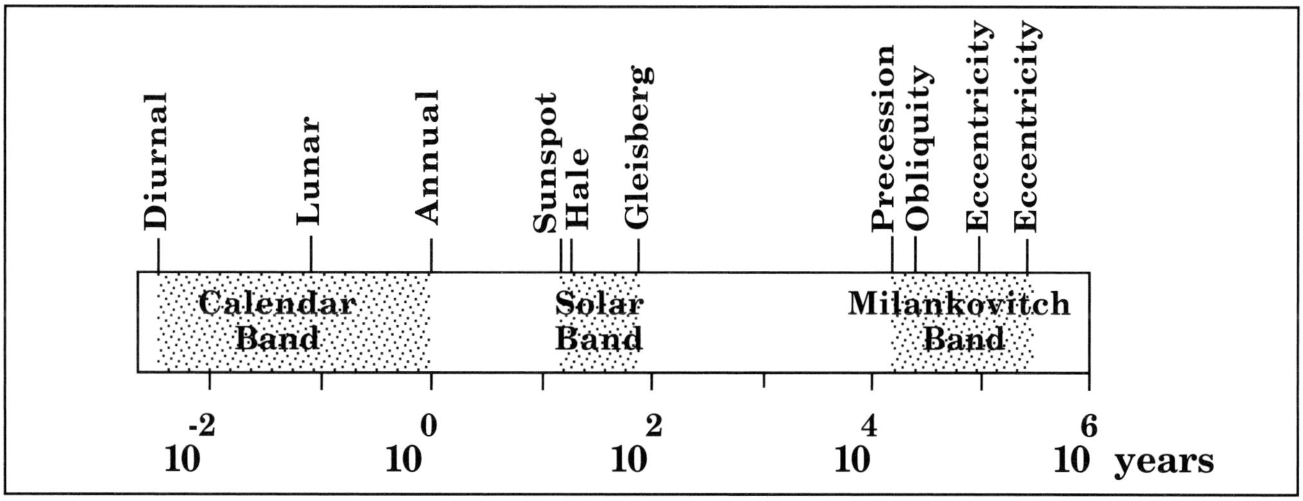

Figure 11. Spectrum of terrestrial and solar rhythms in geological time with periods of less than a million years. Adapted from Fischer and Bottjer (1991).

then be generated by the Fast Fourier Transform (FFT) introduced by Cooley and Tukey (1965). The computer requirements are greater than those used in the Fourier analysis of discrete harmonics. Basically, the Fourier transform is expressed in an exponential form that incorporates imaginary components (functions of the square root of minus one). The FFT then finds the complex coefficients of all wavelengths down to the Nyquist frequency. A limiting condition of the FFT operation is that it must be applied to a record whose number of sample points is a power of two.

The Fast Fourier Transform is used routinely in seismic processing by geophysicists, who find it convenient, or even preferable, to operate in the frequency domain. Most geologists feel more comfortable in the time domain of the stratigraphic framework. However, the use of spectral analysis by geologists has become increasingly common. In part, this is because notions of cyclicity in the rock record have a long history and geologists are making increasing use of the computer. In addition, recent models of climate forcing by cyclic astronomical phenomena have been a major stimulus to the search for systematic periodic components that can be linked with correlatable time events.

ORBITAL FORCING MODELS FOR SEDIMENTARY RHYTHMS

Up to this point we have discussed various mathematical techniques to extract potential systematic geological trends and rhythms from wireline logs. However, these empirical results must be explained in terms of some causative process. One hopes that the observations will match clearly the expectations of one of several alternative models. While not proving any given model to be true, a broad consistency between data and theory is a first step towards acceptance. A model will find acceptability if support is

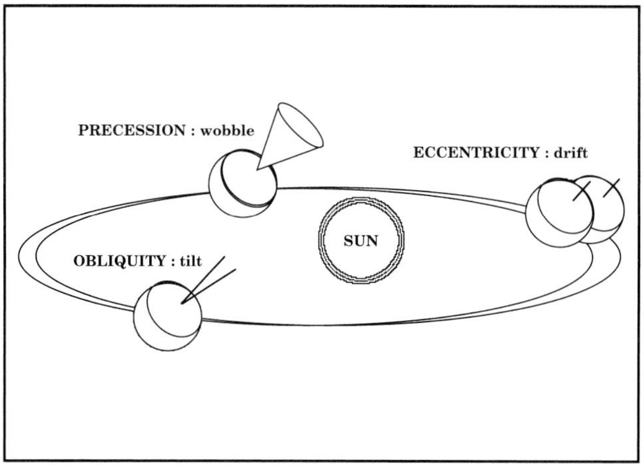

Figure 12. Components of variations in the Earth's orbit (eccentricity) and its angle to the orbital plane (obliquity and precession).

forthcoming from a wide variety of geological sources of evidence. Ultimately, any model "proof" is circumstantial, because the rocks were formed at some remote time in the past. The match of theory to observations is of more than academic interest. Depending on the model, rhythmic patterns may be shown at various scales ranging from localized elements in the architecture of a single reservoir, up through basin-wide phenomena, to events of a truly global scale. The identification and numerical description of an underlying process model has important applications in reservoir engineering, correlation, basin analysis, and exploration.

The roots of modern geology are founded in the uniformitarianism propounded by Lyell. The Earth was considered to be effectively constant through time, so that rocks of the past were formed under similar conditions that prevail today. Regional differences were attributed to tectonic mechanisms; smaller changes could be explained by localized shifts in

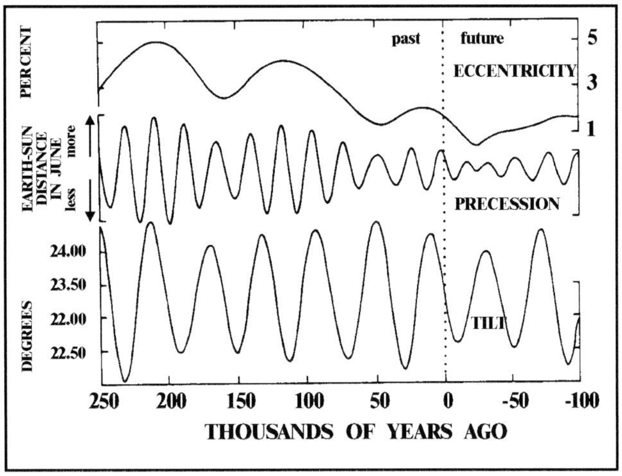

Figure 13. Cyclical patterns of orbital variations of tilt (obliquity), precession, and eccentricity of the Earth in recent history based on astronomical data. From Imbrie and Imbrie (1979).

depositional environment. The discovery of the ice ages required a relaxation of the ultra-uniformitarian viewpoint. Cycles of cooling and warming were linked with periods in the orbital characteristics of the Earth. The mechanism of an "orbital forcing" of climate oscillations was found to be consistent with dates from Pleistocene geology. The strongest link between geology and astronomy came from cyclic phenomena observed in Pleistocene sediments from deep-sea cores. Gilbert (1900) was probably the first to suggest that the results of orbital forcing should be apparent in rhythmic patterns within rock sequences from non-glacial times. There has always been a strong divergence in opinion on this issue. While most geologists would concede that such orbital forcing cycles probably exist, many consider the stratigraphic record to be far too fragmentary to record these rhythms in a recognizable form. However, although sequences on the continents are generally shot through with gaps, deep-sea sediments are more likely to retain a complete record and have been the special focus of studies in cyclicity.

There is a hierarchy of processes that have been recognized to be cyclic in nature and that have radically different periods measured over geologic time (Figure 11). The calendar band of frequencies picks up effects of a year or less duration that are generated by the Earth–Moon–Sun orbital system. The solar band registers climatic phenomena triggered by sunspot cycles and other solar rhythms. The extremely short periods of both the calendar and solar frequency bands mean that their potential records in the rocks will generally be at the scale of laminations. Consequently, their effects are impossible to see from the traces of common well logs other than (perhaps) electrical borehole imaging logs.

The rhythmic features most likely to be perceived on typical well logs are those that may have been induced by processes that occur within the Milankovitch frequency band of time. Milankovitch cycles bear the name of Milutin Milankovitch (1941), who calculated the periods of climatic forcing generated by orbital parameters with durations mostly between 10,000 and 100,000 years. The three principal causes of these cycles are the eccentricity of the Earth's orbit, the Earth's obliquity, and precession (see Figure 12). The precessional cycle tracks the cone that is traced out by the rotation of the axis relative to the orbital plane and currently has two principal modes at about 19,000 and 23,000 years. The obliquity picks up changes in the tilt of the axis with a total variability of 3.5 degrees and periods of about 41,000 and 54,000 years. Finally, the eccentricity of the Earth's orbit changes rhythmically with time with principal cycles concentrated around periods of 95,000, 123,000 and 413,000 years. The cyclic patterns in orbital characteristics that are generated by these processes are shown on Figure 13.

Because they are both lengthy and continuous, wireline logs are an excellent medium to study potential Milankovitch cycles in the sedimentary record. Such logs contrast with core samples, which tend to be short and fragmented. In practice, both sources of information are required to validate rhythmic patterns in terms of reasonable geological models. In particular, core data are needed to supply ages and sedimentation rates, estimated from zone fossils and projections of radiometric ages. This is because the Milankovitch cycles are astronomical phenomena measured out in clockwork time, while the depth dimension of the log is an indirect and monotonic function of time. The depth-to-time conversion is complicated by variations in sedimentation rate and any disruptions that are caused by events of erosion or nondeposition.

The holes drilled by the Deep Sea Drilling Program (DSDP) have penetrated stratigraphic sections that are excellent for testing these ideas. The record of oceanic sedimentation is more complete than comparable successions at continental locations; lengthy cores and logs have been taken throughout the program. The lithological properties that are keyed to climate change appear to be changes in clay mineralogy and clay content, grain size, and types and abundances of planktonic fossils. The common logs are indirectly sensitive to the variations in these properties, which reflect cycles of temperature and aridity/humidity driven by changes in solar energy. Detection of potential cycles is made by Fourier transformation of the logs to amplitude spectra, and conversion of depth to time scales deduced from fossil evidence. The Milankovitch cycles are most easily detected in Pleistocene deep-sea sediments because of the combination of marked climate fluctuations between glacial and interglacial periods, precision dating from cores, and the knowledge that orbital forcing periods should be similar to those of today. All three of these factors become progressively weaker in older sequences, but some DSDP case studies suggest that the Milankovitch cycles can still be detected in a convincing manner.

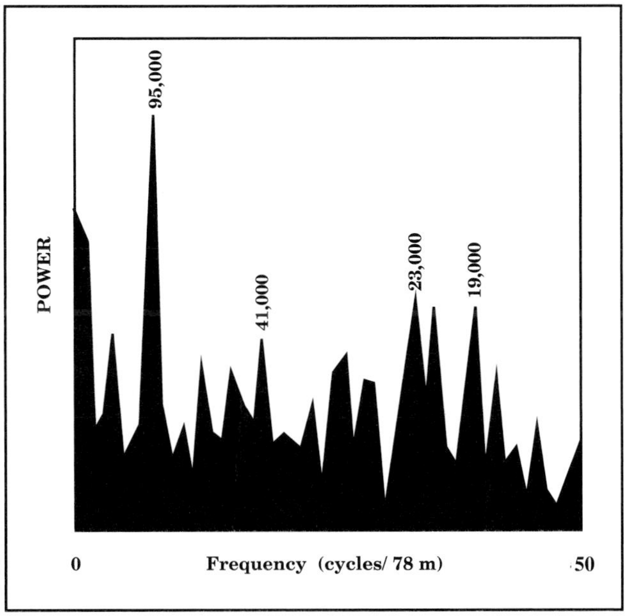

Figure 14. Power spectrum of calcium/silicon elemental abundance ratios from geochemical logs in Tertiary sediments of the Labrador Sea (ODP Leg 105, Site 646). Abundance peaks can be matched with 95,000-, 41,000-, 23,000-, and 19,000-year Milankovitch cycles (from Jarrard and Arthur, 1989).

Jarrard and Arthur (1989) described their analysis of periodicities observed on resistivity and acoustic velocity logs recorded in the Baffin Bay and Labrador Sea sites of Ocean Drilling Program (ODP) Leg 105. The succession consists of silty clays and clayey silts deposited over the last 24 million years. Cyclicity on the logs appears to be strongly linked with variations in clay content and porosity. These, in turn, probably reflect changes in the strength of ocean bottom currents. At times of weaker currents, high-porosity clay-rich sediments accumulated; stronger currents may have resulted in the transport and deposition of greater volumes of quartz and other larger grained minerals. Ultimately, the waxing and waning of the bottom currents is controlled by warming and cooling cycles at the ocean surface, which themselves reflect the migration of upwelling zones. A power spectrum of calcium/silicon ratios from a geochemical log run at the Labrador Sea site (Figure 14) shows peak developments that can be attributed to all three Milankovitch-cycle types.

Mwenifumbo and Blangy (1991) computed amplitude spectra for logs from a Miocene to Pleistocene sequence penetrated on the Meteor Rise of the south Atlantic on DSDP Leg 114. The sediments consisted of calcareous and siliceous oozes. Cycles in the calcium log were clearly out of phase with silicon and hydrogen fluctuations. Collectively, these log characters distinguished between sediments dominated by porous, siliceous diatoms and low-porosity calcareous foraminifera. The cyclic alternations were caused by latitudinal movements of an upwelling current that was powered by Milankovitch climatic cycles. In periods when the upwelling current was located at the Meteor Rise site, conditions were favorable for diatom production. When the current moved away, calcareous foraminifera became abundant. Sedimentation rates were estimated from the fragmentary core recovery to effect the conversion from depth to time. The predominant spectral peaks could be identified readily with cycles of obliquity and eccentricity. The calculated eccentricity period of 77,000 years was significantly less than the commonly accepted figure. This led Mwenifumbo and Blangy (1991) to conclude that sedimentation rates deduced from core were overestimated by about 10%.

There will be many instances where realistic estimates of sedimentation rates will be as elusive. Traditional methods rely on the computation of elapsed time between biostratigraphic events on the section, which are themselves keyed to some absolute time scale. Hiatuses, unfossiliferous intervals, sporadic core recovery, and marked changes in sedimentation rates can all compound the problem of arriving at usable figures. The recognition of Milankovitch cycles therefore offers an "internal clock" to date sedimentary sections. Sedimentation rates deduced from core are used as initial estimates for a preliminary depth-to-time transformation. Spectral peaks from logs are then identified with the orbital parameters of the Milankovitch band. Then the periods of the peaks are adjusted to their orbital parameters. Finally, the adjusted spectrum is used to fine-tune both the sedimentation rate and the ages of section between the biostratigraphic markers.

Clearly, this ambitious strategy requires that several conditions be met. How far can the current Milankovitch periodicities be extended back into geological time? Berger et al. (1989) suggested that the periods have probably only lengthened by a minor amount since the Triassic. What are the minimum sedimentation rates at which we can presume to see Milankovitch cyclicity on wireline logs? Worthington (1990) examined this problem in detail by modelling cyclic sequences with differing sedimentation rates and convolving these with the response functions of common tools. He suggested employing the following formula for a "log association parameter" *(LAP)* as a useful guide:

$$LAP = \left(\frac{s\tau}{2h}\right) \cdot 10^{-6}$$

where s is the sedimentation rate (meters per million years), τ is the cyclic period (years), and h is the log sensitivity (meters). The quantity LAP is dimensionless and should have a value of at least one in order to resolve cycles of periodicity τ. Based on experience with field examples, Worthington (1990) concluded that a minimum value of two would be more realistic for effective resolution. Molinie and Ogg

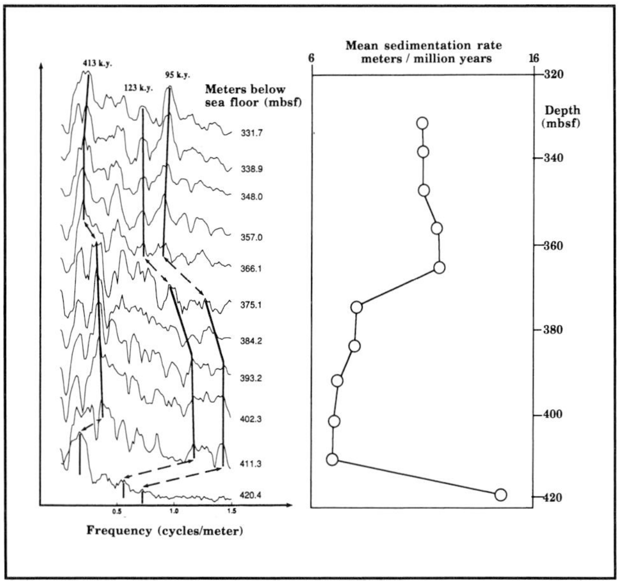

Figure 15. Power spectra computed by sliding-window analysis of a gamma-ray log of Upper Jurassic–Lower Cretaceous radiolarian mudstones from the Equatorial Pacific at ODP Site 801B. At the right is a relative sedimentation rate curve calculated from correlation of peaks on the spectra. Adapted from Molinie and Ogg (1990).

(1990) considered that a resistivity tool with a resolution of 1 m would require a sedimentation rate of at least 20 m per million years to detect eccentricity cycles, and at least 100 m per million years to discriminate precession cycles. The finer resolution of a gamma-ray tool (approximately 0.3 m) would allow the recognition of eccentricity cycles at sedimentation rates exceeding 6 m per million years. Finally, all Milankovitch cycles would be well within the detection limits of the millimeter resolution of electrical borehole imaging devices.

The computation of amplitude spectra for entire successions presupposes that the rate of sedimentation is effectively constant. If, as is likely, the rate varies with time, then the spectral peaks of any Milankovitch periodicities will tend to be smeared across a frequency range. Again, this is because the amplitude spectra are referenced to depth rather than absolute time. A way to accommodate this problem is to compute a series of spectra for successive positions of a moving window of depth. The dimension of the window is determined by the length of the longest period to be detected and the likely change in sedimentation rate. Narrow windows will exclude cycles with long periods; broad windows may obscure depths at which there are breaks in sedimentation rate. Molinie and Ogg (1990) ran a sliding-window spectral analysis of gamma-ray logs of Upper Jurassic–Lower Cretaceous radiolarian mudstones from the Pacific ODP site 801B. The plots of wavelength versus peak intensity (Figure 15) show the progressive shift of peak position with depth that reflect changes in sedimentation rate. Estimates of sedimentation rate were based on three peaks of the eccentricity cycle and are plotted as a function of depth in Figure 15. While the gamma-ray log could discriminate eccentricity cycles, its resolution was insufficient to discriminate cycles with shorter periods, such as precession cycles. The sharp discontinuity in sedimentation rate is matched with an abrupt change in the chert/shale ratio, also observed on both geochemical and electrical borehole imaging logs. The spectral analysis was a successful application, because the radiolarian biostratigraphy and poor core recovery precluded the recognition of this discontinuity.

The ODP examples have been drawn so far from successions that are fairly monotonous oozes, distinguished primarily by their rhythmic character. By contrast, the geochemical logging run from ODP Leg 115 in the Indian Ocean (Figure 16) provides a section of spectacular geology (Lamont-Doherty Borehole Research Group, 1990). The logs show a clear historical picture of a volcanic island which subsequently subsided with the development of a reef in lower Eocene time. Cores from the volcanic sequence exhibit mostly vesicular olivine basalt flows with weathered zones succeeded by plagioclase basalt. The basaltic composition is shown by the relatively high amounts of iron, aluminum, and silicon. The high aluminum spikes are coincident with weathered "soil" horizons between the flows. A thin calcarenite zone (interpreted from core as a beach deposit) is succeeded by a distinctive titanium-rich basalt which marks the termination of volcanic activity. Core recovery in the overlying reef was only 5%, but indicated an upward transition from grainstones to packstones and faunal changes which collectively mark a progressive deepening of water. The reef limestone is contrasted starkly with the volcanic basalt by low iron, aluminum, and silicon contents, but high calcium content. The sulfur curve is of particular interest as it reveals zones of high concentration within the reef, whose fluctuations may be cyclic. The sulfur has been interpreted to reflect sulfate content associated with evaporite zones. No evaporites have been observed in the limited core available (5% recovery). If the high-sulfur zones signify evaporites, then the log may be a depth record of eustatic changes in sea level, with low stands marked by sulfur anomalies. Amplitude spectra from the sulfur trace (Figure 17) show distinctive peaks at wavelengths of 25 and 50 ft, suggesting a cyclic pattern which may be related to the Eocene low stands of 36, 40, 42, 49, and 54 million years of the Vail eustatic curve.

The emergence of the concepts of Milankovitch cycles and sequence stratigraphy has also encouraged spectral analyses and cyclic interpretations of logs at sites onshore. Examples of recent studies along these lines include the analysis of North Carolina Upper Triassic lacustrine beds using gamma-ray logs (Hu et al., 1990), Argentinean Lower Cretaceous highstand-period deposits using a combination of resistivity, sonic, and gamma-ray logs (Spalletti et al., 1990), and English Upper Jurassic sequences using filtered sonic

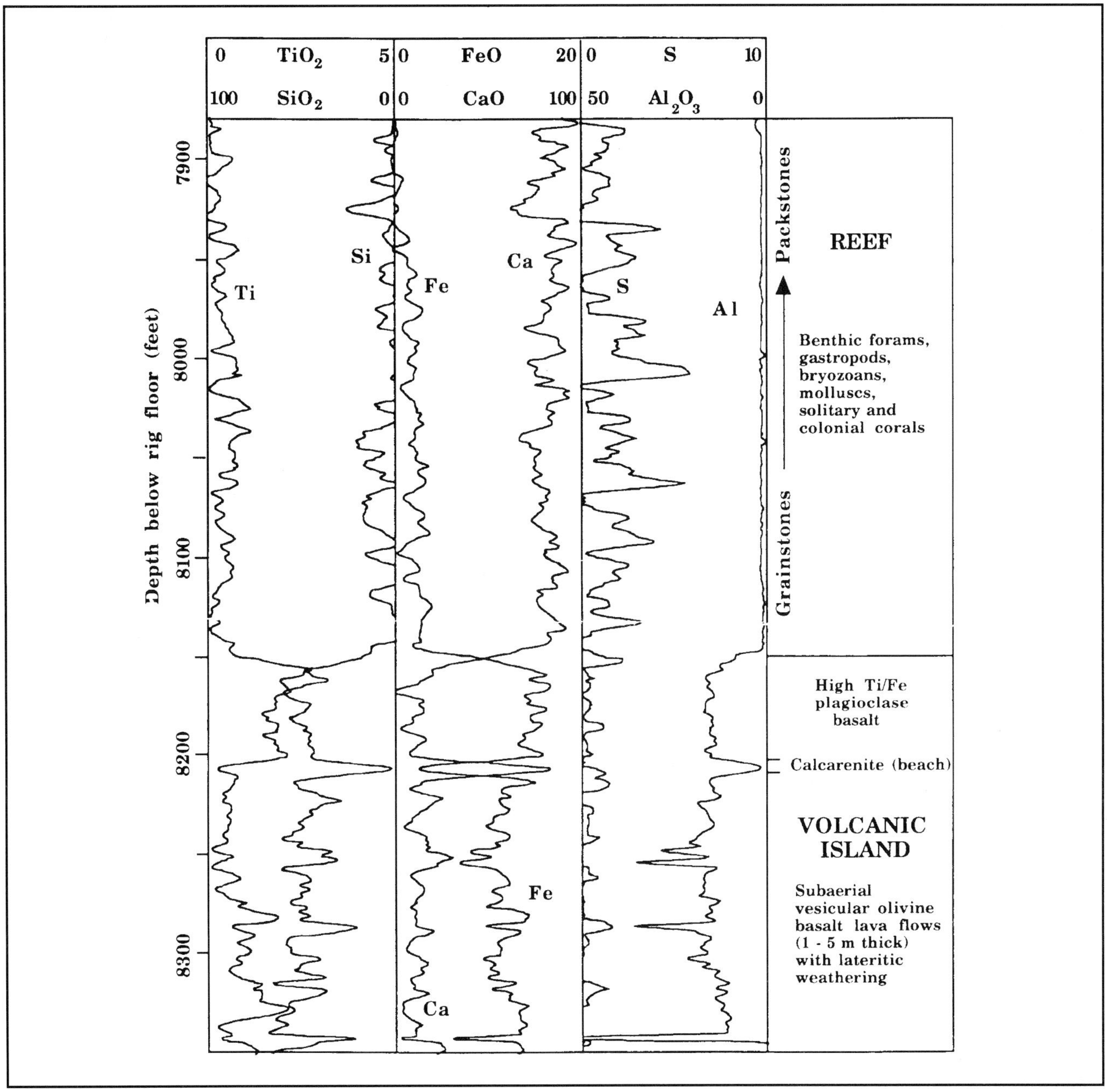

Figure 16. Geochemical logs of a Lower Eocene section from the Maldives Ridge of the Indian Ocean, ODP Leg 115, Site 715 (from Lamont-Doherty Borehole Research Group, 1990).

and gamma-ray logs (Melnyk, 1990). In contrast with deep-sea sediments, these successions would be expected to be more challenging for the identification of plausible Milankovitch periodicities. As discussed earlier, multiple episodes of transgression and regression could obscure periodic patterns irretrievably. This fact of life is recognized by many geologists who work with interpretations of Milankovitch cycles in the sediment record. So, for example, Fischer and Bottjer (1991) stated that lacustrine facies, evaporite sequences, and deposits from carbonate platforms have provided the clearest examples of rhythms induced by orbital forcing.

Borer and Harris (1991) interpreted the Upper Permian Yates Formation in the Permian basin of Texas as the product of orbitally forced cycles. The faster spin of the Earth at that time, coupled with changes in the distance between the Earth and the Moon would have caused shorter periods for both precessional and obliquity cycles. However, it is considered that eccentricity cycles have maintained a fairly constant duration throughout time (Kominz and Bond, 1990). Although numerical estimates can be made for the periods based on astronomical calculations, they are partly academic, because the chronostratigraphy of the Permian basin is weakly defined. Poor time

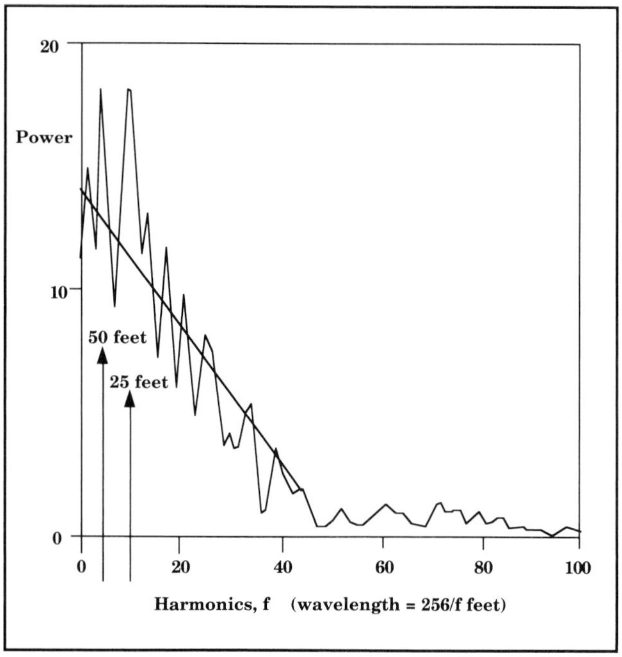

Figure 17. Power spectrum of geochemical log sulfur concentration in an Eocene reef section from the Maldives Ridge of the Indian Ocean, ODP Leg 115, Site 715 (from Lamont-Doherty Borehole Research Group, 1990).

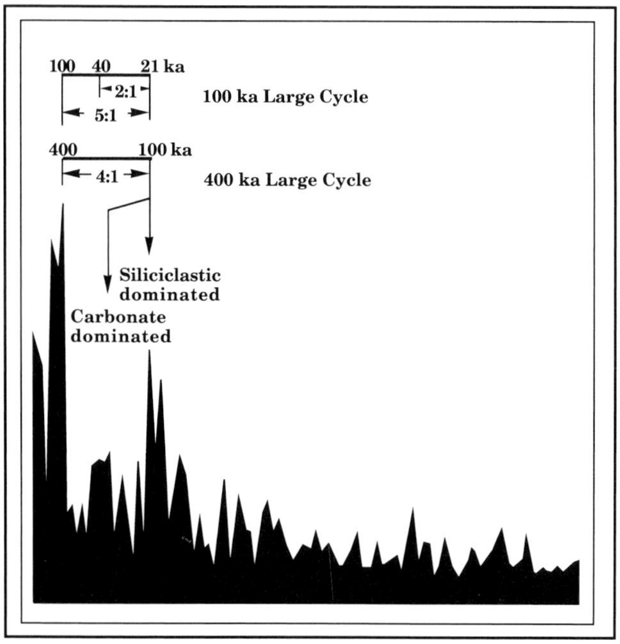

Figure 18. Fast Fourier transform of a gamma-ray log from the Yates Formation of the Permian basin, Texas, marked with two alternative interpretations of Milankovitch cyclicity pattern, based on bundling ratio. Adapted from Borer and Harris (1991).

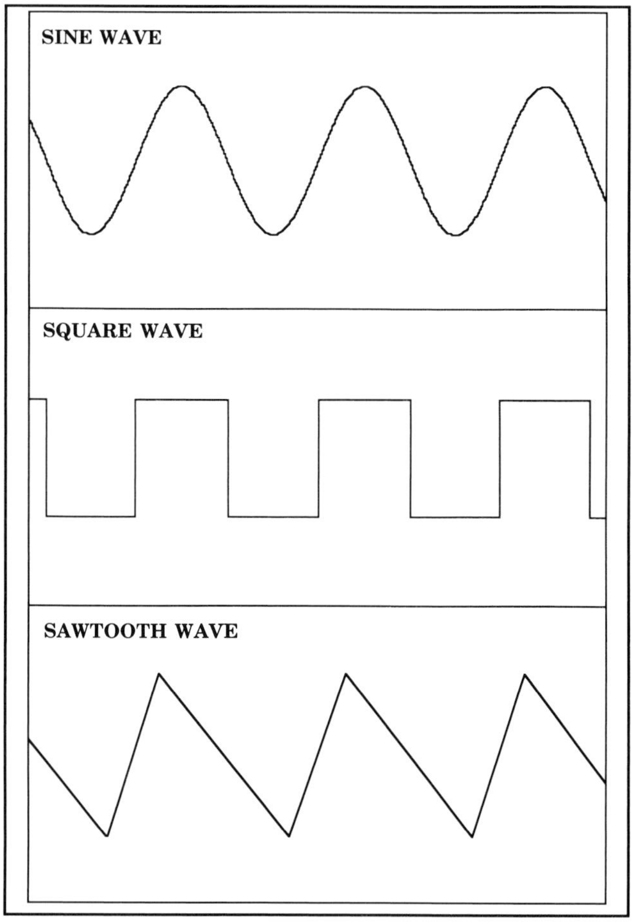

Figure 19. Simple sine, square, and sawtooth waveforms.

constraints within ancient sequences are a common problem. Consequently, a different approach to cycle recognition is commonly used. The hierarchy of Milankovitch cycles generates characteristic ratios between them. The ratio of precessions to the short eccentricity periods is approximately 5, while the ratio of the short to the long eccentricity cycle is about 4. It has been suggested that these ratios have been roughly the same from the Mesozoic onwards (Fischer and Bottjer, 1991). "Schwarzacher bundling" of a ratio of 5:1 has been commonly observed in the stratigraphic record and suggests that the eccentricity–precession syndrome may have been the principal agent of orbital forcing.

Borer and Harris (1991) computed a Fast Fourier Transform (FFT) of a gamma-ray log from a well located on the outer shelf (Figure 18). The gamma-ray log makes a useful distinction between thick alternating intervals of low-radioactive dolomites and the more radioactive arkosic siltstones and sandstones. They concluded that the sequence was controlled by a fundamental eustatic oscillation with a period of about 400,000 years. Siliciclastics were deposited during times of lowstand, while carbonates accumulated at times of highstand. However, the asymmetry of these cycles, weak time constraints, and restricted interval caused ambiguities in the interpretation of the Fourier transform. The ratios of spectral peak could be used to satisfy two alternative cycle

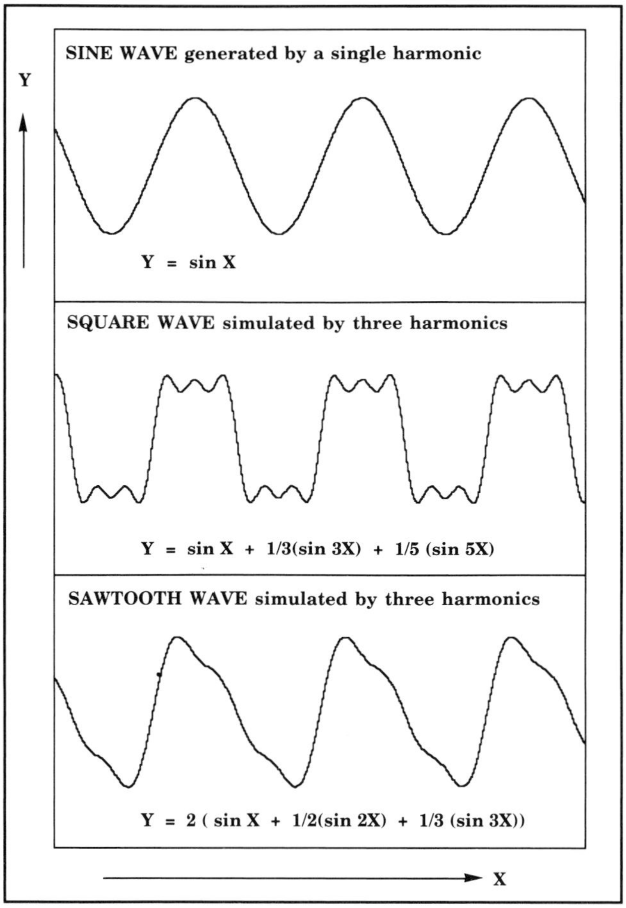

Figure 20. Sine, square, and sawtooth waveforms generated by simple harmonic series of sine waves.

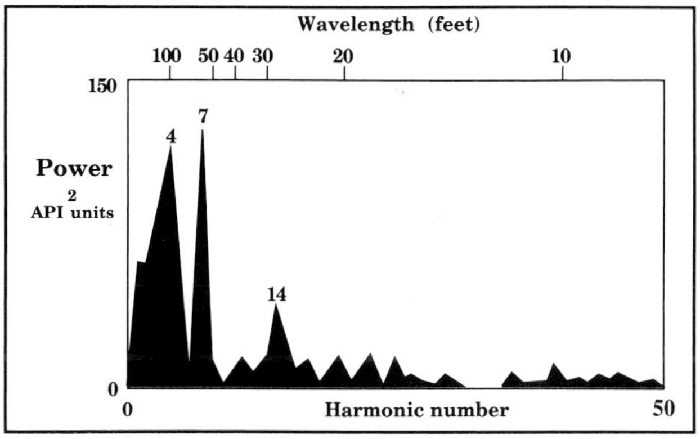

Figure 21. Periodogram of the gamma-ray log (see Figure 22) of a Lower Cretaceous clastic succession in central Kansas.

bundling models (Figure 18). In further modelling, they were able to discriminate carbonate cycles from siliciclastic-dominated cycles. While both appeared to have periods of about 100,000 years, accumulation within the carbonate cycles was between two and three times greater.

The case study of Borer and Harris (1990) is a thorough and well-documented example of cyclostratigraphy. It shows the potential of the orbital-forcing model for describing the vertical and lateral geometry of lithofacies at a variety of scales. At the same time, it reveals potential pitfalls and ambiguities in interpretation that must be addressed. Even if the orbital forcing cycles are represented by compound sinusoidal functions, does the sedimentary record of the continents take the form of sinusoidal cycles? Most depositional models call for marked and characteristic changes of sedimentation rate. The "sequences" of sequence stratigraphy are thought to be produced by changes in base level. Sequences are bounded by time-correlatable boundary event surfaces. Each surface can represent a disconformity at the time of relative sea-level fall with either erosion or nondeposition. Alternatively, the surfaces can be generated by condensed sections of very slow deposition during sea-level rise. Sedimentation between these surfaces has proceeded at variable rates. The sum effect is that the record of time in the rocks of these successions is a highly distorted and disrupted version of the cyclical time measured out evenly by the orbital parameters. Thanks to the classic work of Jean-Baptiste Fourier, almost all frequency domain analysis of rhythms is predicated on the circle. In the next section, we will examine some of the effects that we may expect to see when sedimentary cycles are not sinusoidal.

PROPERTIES OF NONSINUSOIDAL SEDIMENTARY CYCLES

Cyclical properties in time are most commonly modelled by a sine wave. The sine wave has useful mathematical properties which make the Fourier transform a particularly powerful means to view records such as logs in terms of their frequency content. However, the sine wave convention is most commonly accepted because it is seen to be the actual representation of physical phenomena. Light waves and other wave forms from the electromagnetic spectrum are often cited as classic examples of naturally occurring sinusoidal forms. Harmuth (1977) pointed out that this is not necessarily so. Prisms and diffraction gratings are merely devices that decompose light into a set of sine waves. The mathematics of the Fourier transform do the same operation. Regardless of whether or not a log actually has sinusoidal components, the transformation results in its representation by a series of uncorrelated sine waves spread over a range of frequencies.

Examples of regular nonsinusoidal waves are common. The sawtooth and square wave (Figure 19) can be heard in any orchestral concert (Bone, 1982). The sawtooth is a "bright" sound produced by an oboe or other double-reed instrument. The square wave sounds "woody" and can be made by a clarinet. The sine wave is created by the pure tone of a flute. The Fourier transformation of these nonsinusoidal waves results in a train of characteristic harmonics. These

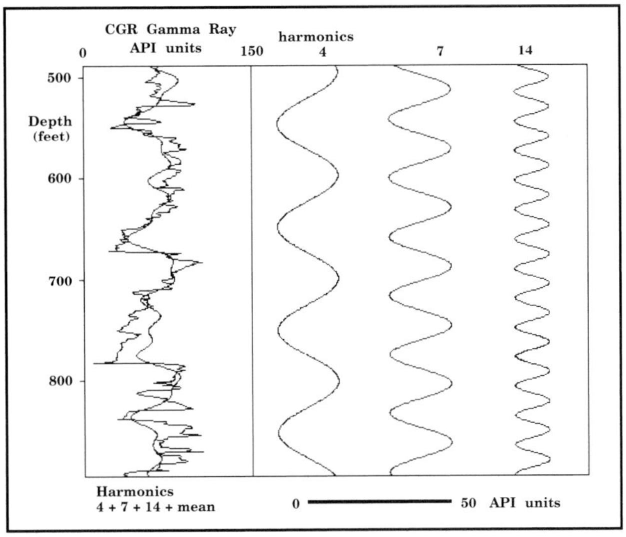

Figure 22. Gamma-ray log of a Lower Cretaceous clastic succession in central Kansas, fitted with sine waves of the fourth, seventh, and fourteenth harmonics.

harmonics are integer multiples of the basic repetition, so that a frequency of f may be echoed by harmonics at $2f$, $3f$, $4f$ and so on. When enough of the harmonics are summed together, they can make a close approximation to the original wave form (Figure 20). A square wave can be fitted closely by the simple function of odd harmonics:

$$Y = \sin X + \frac{1}{3}(\sin 3X) + \frac{1}{5}(\sin 5X) + \ldots$$

when taken to enough terms. A sawtooth wave can be built from both odd and even harmonics using:

$$Y = 2\left(\sin X + \frac{1}{2}\sin 2X + \frac{1}{3}\sin 3X + \ldots\right)$$

In the analysis of conventional time series, the presence of harmonics in an amplitude or power spectrum is most commonly interpreted to suggest that the fundamental cyclic pattern is nonsinusoidal (Chatfield, 1975). This observation is of more than passing interest, because cyclic bundling is often seen as a method to differentiate frequencies within the Milankovitch band. Furthermore, while the orbital cycles may be symmetric in time, waxings and wanings of depositional rates can generate sawtooth records of sediment accumulation.

A practical illustration is given by the periodogram of the gamma-ray log from a Lower Cretaceous clastic succession in central Kansas (Figure 21). The periodogram is marked by three strong peaks at the fourth, seventh, and fourteenth harmonics. The series is eerily close to simple integer multiples of a fundamental harmonic. Using the methods described earlier, the harmonics can be transformed back to the time domain. When compounded, they show a good generalized match with the original gamma-ray log (Figure 22). The harmonic pattern appears to have been triggered by the sawtooth motif of the log profile. The shapes are consistent with the origin of the thick sandstones in channels, with sharp erosional bases and fining-upward profiles. While there seems to be a fundamental repetition with a wavelength of about 100 ft, the multiple harmonics are more likely to be caused by a nonsinusoidal character, rather than additional cyclic processes.

Harmuth (1977) credited Plato with the creation of the "dogma of the circle." Ultimately, the belief in the circle as the fundamental descriptor of nature led to the Ptolemaic system of planetary orbits. Each planet was considered to orbit on a primary circle called a deferent, while superimposed on each deferent was a nested hierarchy of smaller circles called epicycles. Even after Copernicus had relocated the Sun at the center of the solar system, orbits continued to be represented by the superposition of circles. These were increased in number to keep pace with the improved precision of observations. Mercury required a system of eleven circles, Earth and Venus were described by nine, and the outer planets each took five. Finally, it all ended when Kepler showed that elliptical orbits were both simpler and better. It is interesting to note that an elliptical orbit can be represented quite adequately by a sufficient number of circles, but this is an unnecessarily complex description of a simple pattern and gives no clue as to the real system where gravity is a driving force.

The epicylic system is uncomfortably close to a multiple harmonic descriptor of a nonsinusoidal carrier wave. However, rather than demolish the Fourier transform representation of the time domain, the analogy should serve to sharpen our awareness of the multiple possible implications from a power spectrum. In some cases, harmonics will be demonstrable Milankovitch cycles. In others, they may be artifacts caused by asymmetry or nonstationarity of the sedimentary signal.

Sawtooth patterns of sedimentation through time should be a common feature of processes that alternate between periods of erosion/nondeposition and accelerated deposition. The alternation should result in markedly asymmetric cycles matched by a hierarchy of sine wave harmonics. There is no information on the orientation of the teeth in the power spectrum. This is because the power spectrum is a graph of the relative amplitudes and wavelengths, which would be the same for the record turned upside down. However, the necessary information is contained in the train of phase angles for the harmonics, since these give relative displacements with respect to a common origin. In practice, the distinction between the two possible directions of the sawtooth pattern can be difficult to make in geological sequences of any real complexity. Some preliminary investigations along

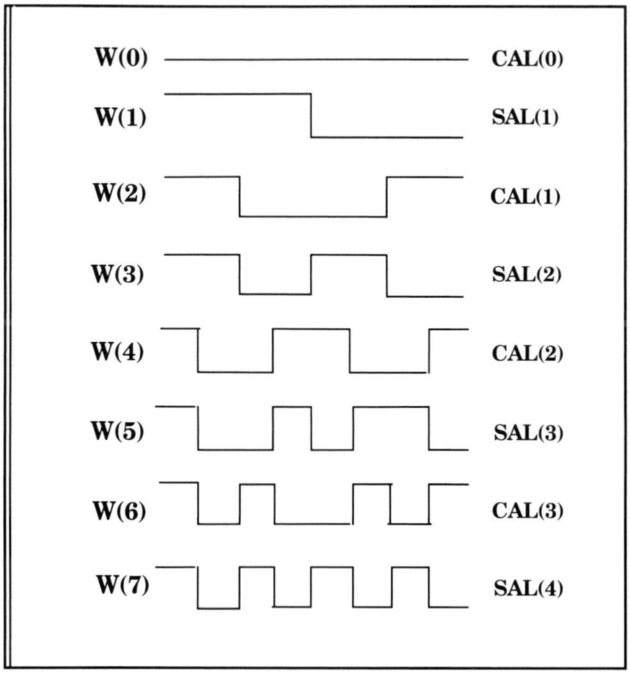

Figure 23. Square waves of the first eight Walsh functions W(0) to W(7). Even-numbered functions are CAL; odd-numbered are SAL functions.

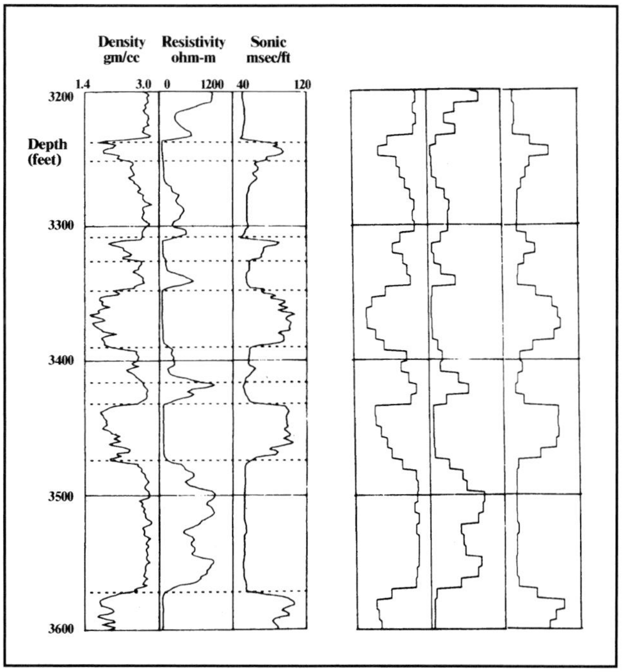

Figure 24. Density, resistivity, and acoustic velocity logs (left) of basalts of the Columbia River plateau marked with rock boundaries picked using the Walsh low-pass version of the logs (right).

these lines were made by Martini et al. (1978), who attempted to isolate patterns of upward thickening and thinning in Cretaceous turbidites from the Apennines, based on spectral characteristics.

Considerably more is known about the properties of square waves. Although several alternative systems of rectangular wave forms have been proposed, the most widely used are square waves that take the value of either +1 or -1. A hierarchy of independent square waves is described by a set of Walsh functions (Figure 23). The concepts and terminology of Walsh functions have much in common with the Fourier series. Instead of frequency, a Walsh function is characterized by a sequency, which is the number of zero crossings per time unit divided by two. The even-numbered functions are symmetric and are called CAL functions because they are analogous to cosine waves. The odd-numbered, asymmetric members are SAL functions, squared versions of sine waves (see Figure 23). The mathematics of Walsh functions are simpler than the equivalent trigonometric Fourier operations, so that a Fast Walsh Transform (FWT) runs much faster than a Fast Fourier Transform.

A conversion to a set of Walsh functions is a transformation to the sequency domain. As with the Fourier transform, the original signal can be recreated exactly from its Walsh transform. Alternatively, low-pass, high-pass, or band-width versions of the signal can be generated by selective summation of Walsh functions of appropriate sequency. Lanning and Johnson (1983) used this property as a means to block well logs in a consistent manner. By excluding Walsh functions with sequencies that exceeded a "minimum resolvable layer thickness," they used the Walsh transform as a low-pass procedure to block well logs as stepped functions (Figure 24).

The application of Walsh functions to logging traces is reasonable since many rock sequences are made up of beds that are fairly homogeneous, but differentiated from one another by sharp boundaries. The basic logging concept of zones is a model that seeks to recover a stepped sequence of layers from the continuous log trace. Ideally, the result will emulate a succession of beds seen in an equivalent outcrop. The discrete nature of many bedding sequences has encouraged the use of Walsh spectra in preference to Fourier transforms (Weedon, 1989). In addition, Walsh spectra are less sensitive to abrupt changes at hiatuses, which tend to generate high frequency Fourier components. By the same token, the abrupt ramping of the lower sequency Walsh functions may cause them to be poor descriptors of longer term gradual changes.

Ultimately, the choice of Walsh or Fourier transform should be made with regard to the nature of the process model. Each transform type has its strengths and limitations. Just as the Fourier transform makes an overly complex description of a blocked function, the Walsh transform requires multiple square waves to approximate a sine wave. In addition, the representation in the sequency domain is not time invariant, as is the case with the frequency domain. Shifts in the starting point of the series can cause noticeable changes in the Walsh spectrogram. However, this shortcoming can be circum-

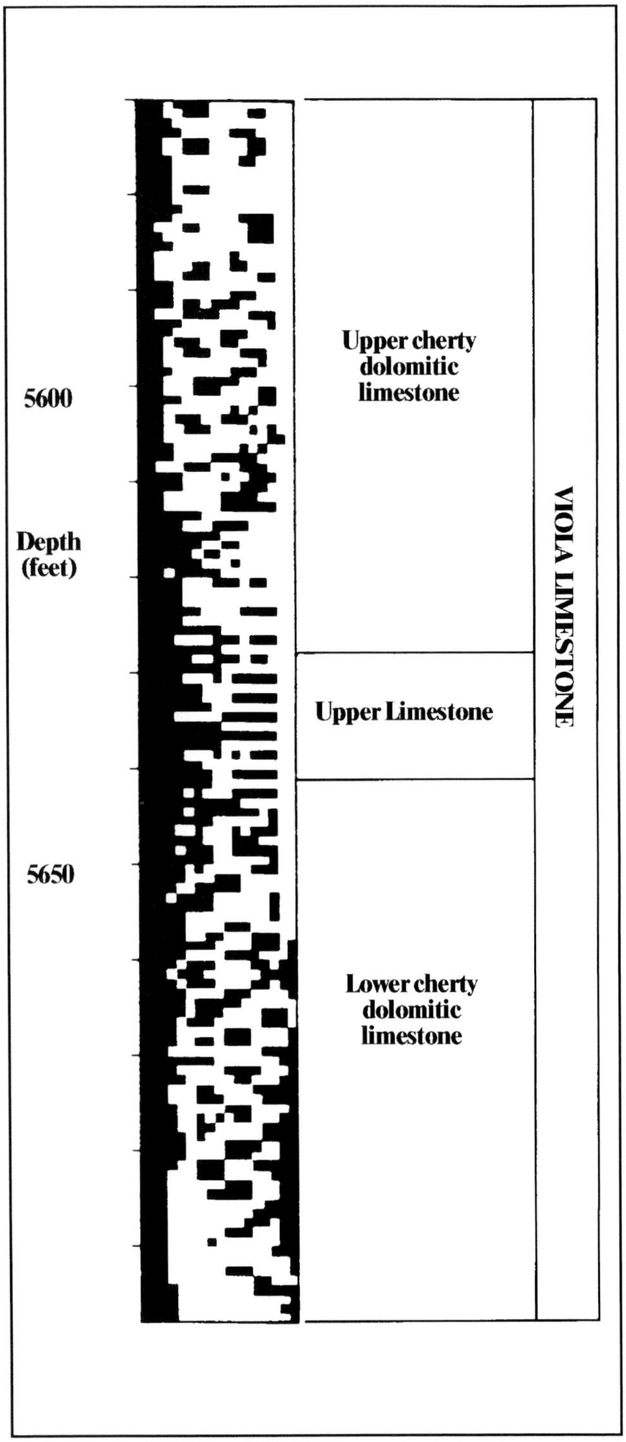

Figure 25. Moving Walsh spectrogram of calcite variation in the Viola Limestone section. Sequency increases from left to right. Modified from Doveton (1986). Reprinted by permission of John Wiley & Sons, Inc., copyright ©1986.

vented by producing an averaged Walsh spectrum (Beauchamp, 1984). Clearly, some intelligent choices should be made concerning the purpose of the analysis and the potential form of the signal at the scale of interest.

Doveton (1986) provided an example of a Walsh transform computed for a short sequence segment. The computations are very simple and form the basis for a sliding window filter program (Ahmed et al., 1976) to generate a moving Walsh spectrogram (Figure 25). This spectrogram graphs sequency amplitudes as a function of depth. Sequency increases from left to right and the spectrum is clipped to differentiate high amplitudes (black) from low amplitudes (white). Essentially, the spectrogram shows changes in the scales of log activity with depth. In this Viola Limestone example, the log records calcite content computed from a compositional analysis based on neutron, density, and sonic logs. Among other things, the calcite log gives indications of the relative thickness of limestone beds within the Viola Limestone section. The spectrogram (Figure 25) shows how the patterns of thickness distribution change with stratigraphic level. Notice, for example, how the lower cherty dolomitic limestone is marked particularly by high-sequency (thin) features, and is contrasted with the low- and intermediate-sequency character of the upper limestone. When applied to a gamma-ray log, the spectrogram typically shows shale interbedding characteristics. However, the same technique can be used to analyze reservoir structure through the transformation of a porosity log.

The preceding discussion is only an introductory treatment of the various techniques used to generate and interpret frequency domain transformations of time domain logs. There are many possible applications and valuable insights to be gained. However, the user should constantly be vigilant for artifacts that are created by conflicts between assumptions of the descriptive model and the reality of the data. The most dangerous artifacts are those that appear to give a close match with theoretical expectations, but in fact are false. Oscar Wilde once said that the poet Wordsworth found in the stones the sermons that he himself had hidden there. We should avoid this trap by being skeptical in our interpretations and calling on independent lines of supporting evidence wherever possible.

PROFILE ANALYSIS OF LOGS

In the review of spectral analysis techniques we have considered logs from the viewpoint of the frequency domain. The frequency representation partitions a log trace into components at different scales of depth resolution. These may be played back selectively as low-pass, high-pass, or band-width filtered log components. A major aim of spectral analysis in stratigraphic applications is the recognition of fundamental repetitive patterns that can be linked with process mechanisms. In the time domain the axis of depth—and therefore time—is retained, so that we can locate the occurrence of events and trends that are obscured in the frequency domain. Since the earliest days of wireline logging, geologists have inter-

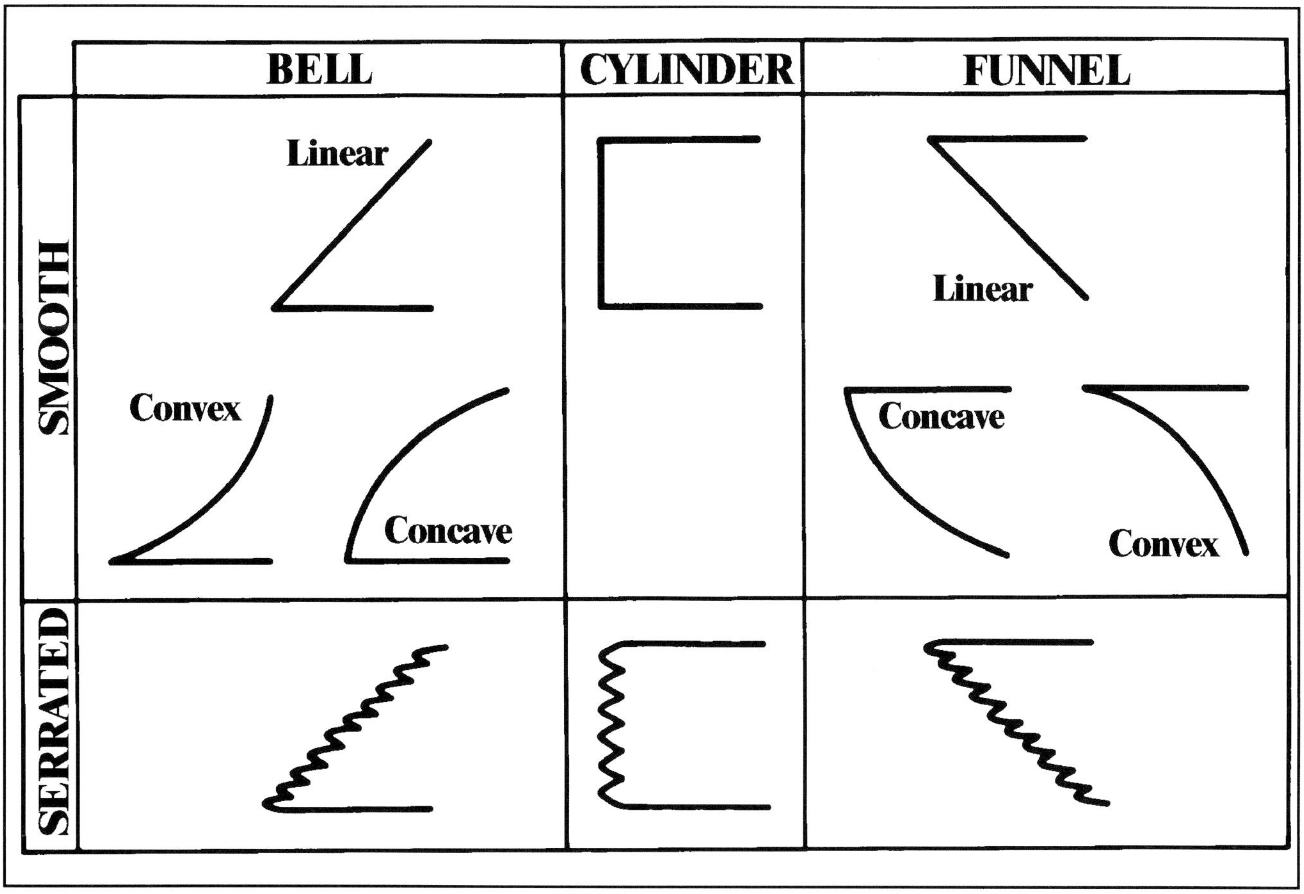

Figure 26. Classic terminology of basic shapes used to describe sandstones on SP and gamma-ray logs.

preted patterns of vertical change in terms of sedimentary geology and stratigraphy. Log profiles contain useful diagnostic information at all scales. Profile shapes of individual units are caused primarily by depositional and erosional processes in individual environments. Profile trends at larger scales reflect long term changes in factors such as grain size and sediment supply. In this section, we will discuss computer techniques developed for profile analysis at all scales. The results are repeatable and consistent between logs, unlike the qualitative interpretations of geologists. Also, the methods can be designed to accentuate features that may not be obvious on the original log trace. However, interpreting the essential meaning of any profile patterns is still the responsibility of the geologist.

Serra and Sulpice (1975) credited engineers within Shell Oil Company in 1958 as the first to classify log profile shapes by simple descriptive motifs. Their terms are still in use today and are linked by a scheme which is readily understood (Figure 26). The log-response shapes are first described by the nature of their upper and lower contacts as either sharp or gradational. Three fundamental motifs are distinguished as "bell" (contacts: sharp lower, gradational upper), "cylinder" (sharp lower, sharp upper), and "funnel" (gradational lower, sharp upper). A fourth motif, the "egg" (gradational upper and lower contacts), appears to have fallen into disuse. At a secondary level, the relative curvature of the gradational limb allows the distinction of a convex, concave, or linear trend. Finally, the shape motif is classified as either "smooth" or "serrated" in character.

Although the descriptive scheme used to classify log traces is purely geometric, the shapes can be interpreted in terms of sedimentology. Visher (1965) introduced the notion of vertical profile analysis, which relates the succession of sedimentary properties observed in outcrop and core to sequences expected as products of different depositional models. The interpretation procedure has been extended to the analysis of profiles of physical properties recorded by logs. Visher (1969) described SP shapes which typify sandstone facies while Saitta and Visher (1968) studied hundreds of wells for variations in SP shape that characterized different deltaic environments. Serra and Sulpice (1975) showed an excellent agreement between gamma-ray logs and grain-size profiles, suggesting that gamma-ray log shapes could be used for the same purpose. Selley (1976) made extensive use of the gamma-ray log in conjunction with carbonaceous matter and glauconite content of cuttings in the genetic interpretation of clastic successions from the North Sea subsurface. The basic shapes of either SP or gamma-ray logs can be considered to reflect transgression or aggradation (bell), regression or progra-

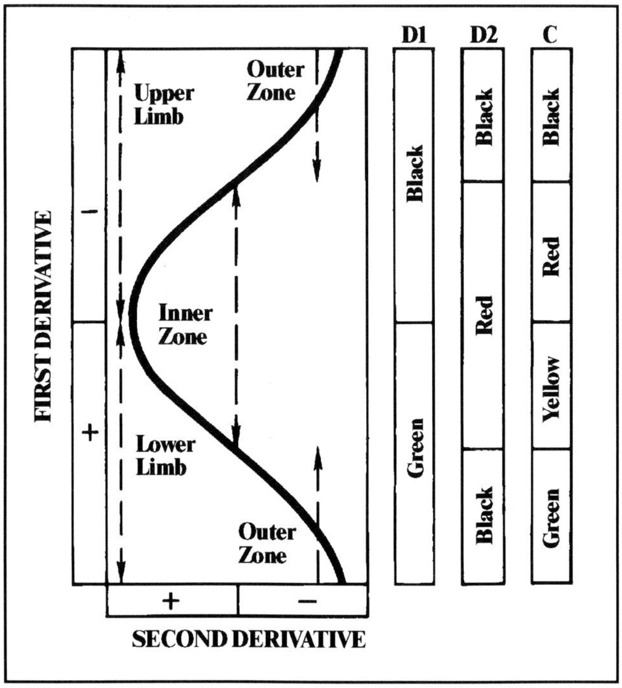

Figure 27. Anatomy of a sine-wave sandstone event on a log with color-characterization of first derivative (D), second derivative (D2), and their composite (C).

dation (funnel), or stability (cylinder) as discussed by Pirson (1970) and Magara (1979).

The log-shape analysis of sandstones is an intuitively simple procedure but has several drawbacks in practical application. Actual log shapes are highly variable and are often difficult to classify in an objective or even consistent manner. Current methods are manual, labor-intensive operations which are time-consuming for work with lengthy sections. Finally, different shapes can often be seen at a hierarchy of scales; for example, a sequence of thin cylinder shapes may be observed to trace out a funnel-shaped trend at a larger scale of resolution. Magara (1979) introduced a computer technique to resolve some of these problems, but the method is not entirely automated and can be cumbersome to apply to complex successions. Collins and Doveton (1988) described a simpler approach which transforms a log trace into a color image. These images are based on the arithmetic sign of the log-curve derivatives and accentuate basic shape character at a variety of scales, while preserving nuances of shape.

A graph of the first derivative of a log with respect to depth is a continuous profile of log slope which oscillates between positive and negative values. The zero crossings of this profile mark positions at which the original log shows the extreme of either a peak or a trough feature. The second derivative measures the rate of change of slope, and when plotted as a continuous function, also varies between positive and negative values, with zero crossings at positions of curve inflection on the original log. The geometry of first- and second-derivative logs can be understood when reference is made to Figure 27.

All wireline logs are smoothed representations of actual physical property variations which are averaged by the vertical resolution of the logging tool. Consequently, log analysts pick the depth boundaries of subsurface zones at the inflection positions of curve traces. The second derivative log provides an automatic zonation device where the depths of zone boundaries are marked at zero crossing points. On both the SP and gamma-ray logs, sandstone zones will be registered as trough features with positive second derivative. These contrast with shale peak features which have negative second derivatives. If positive values are assigned the color red and negative values are matched with black, the analog trace of the second-derivative log is transformed into a simple color-banded image. Red bands will discriminate sandstones and black bands distinguish shales (Figure 27).

A similar scheme of representation can be applied to the first-derivative trace. The first derivative will be positive within segments trending from sandstone to shale and negative within those trending from shale to sandstone (with increasing depth). These two types of features were characterized by Pirson (1970) as transgressive and regressive, respectively. If positive values are assigned a green color and negative values black, the first-derivative log is transformed into a second color-band image, where the color alternations pick out the sandstone-trending and shale-trending features of the original log (Figure 27).

The log character information contained in the two color transformations can be combined in a composite image using simple color theory. In an additive color scheme, segments which are both red and green are mixed as yellow. Since black is treated as the absence of any color, the remaining possible color composites are green, red, and black. The two derivative images and their composite are shown with reference to a sine wave which simulates a hypothetical sandstone event in Figure 27. The sequence of colors corresponding to depositional order is easily remembered. It obeys the traffic-light convention of green, amber, red, and is terminated by the "null" of black.

The symmetric gradational character of the sine wave results in a color composite image of the sandstone as a sequence of color bands of equal thickness. The basic shape descriptions discussed earlier (Figure 26) can be thought of as distortions of this sine wave model, which translate into different expectations for the color-band display. For a bell-shaped feature, the sharp basal contact results in a marked contraction of the lower limb and consequent thinning of green and yellow banding. A well-developed smooth bell shape will be dominated by red and black bands. If the bell shape shows pronounced convexity, the inflection point in the upper limb will be moved downward, with suppression of the red band in favor of the black. If the bell shape is strongly concave, this inflection point will be moved upward, with accentuation of the red over the black. Similar considerations can

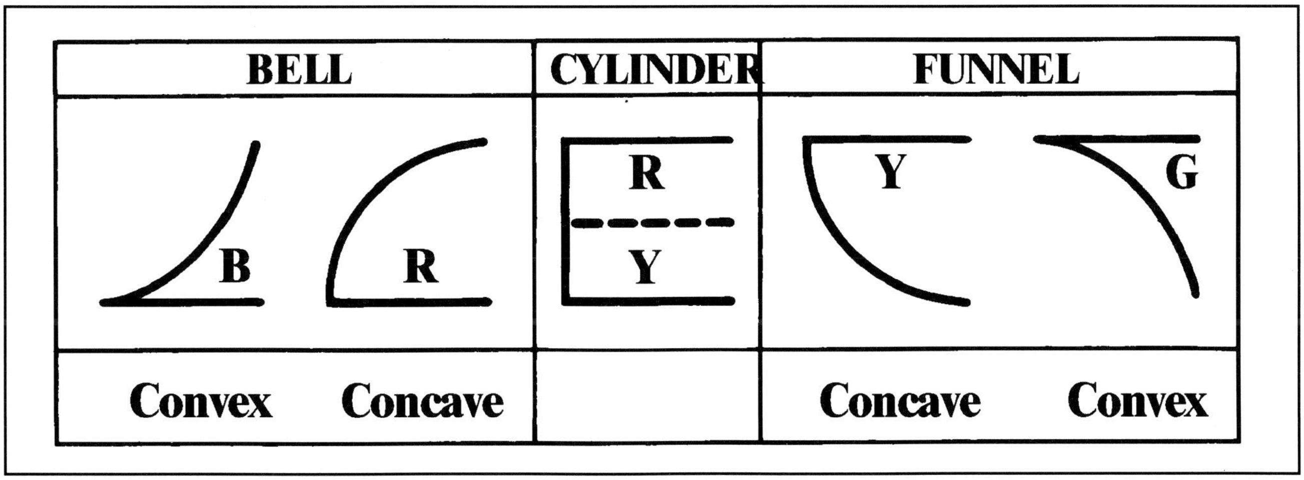

Figure 28. Derivative image colors of basic descriptor shapes. B = black; R = red; Y = yellow; G = green.

be applied to the other shapes to extract the dominant color which will typify any of the shape descriptions. The shapes and their dominant colors are shown on Figure 28. Notice that the relative dominance of colors in a sequence may suggest a fundamental wave form type for any repetitive pattern. Not only can square waves be distinguished from sine waves, but the orientation of a sawtooth wave results in different colors. These displays can therefore be helpful in the interpretation of spectral analysis results.

The computation of the first and second derivatives of a log is a straightforward operation. The differences between adjacent log values on the digitized log string are the first derivatives and represent slopes between successive readings. The differences between these first differences are the second derivatives and show rates of change in slope. The resulting color image transformation will be sensitive to minor fluctuations and smooth major features, but will be blind to shapes which are serrated and to long-term trends in shapes which are punctuated by multiple zones. This is because the slope perception of simple differences is a function of the sample frequency which is conventionally two digitized readings per foot. Consequently, the color image method was modified to mimic the human ability to discern shapes and trends at all scales.

Convolution of a log with a low-pass filter, such as a simple moving average, results in the suppression of high-frequency elements in favor of longer term components. A more sophisticated filter can be designed which fits a polynomial curve by the least-squares criterion to the log segment scanned by the filter window. A cubic polynomial was selected as a useful function which would be adequate to fit peaks, troughs, or shoulders at the scale of the window span. Polynomial curve fitting is a form of multiple regression, but is considerably simplified when applied to data where the independent variable is sampled at a constant rate (the increment of depth digitization). This property allows the design of simple filters which will generate the smoothed value and derivatives of a polynomial best-fit curve at the midpoint of the window, using methods described by Savitzky and Golay (1964).

The principles of polynomial filter design are explained with reference to a cubic function which is specified by the equation:

$$s = a_0 + a_1 d + a_2 d^2 + a_3 d^3$$

where s is the cubic trend in log response and d is depth position. The filter window is made up of $(2m+1)$ elements ranging in depth value from $-m$ to $+m$, with a value of zero at the midpoint. At the location of the window midpoint (where the depth value is zero), differentiation of the cubic equation gives a first derivative of a_1 and a second derivative of $2a_2$. The standard matrix algebra solution for the coefficients of a cubic regression model can be written as:

$$\begin{bmatrix} n & \sum d & \sum d^2 & \sum d^3 \\ \sum d & \sum d^2 & \sum d^3 & \sum d^4 \\ \sum d^2 & \sum d^3 & \sum d^4 & \sum d^5 \\ \sum d^3 & \sum d^4 & \sum d^5 & \sum d^6 \end{bmatrix}$$

where n is the number of filter elements and the summations are over all $(n=2m+1)$ elements (Davis, 1986). The system collapses to a simpler equation set, because all terms that involve the summation of depth values with odd exponents are zero. The solution for the first derivative at the window midpoint is then:

$$\frac{\left(\sum d^4 \sum d^3 s - \sum d^6 \sum ds\right)}{\left(\sum d^4 \sum d^4 - \sum d^2 \sum d^6\right)}$$

TABLE 3. Coefficients of Convolution Filters for (a) Smoothed Estimate, (b) First Derivative, (c) Second Derivative of a Nine-Point Cubic Polynomial.

Filter Element	(a)	(b)	(c)
1	–21	86	28
2	14	–142	7
3	39	–193	–8
4	54	–126	–17
5	59	0	–20
6	54	129	–17
7	39	193	–8
8	14	142	7
9	–21	–86	28
Normalizing Factor	231	1188	462

and the second derivative is:

$$\frac{2\left(n\sum d^2 s-\sum d^2\sum s\right)}{\left(n\sum d^4-\sum d^2\sum d^2\right)}$$

The calculation of each of these derivatives may be recast as a sequence of loadings for elements in a convolution filter. An example of the loadings to create derivatives from a nine-point cubic polynomial filter is listed in Table 3.

By moving these filters down a log, the operation of convolution generates cubic polynomial first and second derivatives as continuous functions (Doveton, 1986, p. 197). The arithmetic signs of these derivatives were processed as color images using the methodology discussed earlier. The colors discriminate features and trends whose dimension is approximately the same scale as the filter window. The multiple application of the cubic filter with progressively increasing window size was used in the analysis of trends on a continuous range of scale.

The hypothetical example in Figure 29 shows how the expanded color image distinguishes a sequence of symmetrical features which coalesce into a generalized bell-shaped trend at a larger scale. The color original of this illustration shows a rapid repetition of

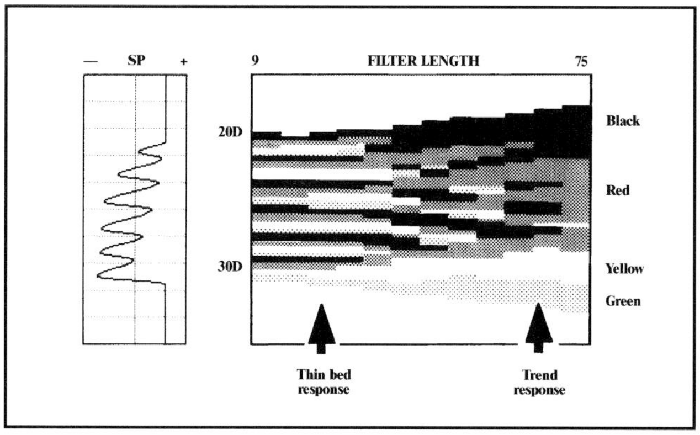

Figure 29. Composite color image from derivatives of a synthetic bell-shaped trend of thin sandstone beds.

traffic-light sequences at short filter lengths of analysis, which merge to a single sequence at longer filter lengths. The dominant colors of red and black at the longer scale classify the trend as bell-shaped while the bias towards red suggests "concave" to be an appropriate modifier. Although the "traffic-light sequence" has been emphasized as the basic color description of a simple sandstone event, real sequences will show "stutters." These are perturbations registered by alternations of green and yellow or red and black bands, and they discriminate shoulder features on the original log.

A color transformation of derivatives from an actual SP log is shown in Figure 30, and is the record of a succession of Tertiary deltaic sandstones and shales from a well in Louisiana. The log sampling rate can be adjusted to be compatible with the scale of features of interest and, in this example, was selected at 20 ft. The original color image contains information of various types. Automatic zonation is shown at a variety of scales on a continuum, which successively clusters thinner units into larger aggregates at coarser scales of resolution. The relative thickness of color bands within each unit is the basis for its description in terms of overall shape. Broad symmetrical units in the center of the succession are contrasted with a funnel-shaped trend at the base and a bell-shaped feature near the top.

This method of transformation is purely a descriptive tool to convert the raw analog trace of a wireline log into a color image that accentuates zonation and shape characterization over a range of scales. These images can be used for a variety of applications. Individual sandstone units or longer term trends can be classified automatically in conventional shape terminology, as specified by the relative thickness of color banding within the basic color sequence. In particular, the image is amenable to interpretation in terms of sequence stratigraphy or allied paradigms because it retains information on breaks that may reflect diagnostic boundaries, as well as gradations linked with fluctuating sedimentation rates.

Entire sequences can be summarized either

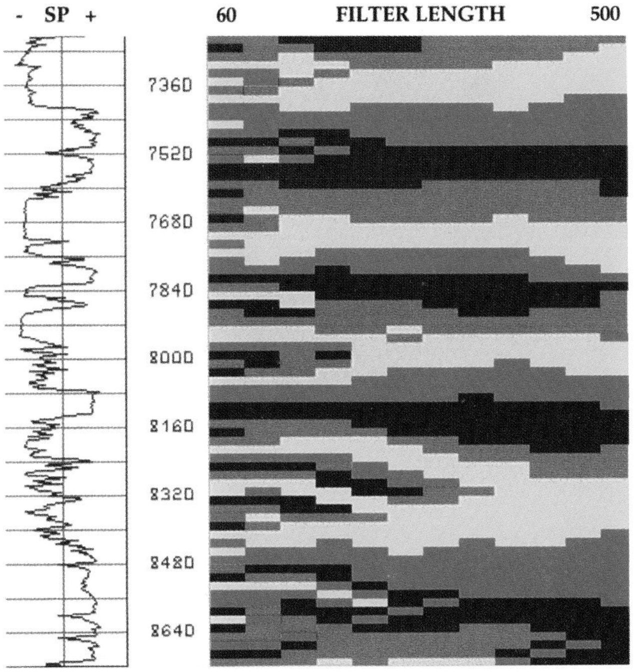

Figure 30. Composite color derivative image of an SP log for a Tertiary deltaic succession from a well in Louisiana.

through cumulative counts of the different types of shape (in a method similar to that used by Magara, 1979) or through aggregation of the total thickness of each color-band type. The color analysis of an individual unit or succession can be expressed numerically at well locations and then interpolated laterally to trace areal changes in shape. Prior to mapping a decision must be made concerning the appropriate filter length to be used, since shape characterization varies with the scale of vertical resolution. The description of topographic surfaces has a similar problem, but is aided by the widely held perception by geomorphologists that landscapes have characteristic scales. There is no intrinsic reason for logs to be self-similar at all scales, so the fractal dimension method for topographic analysis described by Mark and Aronson (1984) would be useful to define distinctive scales of log shapes.

The zonation and blocking techniques discussed earlier produce subdivisions at the finest scale compatible with tool resolution. The multiple-scale derivative approach allows segmentation to be made at a variety of scales. Work along similar lines has been described by Vermeer and Alkemade (1992). They located the edges associated with zero crossings of the second derivative as a method for multiscale segmentation of well logs.

The recognition of characteristic subdivisions on the color image has immediate applications in sequence stratigraphy. Studies have been made to define a hierarchy of stratal units that range upward in thickness from lamina, through lamina set, bed, bed set, parasequence, parasequence set, to sequence. Laminae and lamina sets are generally thinner than the resolution of most wireline logging tools. However, higher order members can often be identified on logs, particularly when supplementary information from facies interpretations of cores or outcrops is available. Upward-coarsening and upward-fining parasequences show distinctive shapes on a variety of scales that can be seen on the common wireline logs (Van Wagoner et al., 1990). Parasequences that coarsen upward are marked by a tendency to increase in thickness in bed sets and decrease in shale ratio. The reverse is true in fining-upward parasequences where bed sets become thinner and there is an increase in shale ratio. These attributes are measured at all scales by the color-image transform and so could be used as a tool for the recognition of parasequences and parasequence sets.

The discussion of potential applications has been restricted to examples involving either gamma-ray or SP logs. There is no intrinsic reason why the same methodology should not be directed to other logs. So, for example, color derivative processing of porosity logs could be used to highlight patterns of porosity development which reflect depositional and/or diagenetic geometries. The resulting images would be a new graphic method to show the internal structure of complex reservoir units.

MARKOV CHAIN ANALYSIS OF ELECTROFACIES SEQUENCES

The vertical arrangement of beds in outcrop successions is a rock type sequence that is ordered in time. If each bed can be assigned to one member of a set of states (lithology, facies, or other descriptor), then the sequence forms a chain of discrete events. The ordering patterns within this sequence can be studied by treating the sequence as a Markov chain. Named after their discoverer, A.A. Markov, whose inspiration was the alternation of vowels and consonants in Pushkin's poem "Onegin," Markov chains are an example of a stochastic process model. Markov chain models take up the middle ground between the extremes of determinism, where every event is exactly determined by its predecessor, and independent events, where there is no relationship between successive events. Instead, the outcomes of a Markov chain are analyzed in terms of probability theory, with a wide range of useful properties and applications. A useful summary of the broader aspects of the model is given by Kemeny and Snell (1960), but we shall restrict our attention to some simple properties of Markov chains that are useful in the analysis of electrofacies sequences.

The zonation of well logs and their assignment to electrofacies types emulates the generation of standard geological descriptions from outcrop or core. The end result has the same form: a coded sequence of events expressed in a relatively small set of electro-

TABLE 4. Tally Matrices of Transitions between Electrofacies of a Lower Cretaceous Sandstone–Shale Sequence.

	OBSERVED TRANSITIONS						TRANSITIONS EXPECTED FROM INDEPENDENTEVENTS MODEL					
	K	C	S	P	D	B	K	C	S	P	D	B
K	–	6	0	0	0	0	–	1.5	1.0	0.7	2.0	0.8
C	3	–	1	2	4	0	1.5	–	1.8	1.3	3.9	1.5
S	1	1	–	0	3	2	1.0	1.8	–	0.8	2.4	1.0
P	0	0	1	–	2	2	0.7	1.3	0.8	–	1.7	0.7
D	2	2	5	1	–	2	2.0	3.9	2.4	1.7	–	2.0
B	0	0	0	2	4	–	0.8	1.5	1.0	0.7	2.0	–

KEY: K= coarse sandstone; C = fine sandstone; S = thinly bedded sandstone; P = proximal flood-plain shale; D = distal flood-plain shale; B = bay shale.

facies types. The information is discrete and not amenable to the continuous time series techniques described up to this point. An example of an electrofacies sequence is given by the cluster analysis of zones from the Lower Cretaceous sandstone–shale sequence described in Chapter 4. The clustering partition at a level of six electrofacies for three sandstones and three shales appeared to discriminate between flood-plain, bay, paralic, and channel environments. In the following treatment the six electrofacies are coded as:

K=coarse sandstone, typically in the lower part of a channel;

C=fine sandstone, commonly in the upper part of a channel;

S=thinly-bedded sandstones representing both crevasse-splay and paralic units;

P=proximal flood-plain shale;

D=distalflood-plain shale;

B=both brackish and fresh-water bay facies (shale).

The relationship between adjacent events in the sequence can be summarized in a transition tally matrix. The matrix is a contingency table whose cells sum to the number of times that one state (identified by the row) is followed by another (identified by the column). The transition tally matrix for the Lower Cretaceous example is shown in Table 4. The individual tallies suggest systematic patterns of ordering—but, which transitions occur more often than might be expected and which may be due simply to chance? Some conclusions can be drawn by comparing the observed tally with that which would be expected for a situation of independent events. This expectation is given by the lower tally matrix of Table 4, generated using the procedure described by Davis (1986).

Basically, the expected number of transitions from state i to state j is given by:

$$e_{ij} = \left(\frac{n_i}{N}\right)\left(\frac{n_j}{N}\right) N = \frac{n_i n_j}{N}$$

where N is the total number of events in the sequence and n_i and n_j are the number of occurrences of state i and state j. The application of this formula is not so straightforward in this application, because it is an example of an "embedded" Markov chain. In following the electrofacies sequence, we have counted the transitions between them, but have no numbers for possible transitions of an electrofacies to itself. This design leads to zero tallies on the main diagonal of the transition matrix, and is the result of an embedded Markov chain. The equivalent randomized expectation is then actually one of pseudo-independent events, because although transitions between states are random, there is a constraint that precludes two successive events from taking the same state.

If the sample were larger, the sequence could be tested for a systematic Markov property through a chi-square contingency test. The test would assess whether the observed transition tally matrix was significantly different from the pseudo-independent events transition tally matrix. When the null hypothesis is rejected, then the alternative of a first-order Markov property will be accepted. This property is sometimes called the memory, and is first-order,

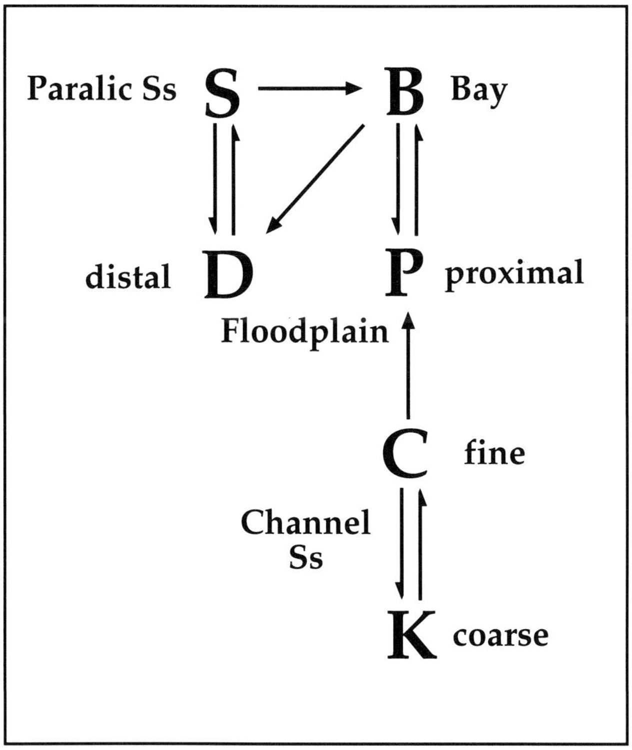

Figure 31. First-order Markov preferred transition path of Lower Cretaceous deltaic sandstone–shale sequence, drawn from the comparison of observed transition frequencies with those predicted by a model of pseudo-independent events.

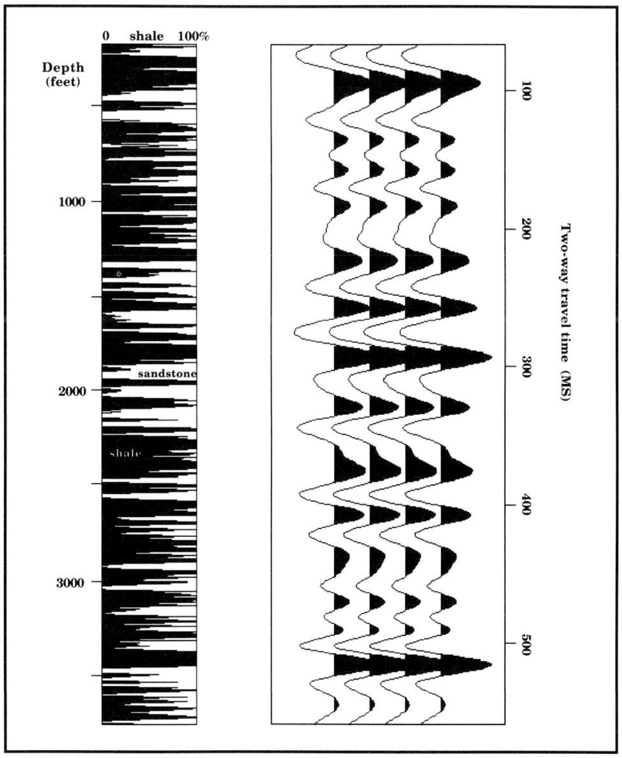

Figure 32. Markov chain simulation of a sandstone–shale sequence (left) convolved with a 30-hertz Ricker wavelet to produce a synthetic seismogram (right). The seismogram has been depth-shifted to match maximum energy of the wavelet with the reflecting surface.

because any systematic property in this model shows a link between immediately successive events. Higher order memories are possible, but become progressively less likely in actual rock sequences.

Any simple underlying pattern can be found through a comparison between the observed transition tallies and those expected from independent events. The network of preferred transitions is shown in Figure 31. The network appears to split between transitions that link channel sandstone facies with proximal flood plain and a second association between thin sandstones and distal flood plain and bay facies.

A preferred transition network generalizes the nature of any simple repetitive schemes for the ordering of electrofacies. It may clarify ideas concerning depositional or diagenetic processes. However, the network is a deterministic simplification of the full network of probabilities shown by the transition probability matrix. This matrix is computed from the transition tally matrix by dividing each cell tally by its row total. These conditional probabilities can be used as additional information in the prediction of electrofacies as described in Chapter 4. In addition, vertical adjacency of rock types translates to lateral adjacency through Walther's Law, as discussed in that chapter. Transition probabilities can therefore be used to aid stratigraphic correlations between wells (see Chapter 6).

In conventional Markov chain analysis, observations are made at equal intervals of time. The resulting Markov chain is not embedded, because successive events can take all possible outcomes, including the same state. The transition tally matrix has values in the main diagonal cells. It is difficult to choose a sampling interval that is both meaningful and practical in many geological applications. If too coarse an interval is selected, then thin beds may be missed. If the interval is too fine, then counts within the main diagonal cells may vastly exceed transitions between lithologies. In this situation, the process is strongly Markovian: states tend to be succeeded by themselves. So, when the objective is to isolate systematic tendencies in the transitions between states, analysis is generally made with an embedded Markov chain model (Doveton, 1970).

Simulated sedimentary successions can be generated from a transition probability matrix. An arbitrary state is selected as the first event. A random-number generator then supplies a value that selects a cell in the row of the first state; the value in the cell determines the state of the second event. The process is repeated as many times as is required to form a sequence of events. If the transition matrix is embedded, then a random selection must also be made from a thickness distribution to set the thickness of each event. If the matrix is not embedded, then the simulated sequence will generate thicknesses automatically. The scale of each step is then the same as the sampling interval of the original observations. The simulations often appear to be quite realistic. However, it should always be remembered that their basis is quite myopic, since they are based solely on conditional probabilities that link successive lithologies. The Markov model can be a useful descriptor of sequences that appear to be stacked repetitions of relatively simple units. Possible examples include turbidite successions or sediments formed on a delta-plain that is dominated by autocyclic processes.

An interesting practical application of Markovian simulated successions to hydrocarbon exploration was described by Sinvhal and Khattri (1983). If acoustic transit times can be assigned to the various states in a Markov chain, then a blocked acoustic velocity log can be created for a simulated succession. When the acoustic log is converted to a reflection coefficient series and convolved with a seismic impulse response, the result is a synthetic seismogram (Figure 32). In itself, the seismogram is only of academic interest, because it is literally one of an infinity of seismograms that could be created from the same transition probability matrix. However, if the process is repeated to generate a sample of synthetic seismograms, then their average characteristics should differentiate them from those generated by a markedly different transition probability matrix.

Sinvhal and Khattri (1983) pursued this idea in a seismic exploration study in the Tertiary of Western India, where dominantly sandy hydrocarbon-bearing successions are contrasted with barren successions that are more shaly. Transition tally matrices were aggregated for sections of each facies type from the logs of a number of wells, sampled at 4-m spacing. A total of 508 synthetic seismograms were generated from the simulations created from the two transition probability matrices. Seismic lines in the vicinity of the well control were used to provide 387 real seismograms in the comparative study that followed. Seismic attributes from the autocorrelation function and power spectra of these seismograms were compiled for each facies type. The attributes were then entered in a discriminant function analysis to differentiate productive from barren successions. The analysis was particularly helpful in showing which of the seismic attributes contributed most strongly to discrimination. The attributes were reduced in number to include only the most effective discriminators, and the ability of the function to classify correctly a validation set of seismic traces was tested. The validation test result, a performance of 73%-correct classification, can be considered moderately successful.

This seismic case-study is an interesting example of how transition probabilities of electrofacies can be used in a practical manner. However, Markovian models should have a wide range of additional applications, because they provide simple and useful descriptors of the arrangement of electrofacies with respect to depth. In addition, in accordance with Walther's Law, vertical ordering implies analogous patterns of arrangement in the lateral dimension. Consequently, transition probabilities contain information on variability in all three spatial dimensions.

REFERENCES CITED

Ager, D. V., 1973, The nature of the stratigraphical record: New York, John Wiley & Sons, 114 p.

Ahmed, N., T. Natarajan, and H. R. Rainbolt, 1976, On generating Walsh spectrograms: IEEE Transaction on Electromagnetic Computability, v. EMC-18, no. 4, p. 198-200.

Beauchamp, K. G., 1984, Applications of Walsh and related functions: London, Academic Press, 308 p.

Berger, A., M. F. Loutre, and V. Dehant, 1989, Influence of the changing lunar orbit on the astronomical frequencies of pre-Quaternary insolation patterns: Paleoceanography, v. 4, no. 4, p. 555-564.

Bone, D., 1982, Music and the circular functions: The UMAP Journal, v. III, no. 3, p. 1-12.

Borer, J. M., and P. M. Harris, 1991, Lithofacies and cyclicity of the Yates Formation, Permian Basin: Implications for reservoir heterogeneity: AAPG Bulletin, v. 75, no. 24, p. 726-779.

Chatfield, C., 1975, The analysis of time series: Theory and practice: London, Chapman and Hall, 263 p.

Collins, D. R., and J. H. Doveton, 1988, Color image transformations of wireline logs as a medium for sedimentary profile analysis: Bulletin of Canadian Petroleum Geology, v. 36, no. 2, p. 186-190.

Cooley, J. W., and J. W. Tukey, 1965, An algorithm for the machine computation of complex Fourier series: Mathematical Computation, v. 19, p. 297-301.

Cox, D. R., and P. A. W. Lewis, 1966, The statistical analysis of series of events: London, Methuen, 283 p.

Davis, J. C., 1986, Statistics and data analysis in geology: New York, John Wiley & Sons, 646 p.

Doveton, J. H., 1970, An application of Markov chain analysis to the Ayrshire Coal Measures succession: Scottish Journal of Geology, v. 7, no. 1, p. 11-27.

Doveton, J. H., 1986, Log analysis of subsurface geology—concepts and computer methods: New York, John Wiley & Sons, 273 p.

Duff, P. McL. D., A. Hallam, and E. K. Walton, 1967, Cyclic sedimentation: Utrecht, Elsevier, 280 p.

Fischer, A. G., and D. J. Bottjer, 1991, Orbital forcing and sedimentary sequences: Journal of Sedimentary Petrology, v. 61, no. 7, p. 1063-1069.

Gilbert, G. K., 1900, Rhythms and geologic time: American Association for the Advancement of Science Proceedings, v. XLIX, p. 1-19.

Granger, C. W. J., 1966, The typical shape of an econometric variable: Econometrica, v. 34, p. 150-161.

Haq, B. U., J. Hardenbol, and P. R. Vail, 1988, Mesozoic and Cenozoic chronostratigraphy and eustatic cycles, in C. K. Wigus, et al., eds., Sea-level research: An integrated approach: SEPM Special Publication 42, p. 71-108.

Harmuth, H. F., 1977, Sequency theory: Foundations and applications: New York, Academic Press, 505 p.

Hu, L. N., D. A. Textoris, and J. K. Filer, 1990, Cyclostratigraphy from gamma-ray logs, Upper Triassic lake beds (Middle Carnian), North Carolina, U. S. A. [Abstract]: 13th International Sedimentological Congress Proceedings, p. 549.

Imbrie, J., and K. Imbrie, 1979, Ice ages: Solving the mystery: Cambridge, MA, Harvard University Press, 224 p.

Jarrard, R. D., and M. A. Arthur, 1989, Milankovich paleoceanographic cycles in geophysical logs, Leg 105, Labrador Sea and Baffin Bay, Section 38, in S. K. Stewart, ed., Scientific results: Texas A & M University, Proceedings of the Ocean Drilling Program, v. 105, p. 757-772.

Kemeny, J. G., and J. L. Snell, 1960, Finite Markov chains: Princeton, Van Nostrand, 210 p.

Kerford, S. J., and D. T. Georgi, 1990, Application of time-series analysis to wireline logs: The Log Analyst, v. 31, no. 3, p. 150-157.

Kominz, M. A., and G. C. Bond, 1990, A new method for testing periodicity in cyclic sediments: Application to the Newark Supergroup: Earth and Planetary Science Letters, v. 98, no.2, p. 233-244.

Lamont-Doherty Borehole Research Group, 1990, Wireline logging manual, Ocean Drilling Program: Palisades, NY, Lamont-Doherty Geological Observatory, 417 p.

Lanning, E. N., and D. M. Johnson, 1983, Automated identification of rock boundaries: An application of the Walsh transform to geophysical well-log analysis: Geophysics, v. 48, no. 2, p. 197-205.

Magara, K., 1979, Identification of sandstone body types by computer: Mathematical Geology, v. 11, no. 3, p. 269-283.

Mark, D. M., and P. B. Aronson, 1984, Scale-dependent fractal dimensions of topographic surfaces: An empirical investigation, with applications in geomorphology and computer mapping: Mathematical Geology, v. 16, no. 7, p. 671-683.

Martini, I. P., M. Sagri, and J. H. Doveton, 1978, Lithologic transition and bed thickness periodicities in turbidite successions of the Antola Formation, Northern Apennines, Italy: Sedimentology, v. 25, p. 605-623.

Melnyk, D. H., 1990, Cyclicity in filtered wireline logs, Kimmeridgian through Portlandian stages, Wessex Basin, U. K. [Abstract]: 13th International Sedimentological Congress Proceedings, p. 354

Miall, A. D., 1992, Exxon global cycle chart: An event for every occasion?: Geology, v. 20, no. 9, p. 787-790.

Milankovitch, M., 1941, Kanon der Erdbestrahlung und seine Anwendung auf das Eiszeitenproblem: Belgrade, Serbian Academy of Science, v. 133, 633 p.

Molinie, A. J., and J. G. Ogg, 1990, Sedimentation-rate curves and discontinuities from sliding-window spectral analysis of logs: The Log Analyst, v. 31, no. 6, p. 370-374.

Mwenifumbo, C. J., and J. P. Blangy, 1991, Short-term spectral analysis of downhole logging measurements from site 704, Chapter 30, in P. F. Ciesielski, Y. Kristoffersen, et al., eds., Scientific results: Texas A & M University, Proceedings of the Ocean Drilling Program, v. 114, p. 577-585.

Pirson, S. J., 1970, Geologic well log analysis: Houston, Gulf Publishing, 370 p.

Rawson, R. R., 1980, Uranium in the Jurassic Todilto Limestone of New Mexico—an example of a sabkha-like deposit, in C. E. Turner-Peterson, ed., Uranium in sedimentary rocks: Application of the Facies Concept in Exploration: SEPM Short Course Notes, SEPM, Tulsa, Oklahoma, p. 127-147.

Saitta, S., and G. S. Visher, 1968, Subsurface study of the southern portion of the Bluejacket-Bartlesville sandstone, Oklahoma, in G. S. Visher, ed., A guidebook to the geology of the Bluejacket-Bartlesville Sandstone: Oklahoma City Geological Society, p. 32-51.

Savitzky, A., and M. J. E. Golay, 1964, Smoothing and differentiation of data by simplified least-squares procedures: Analytical Chemistry, v. 36, no. 8, p. 1627-1639.

Selley, R. C., 1976, Subsurface environmental analysis of North Sea sediments: AAPG Bulletin, v. 60, no. 2, p. 184-195.

Serra, O., and L. Sulpice, 1975, Sedimentological analysis of shale-sand series from well logs: Transactions of the SPWLA 16th Annual Logging Symposium, Paper W, 23 p.

Sinvhal, A., and K. Khattri, 1983, Application of seismic reflection data to discriminate subsurface lithostratigraphy: Geophysics, v. 48, no. 11, p. 1498-1513.

Solterer, J., 1941, A sequence of historical random events: Do Jesuits die in threes?: Journal of the American Statistical Association, v. 36, no. 5, p. 477-484.

Spalletti, I., A. Del Valle, and A. Kielbowicz, 1990, An orbital induced cyclic succession in the early Cretaceous of western Argentina [Abstract]: 13th International Sedimentological Congress Proceedings, p. 519.

Van Wagoner, J. C., R. M. Mitchum, K. M. Campion, and V. D. Rahmanian, 1990, Siliciclastic sequence stratigraphy in well logs, cores and outcrops: AAPG Methods in Exploration Series, No. 7, 55 p.

Vermeer, P. L., and J. A. H. Alkemade, 1992, Multiscale segmentation of well logs: Mathematical Geology, v. 24, no. 1, p. 27-43.

Visher, G. S., 1965, Use of vertical profile in environmental reconstruction: AAPG Bulletin, v. 49, no. 1, p. 41-61.

Visher, G. S., 1969, How to distinguish barrier bar and channel sands: World Oil, v. 168, no. 6, p. 106-113.

Weedon, G. P., 1989, The detection and illustration of regular sedimentary cycles using Walsh power spectra and filtering, with examples from the Lias of Switzerland: Journal of the Geological Society of London, v. 146, no. 1, p. 133-144.

Worthington, P. F., 1990, Sediment cyclicity from well logs, in A. Hurst, M. A. Lovell, and A. C. Morton, eds., Geological applications of wireline logs: Geological Society of London, Special Publication 48, p. 123-132.

Zeller, E. J., 1964, Cycles and psychology: Kansas Geological Survey Bulletin 169, p. 631-636.

Chapter 6

Lateral Correlation and Interpolation of Logs

It is ironic that petrophysicists are only minority users of wireline logs. They are easily outnumbered by geologists who apply logs routinely as frameworks for lithostratigraphic correlation (Lang, 1986). By linking tops of correlative units, geologists trace the structure and thickness changes of formations across fields or even basins, through correlation profiles and maps of surfaces. However, even today, the task of correlation is usually a laborious manual process. Although many automated methods have been proposed, their routine use in field applications still lies in the future. This changeover is only a matter of time, for much the same reasons that automated contouring has made such substantial inroads into manual contouring. Manual methods for both contouring and correlation are slow, highly labor-intensive, inconsistent, and expensive.

The major rationale for decisions by a human operator is that stratigraphic correlation is fraught with ambiguities and requires the pattern recognition skills of a geologist. However, correlations of a set of logs made by several geologists often differ substantially. If geologists cannot find a common correlation, perhaps they can at least agree on a set of procedures and criteria for the practice of correlation. A computer method is the obvious practical means to codify and apply these rules. The results of any such approach have the immediate advantage of consistency. The answer is the same each time the method is applied to the same logs using the same rules. However, consistency is only a step on the road to the goal of a unique and correct correlation. If a true correlation is known for at least some of the logs, then the system of rules can be adjusted until the results are not only consistent, but correct. The task may be difficult but, hopefully, not impossible.

HISTORICAL PERSPECTIVE

Automated correlation was one of the first applications of computers to wireline logs. Moran et al. (1962) described a simple computer procedure to correlate microresistivity curves recorded on the pads of a dipmeter. By this means, the excruciating manual chore of calculating dips and strikes of bedding planes over long intervals could be delegated to a machine. Although microresistivity tools record logs with very fine vertical resolution, changes and shifts between them across the width of a borehole are generally fairly minor, unless the formation is markedly heterogeneous or dip angles are steep. However, even in this simple setting, dipmeter processing algorithms have been refined progressively over the years (see Kerzner, 1986).

The problem of correlating logs over the typical spacing between wells is generally much more complex. While some beds are laterally continuous over great distances and maintain essentially constant log characteristics, others show marked changes or pinch out completely. Cyclic repetitions of lithologies often make it hard to make unique assignments of correlative units. Unconformities, hiatuses, faults, and other discontinuities can disrupt correlations in many different ways. All these factors are well known, although their complex role in the conceptual world of the subsurface can be underestimated. Outcrops show that while some rock units show remarkable continuity over long distances, many others have little lateral persistence. Jordan et al. (1993) discussed some useful practical experiences with correlation problems based on wireline logs of quarry walls. The example in Figure 1 shows three gamma-ray logs of the Pennsylvanian Jackfork Group from a quarry in Arkansas. The sandstones and shales were deposited as turbidites and debris flows in a slope setting. The discontinuous nature of these rocks caused a drastic revision of the simple log correlation that was made before examination of the outcrop. The actual disposition of the succession is that of a thick slumped sandstone, onlapped by thinner, lenticular sandstones. Subsurface correlation problems are scaled-up versions of these learning examples, where the spacing between typical well control will exceed the lateral extent of many rock units.

Subsurface correlation is most commonly based on logs, but geology, insofar as it is known from core and cuttings, is used as the arbiter of the validity of the correlation result. This is appropriate, because the wireline log is a remotely sensed measurement of a physical property that is only indirectly related to diagnostic compositional and textural rock properties. It also means that any successful correlation is one of lithostratigraphy that will usually not match coeval units of chronostratigraphy. Instead, the correlation will track lithofacies in their migrations through space and time. Surprisingly little attention

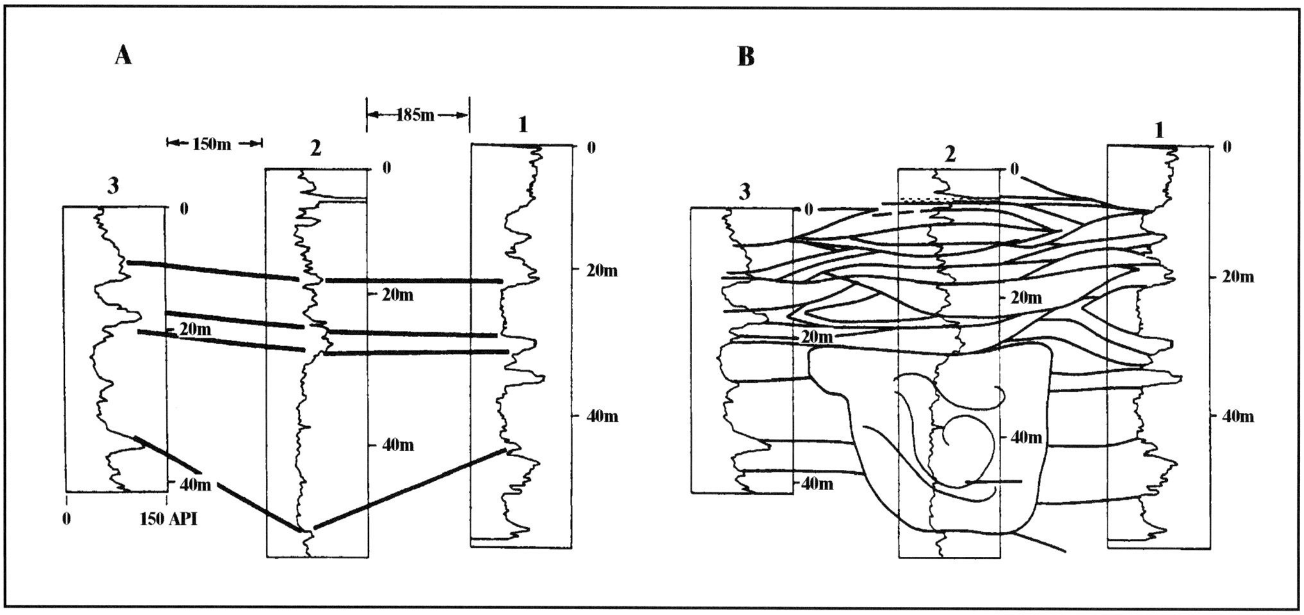

Figure 1. Correlations of Pennsylvanian Jackfork Group sandstones and shales from gamma-ray logs of a quarry wall, (A) before and (B) after examining the outcrop. From Jordan et al. (1933).

has been paid to the type of log that is best suited for a given stratigraphic correlation problem or the appropriate scaling of measurement units. Although all logs record rock properties, do the properties of a given log match the geological criteria set for correlation? If they do not, then the most sophisticated computer procedure is doomed to failure.

A simple example of the wrong log would be the use of a gamma-ray trace in sections of a thick, shale-free carbonate sequence. If the lithostratigraphic subdivision and correlation were based on textural and carbonate mineral changes, then the gamma-ray log might well be ineffectual for correlation. Success cannot be ruled out, since fortuitous changes in natural radioactivity might be caused by uranium concentrations that were preferentially partitioned between distinctive lithofacies. However, human nature being what it is, failure would be blamed on logs in general and computers in particular, rather than shoddy thinking by the stratigrapher. A little thought as to the nature of the geological measurement of the wireline log is the essential first step in any automated stratigraphic correlation.

Matuszak (1972) listed the properties of the common logs as they are used for correlation purposes. For the most part, the fundamental properties are shale content and pore fluid volumes. Matches between logs from different wells are then made in a hybrid strategy operating at a variety of scales. Wherever possible, distinctive marker beds such as anhydrite beds or coals are identified and traced between wells. Between these levels, sets of individual peaks and troughs are matched in a process of quiddity—their shapes and arrangement are compared for apparent similarity. If, indeed, they are the same, then the porous or shaly units that they represent must have lateral dimensions that are at least that of the well spacing. If they are not, similar but different units have been mistakenly linked. At a larger scale, entire packages are compared in terms of overall trends and shapes.

The logs that are used most commonly for correlation, even today, are the SP and resistivity logs. This practice is a result both of ingrained tradition and the simple fact that they are so much more widely available than any other log combination. However, Lang (1984) asserted bluntly that they would probably not have been used as a correlation tool if modern technology had been available earlier. Resistivity logs tend to emphasize porosity fluctuations and shale stringers within sandstones, at the expense of more subtle, but possibly systematic changes within shales. Because the SP log is sensitive to porous, permeable zones, it generally results in a fairly featureless trace opposite shales, even when the shales are quite variable. In clastic sequences, sandstone units are commonly the discontinuous phase, with a better likelihood that the encompassing shales maintain continuity between wells. In these circumstances logs used for correlation should be most sensitive to shale character, with a suppression of laterally ephemeral sandstone features.

In some cases the situation can be improved by changing the scaling convention of the log. An illustrative example is given by the induction log in Figure 2, in which a Cretaceous marine shale overlies a deltaic sandstone. Two traces are shown, but they are the same induction log, scaled in both its original measurement of conductivity units and its transformation to resistivity as the reciprocal of the conductivity. The resistivity log shows very well the thin shales (that are probably laterally impersistent) with-

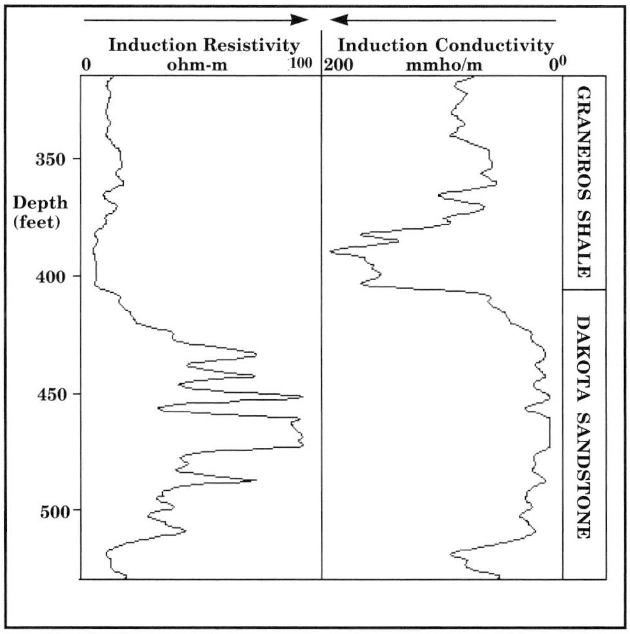

Figure 2. Induction log of a Cretaceous section of shales and sandstones plotted both in resistivity form (left) and original conductivity recording (right). Notice that, although the traces show the same physical measurement, the reciprocal relationship between resistivity and conductivity causes a marked contrast in scaling of measurement units and emphasis of different bed types.

in the deltaic sandstone, but displays no distinctive character in the overlying marine shale. The conductivity log has a radically different appearance, with a suppressed sandstone character and a strong differentiation within the Graneros Shale. The basal, more conductive, and clay-rich shale changes upwards into calcareous shales that underlie the Greenhorn Limestone and mark the passage of a major marine transgression. This example and others suggest that a conductivity log may be preferable to its resistivity reciprocal when lateral continuity is keyed to shale units, rather than to intervening lithologies.

The choice of appropriate scaling can also be helped by thinking through some basic petrophysical concepts. So, for example, the Archie equation is:

$$F = \frac{R_o}{R_w} = \frac{a}{\Phi^m}$$

which, when rearranged, gives:

$$R_o = (a \cdot R_w) \cdot \left(\frac{1}{\Phi^m}\right)$$

The resistivity of a shale-free rock is inversely proportional to the mth power of the porosity. If porosity variations are linked with correlative lithofacies changes, then the use of a resistivity log will give preferential weighting to variations in lower porosity zones. Alternatively, substitution of the square root of the conductivity (if m is taken to be about 2), will give a log that is proportional to porosity throughout the whole range.

Lang (1984) considered the acoustic velocity log to be possibly the best log for subsurface correlation. This log makes useful discriminations of many geologic features, is less affected by borehole conditions or fluids, and is run widely for both geological and geophysical purposes. When scaled as transit times, the log is approximately a linear measure of porosity and also tends to accentuate shale units. The use of acoustic velocity (the reciprocal of transit time) would tend to suppress shale features and accentuate low-porosity fluctuations.

Similar considerations of both scaling and potential geological meaning apply to other logs. If correlation from logs is to be judged against a standard set by more conventional geological observations, then choices of log types and measurement units should be made accordingly. Special attention should be focused on the discrimination of lithologies that are likely to be correlative between wells as opposed to those that are not. In some cases, simple mathematical transforms may be helpful in improving correlation matches. Robinson (1975) demonstrated how root, power, and derivative transformations of logs preferentially accentuated and suppressed features at different scales. Conversion to square- (or other) root values enhances low-amplitude features relative to high-amplitude peaks. When logs are powered, high-amplitude events are further accentuated. Derivative logs are sometimes useful when the original trace shows little character. Zero-crossings of the first derivative locate feature boundaries (see Chapter 5). The derivative log can be useful in correlation as a measure of bed frequency, provided that the noise error of the nuclear logs is first suppressed by smoothing.

STRATIGRAPHIC SEGMENTATION OF WIRELINE LOGS

In Chapter 4 we briefly discussed the subdivision of a continuous analog trace into a sequence of zones. Traditional zonation aims to recover a blocked representation of the log in which each zone is more closely representative of a bedding unit in the borehole wall. The flanks of peaks and troughs are pared away as artifacts of the smoothing by the logging tool response function. Generally, in an attempt to compensate for the averaging effect of the tool, the value of each zone is taken as the extreme peak or trough feature. The vertical resolution of the more common logging tools is such that the scale of the zones corresponds broadly to beds. Zones defined in this manner therefore represent basic bedding units as discriminated by the physical measurement of the tool.

Stratigraphy has a different goal in the subdivision

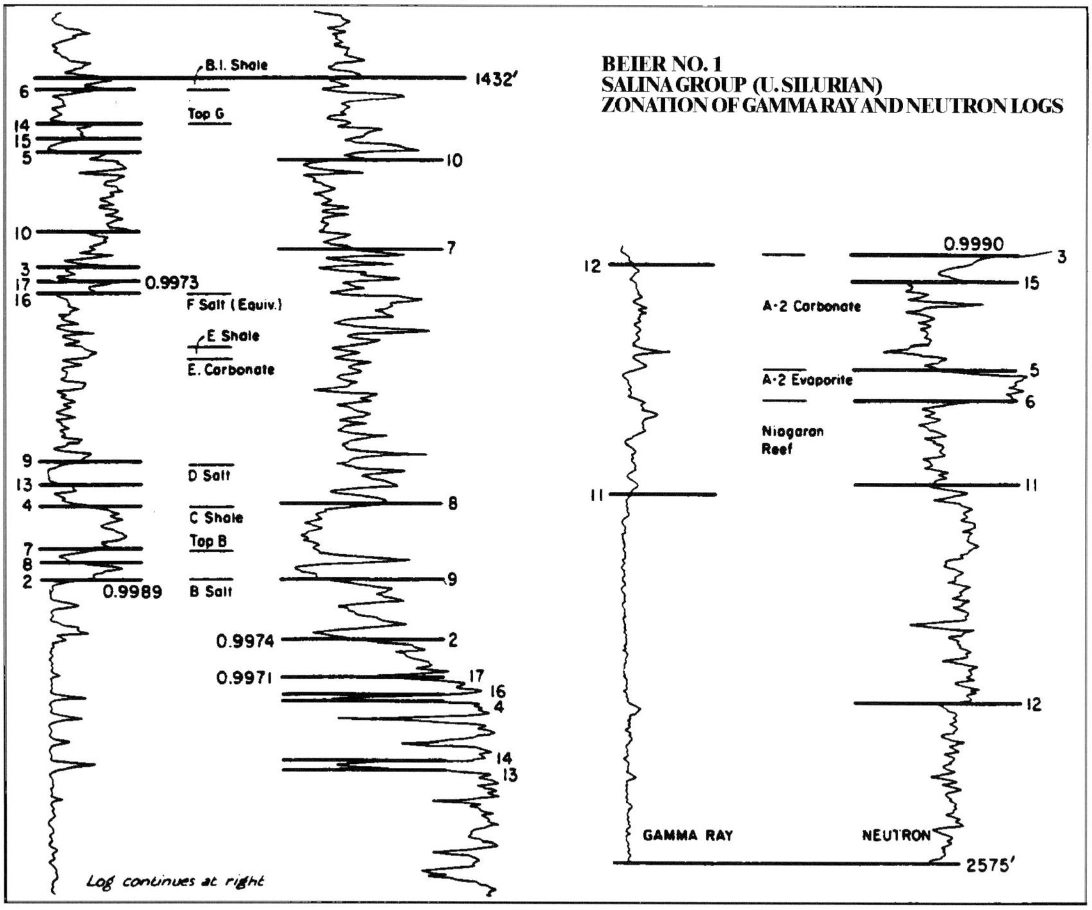

Figure 3. Segmentation of gamma-ray and neutron logs in an Upper Silurian Salina Group succession of carbonates, evaporites, and shales in a Michigan well. The numbers signify the order in which the segments were picked. From Gill (1970).

of a succession into lithostratigraphic units. Rather than discriminate every individual bed, the log should be subdivided into units that aggregate similar consecutive beds and differentiate them from dissimilar contiguous units. The process should isolate major boundaries that have potential significance in lateral correlation. Between the boundaries, each unit should be relatively homogeneous internally and show a marked difference with adjacent units. We will call this process "segmentation" to avoid confusion with bed-discrimination methods that we have termed "zonation."

The most common methods for log segmentation define subdivisions using either a variance or a mean-value criterion. Both Testerman (1962) and Gill (1970) devised computer methods that minimized variance within segments, while maximizing variance between them. Their procedures are adaptations of a single-factor, fixed-effect analysis-of-variance model. On the first pass, a boundary is moved progressively down the succession in an iterative series of trials that contrasts the variance within and between the two resulting segments. The boundary with the best differentiation marks the interface between the first two segments. The operation is then applied to each of the segments, and the process repeated in progressively finer segmentation. Either the number of segments must be known beforehand or some type of variance criterion must be applied to define a stopping point for the procedure.

An example of this method is given in Figure 3 (from Gill, 1970) and shows the segmentation of gamma-ray and neutron logs in an Upper Silurian Salina Group section of carbonate, evaporite, and shale units from Michigan. Notice the numbers marked on the boundaries that register the order of their appearance in the segmentation procedure. This information can be useful in the weighting of the relative importance of the boundaries as markers for stratigraphic correlation. The segmentation follows a different hierarchy for the two logs because of their sensitivity to different petrophysical properties. Gill (1970) noted that the segments defined from the gamma-ray log were compatible with lithostratigraphic units picked manually. However, the association was not so close for segments partitioned from the neutron log, which responds primarily to porosity variability. Either of these segmentation results could be the "correct" result, depending on whether the end product should be a stratigraphic section keyed to lithologies, or a reservoir subdivided into distinctive porosity units.

Alternatively, segmentation can be defined on the basis of mean value rather than variance. Webster and Wong (1969) used the mean value of a sliding window as the test statistic to locate boundaries in data sequences. At each step the sequence within the window was halved and the means of the two subdivisions checked by a t-test to assess any significant difference. The method results in the location of boundaries and segments that are commensurate with the window size. Hawkins and Merriam (1975) made a study of the comparative performance of segmentation defined by variance and mean-value criteria. They concluded that mean-value methods worked best when the data were marked by sharp breaks with intervening plateaus and spikes. Variance techniques were most appropriate when the trace was characterized by changes in gradient or gradational drifts in average value.

The earlier segmentation algorithms were only applicable to single logs, but have been extended to the simultaneous consideration of several logs (Hawkins and Merriam, 1974). Alternatively, a log suite can often be condensed to a single principal component score log with comparatively little loss of useful information. If this is so, then the score log can be segmented by univariate methods. More recently, Gill et al. (1993) have applied adjacency-constrained cluster analysis to the segmentation problem (see Chapter 4). This approach has several novel features that should stimulate renewed interest in automated stratigraphic subdivision. Unlike the divisive strategy of traditional variance techniques, the method is agglomerative in building coarser segments from finer divisions. A cluster dendrogram display is helpful in the search for possible natural levels in the hierarchy that may reflect meaningful geological boundaries (see Chapter 4, Figure 30). Finally, the method can be used with any number of logs, because it operates on similarity measures between adjacent zones or segments.

Some authors, such as Shaw and Cubitt (1979), considered that segmentation of logs into stratigraphic units must be made before analysis by any automated correlation method. Others, such as Olea (1988), suggested that segmentation was of limited use in the development of a viable automated correlation computer program. As yet another viewpoint, Mann (1979) concluded that the location of boundaries by segmentation was probably most useful as a refinement that followed the results of correlation. Clearly, there is a wide diversity of opinion on how segmentation methods can be integrated most effectively with automated correlation.

Also, in the past, the aim of log correlation has often been to emulate classical methods of lithostratigraphy that correlate rock types between outcrops. Field observations of lithology, textural properties, and fossil content are commonly the criteria for these judgments. However, these factors may be of indirect importance in the subdivision of a reservoir unit in the subsurface. The partition of a heterogeneous reservoir into subunits should be keyed to flow units if they are to be used in reservoir engineering decisions. Because the common logs are sensitive primarily to porosity and shale content, they are particularly useful for this purpose. Segmentation and correlation from an engineering perspective will mean a definition of flow units and seals, followed by a trace of their continuities and breaks between wells. Flow units will often map onto conventional lithostratigraphic units; sometimes they will not. Clearly, the goals of both segmentation and correlation must be thought through at the outset. A lithostratigraphic framework may be marginally helpful both to a geologist interested in absolute time lines of chronostratigraphy and an engineer responsible for reservoir management. However, segmentation and correlation based on wireline logs are especially attuned to the needs of the reservoir engineer.

AUTOMATED CORRELATION METHODS FOR LOGS

The most widely used methods of automated correlation of wireline logs have adapted the basic approach of algorithms used in dipmeter processing. The measure of similarity between two log traces is computed by the standard (Pearson) correlation coefficient, r:

$$r = \frac{\mathrm{Cov}\left(L_1 L_2\right)}{s_{L_1} \cdot s_{L_2}}$$

where *Cov* and *s* are the covariance and standard deviations of logs L_1 and L_2. A segment of the first log trace is then moved by small increments past the trace of the second log. At each step a correlation coefficient is computed and plotted as a function of lag in a correlogram (Figure 4). The lag is the relative shift in depth between the two logs at each position of comparison.

The parameters that must be set before running the procedure are the same as those used in dipmeter processing, namely the window length, step length, and search length. The window length is the thickness of the template segment to be extracted from the first log for comparison with the second. A large window length increases the size of the sample, so that correlation coefficients are more stable estimates. However, this ideal must accommodate the reality that longer segments are more likely to be disrupted by heterogeneities, variations in thickness, and discontinuities of one kind or another. Some compromise may be necessary and can often be found by experimentation with different window lengths. Alternatively, rather than make a fixed choice of window length, segments can be drawn from the results of a segmentation procedure.

The step length sets the increment of depth displacement for successive comparisons to be made between the segment and its corresponding window on the second log. A step length that is too fine will be an inefficient use of information and sometimes cause horrendous computer run times. If an interval is too coarse, it degrades the depth resolution of the optimal match, and may even result in the match being missed altogether. Computer time is also a major factor in the selection of the search length. This consideration determines the maximum lag shift to be used as a cut-off for comparison. Generally, some reasonable limits on search length can be set prior to correlation using prior information about ranges of structural dips of local geology. In the worst case of no knowledge, experiments with different search lengths will generally result in a satisfactory solution.

At the completion of a correlation run, the resulting correlogram is examined for the optimum correlation coefficient. The correlation coefficient gives a useful measure that is easily understood. A perfect match of the two traces is given by a positive correlation of unity. A zero correlation means that there is no simple match; negative correlations imply some tendency towards a mirror image. Most commonly, the best match will coincide with a peak correlation value at small lag when the distance between the wells is small and structural dip is not excessive. Secondary peaks are to be expected at other lags as the moving template locks onto similar lithofacies that are repetitions of the target interval. In some cases these false picks may have better correlation coefficients than the true pick (Figure 4). This can be caused by lateral changes in the segment signature or simply by the limitations of the correlation coefficient as a measure of comparative shape. Many ambiguities can be resolved by selecting a search length that excludes matches that are not structurally feasible. Also, because the procedure is applied to many intervals, expert system rules can be applied to eliminate picks that are geometrically inconsistent with the majority (Olea, 1988).

Most practical correlation algorithms use a critical cut-off value for the correlation coefficient to decide if the attempt to match a particular segment should be abandoned. This is appropriate, because some segments will represent intervals that pinch out between the two wells. In other cases, the best value from the correlogram will be so low that visual comparison is unconvincing and the match is probably erroneous. So, the correlation coefficient has a secondary role as a measure of the relative strength of any correlation. Poelchau (1987) applied this property in a novel manner by pointing out that the coefficient could be mapped across a reservoir as a coherence indicator and used to indicate possible flow paths and discontinuities. Unfortunately, the coefficient cannot be mapped in the conventional way because it is a value that relates the similarity of a log segment between two wells, rather than a point value at a single well.

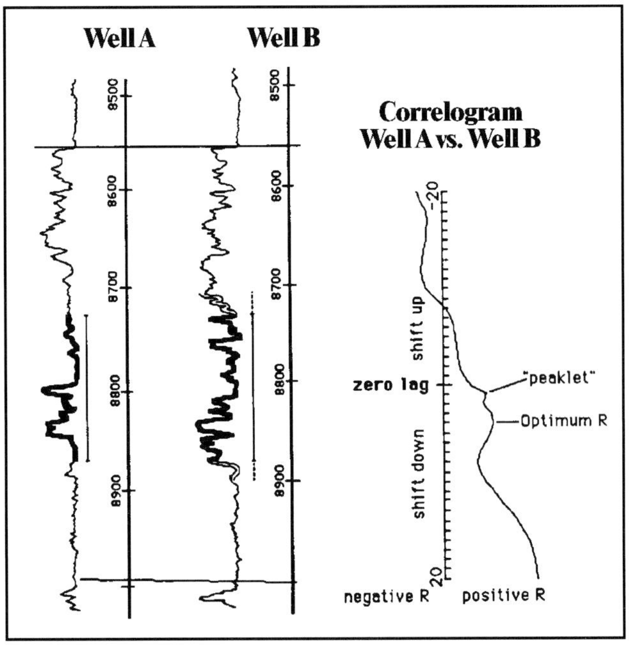

Figure 4. Correlogram produced by cross-correlation of sliding window of data from Well A matched against log of Well B. The optimum was selected as the peak with the maximum correlation and the smallest lag. From Poelchau (1987).

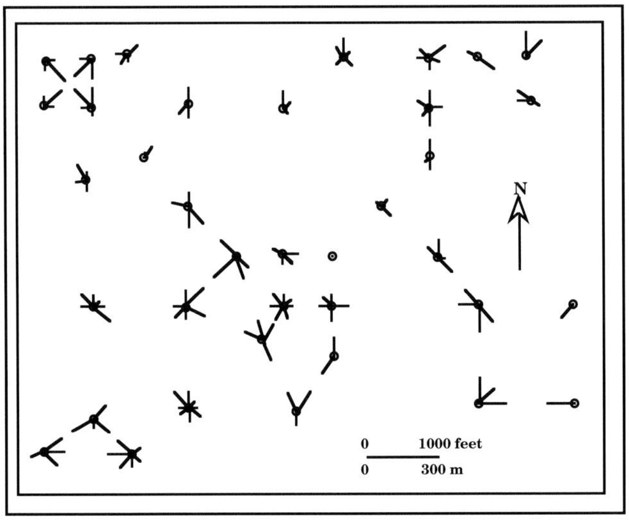

Figure 5. Coherence vector map of correlations between porosity logs of neighboring wells in a carbonate–chert reservoir in west Texas. The length of each vector is proportional to the peak correlations. Correlations tend to be better along strike in a north–south direction and pick up the "grain" of the reservoir. Detail of a larger map published by Poelchau (1987).

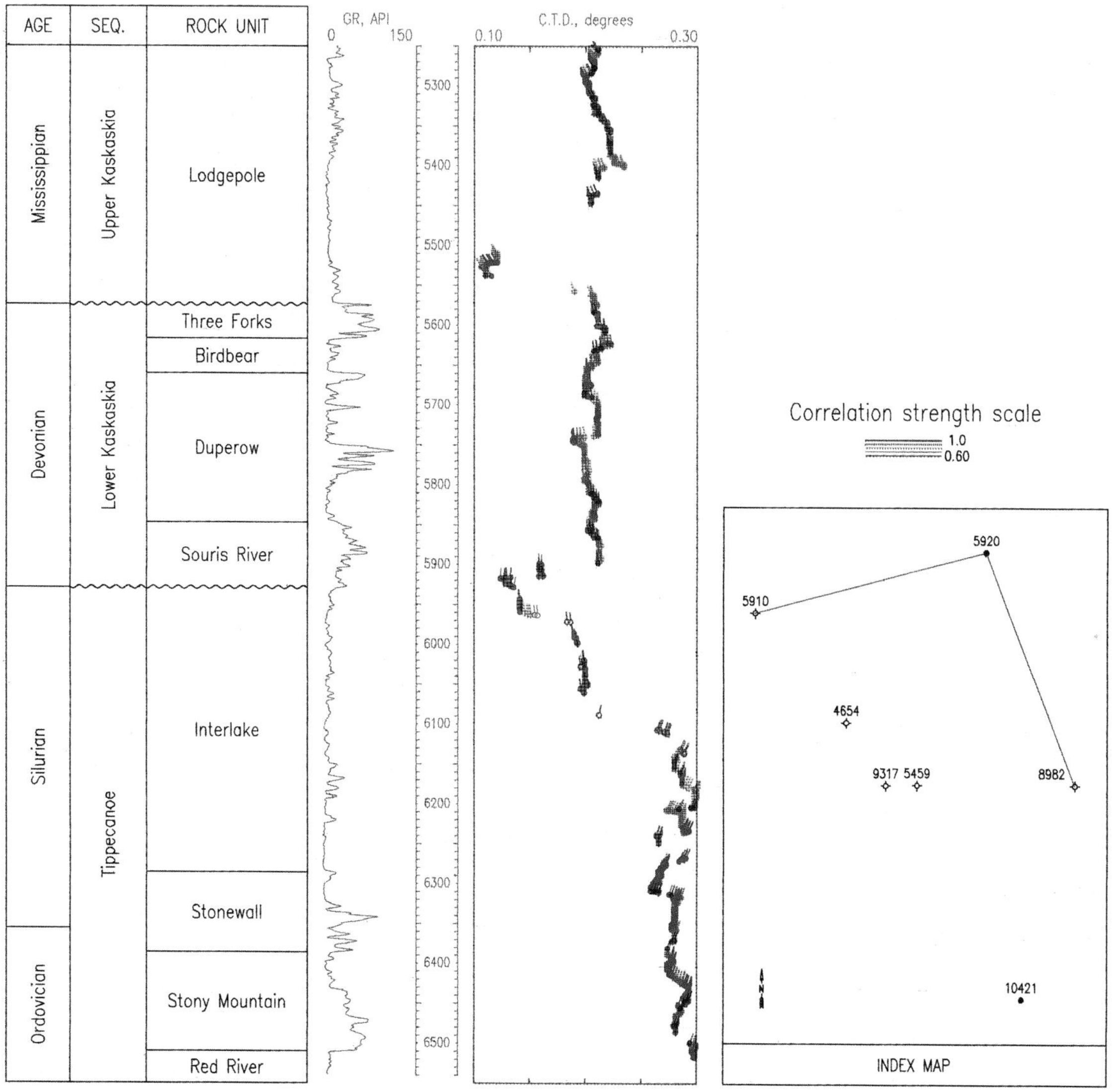

Figure 6. "Tadpole" plot of dip vectors computed from the stratigraphic correlation of gamma-ray logs in three wells in the Cold Turkey Creek field of North Dakota by the CORRELATOR program (Olea, 1993). The vectors are plotted by the same convention as traditional dipmeter plots.

However, one way to present the information is as a vector map that shows the similarity of nearest neighbors. An example of a coherence vector map is shown in Figure 5 (from Poelchau, 1987). The map shows coherence vectors based on porosity logs of a thick reservoir sequence of carbonates and cherts in West Texas. The map was prepared to evaluate the economic feasibility of infill drilling. Correlation between wells was known to range in difficulty, which raised the possibility that the field was compartmentalized rather than continuous. The coherence map therefore had the practical objective of locating the boundaries of these compartments, if they existed.

The correlation coefficient is a good and robust measure of similarity between log traces. It is invariant to scaling changes that involve either the addition of a constant or multiplication of either log by a constant. Minor lateral changes in thickness will degrade the correlation coefficient slightly. Major changes, however, are a problem that can be tackled through selective stretching of one of the logs. So for example, Shaw (1977) advocated stretching the thinner of the two segments to be compared and resampling the stretched curve. This could be done in the time domain by interpolation with B-splines. Other authors have used the frequency domain as the medium in which to stretch logs (e.g. Rudman and Blakely, 1976; Mann and Dowell, 1978; Kwon and Rudman, 1979). Cross correlation of the power spectra of the two logs to be correlated determines the relative stretch factor in the differences in frequency scaling. The idea behind the approach is attractive, but its implementation is computationally complex and rather rigid in its use of a fixed offset and stretch factor for the two logs. Also, the method takes no account of gaps within the sequences.

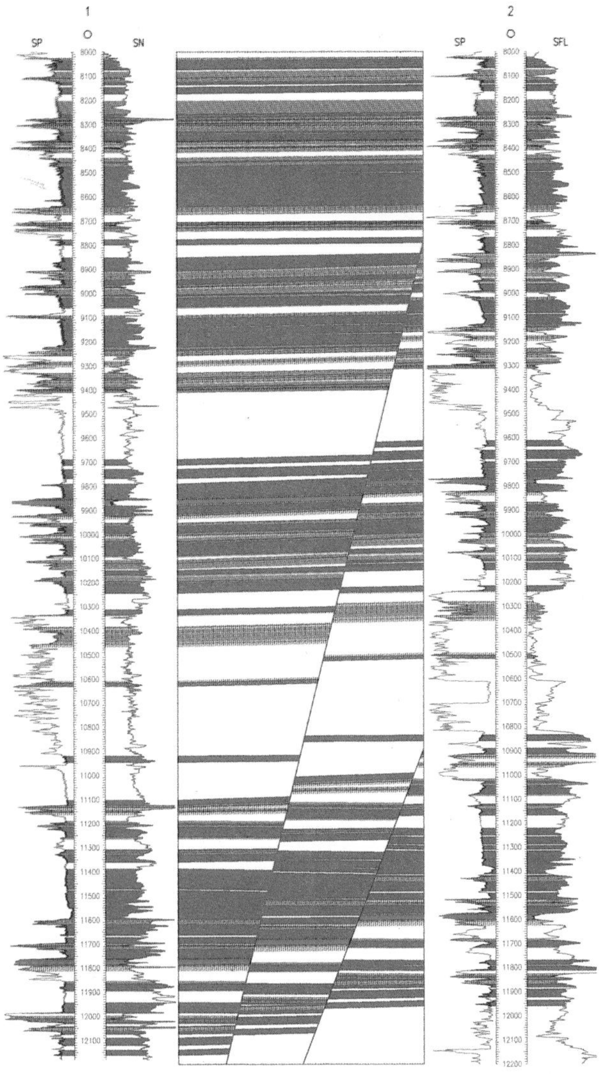

Figure 7. Automated correlation of SP–resistivity logs from two wells of a Gulf Coast Tertiary clastic succession modelled by a double faulted section. The length of each sequence is 4000 ft and there is no vertical exaggeration on the cross-section. Colors represent shale content of correlative intervals, increasing from blue, through yellow, to a maximum of red. Original data provided courtesy of Marathon Oil Company and processed by R.A. Olea.

The development of a practical automated correlation package is now a real possibility thanks to the steady increase in the speed and performance of computer hardware in recent years. Quite complex procedures can now be run interactively, rather than in the inflexible mode of batch processing. This feature allows the human interpreter to monitor results and intervene in the process whenever necessary. Although many correlation problems may be simple enough to be delegated to routine machine processing, many others are challenging both for the human and any computational procedure. While the human often has additional information that can be used to eliminate alternative solutions, it may be difficult to codify.

In some cases, it is possible to formulate both experience and logic in a system of rules that can guide the automated correlation process. Olea (1988) incorporated an expert system within an interactive computer system. The expert system is comprised of a set of production rules, a global database (tables of correlations), and a rule interpreter. The 19 production rules focus mostly on the geometrical consequences of tie-lines computed between the two logs. Crossing lines are flagged as impossible and the likelihood of abrupt changes in dip is checked. The expert system follows a logical sequence that will be reviewed in Chapter 7. Essentially, the process proceeds in a sequential search from the strongest to weakest hypothesis in its evaluations of the bundles of correlations established in the correlation analysis. The user is then alerted to the location of impossibilities or improbabilities and the nature of the rules that are violated or called into question.

Olea's (1988) system evaluates two logs from each well, rather than a single log. The first log (usually either SP or gamma-ray) is used as a shale indicator to segment the sequence into shale and non-shale lithologies. Correlation is then based on the second log, which can reasonably be expected to contain some petrophysical features that are laterally persistent between well control. The cross-correlation coefficient from comparison of the second log between two wells is then weighted by a coefficient of similarity in shale content. The weighted coefficient provides a discrimination based on both logs as to correlative signature and relative shale content.

As with earlier methods, there are ambiguities and problems. However, the addition of the expert system module allows the results of many correlative comparisons to be considered collectively and evaluated in the light of commonsense rules of geometry. The design philosophy of the expert system approach allows easy communication of the program's actions to the user. Any problem situation can be flagged by a "paper trail" of rules that have been triggered and relayed to the user for inspection together with suggested remedial action. The operations form an "editing expert system" that eliminates impossibilities and weighs viable alternatives in the search for an accurate correlation.

In a further refinement, Olea (1994) experimented with elements of a "stratigraphic expert system" to help in the geological modeling of the results of the edited correlations. The linkage of correlative horizons sets the dips and strikes of surfaces between wells. The pattern sequence of these vectors can be thought of as a scaled-up version of vectors processed from dipmeter runs. The sampling space of the dipmeter tool is a matter of inches. Many of the features that are reconstructed from dipmeter logs are very localized, although persistent patterns of common dip and strike commonly reflect regional dip. By contrast, dips and strikes are easily comput-

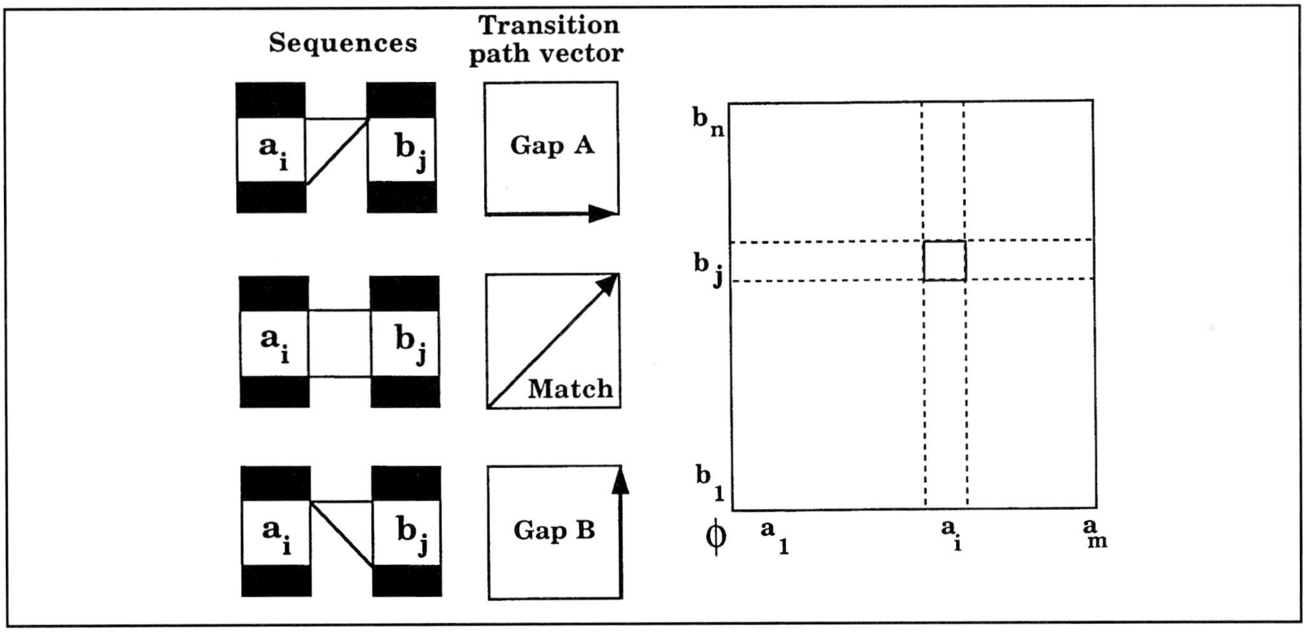

Figure 8. Consequences of gaps or a match of units a_i and b_j as vector directions of a local path (left) in one cell of a dynamic programming framework (right).

ed from matched horizons in automated correlation and pick up stratigraphic and structural features on a larger scale.

The results of automated correlation between three wells can be presented as a vector or "tadpole" plot. This convention has the immediate advantage of familiarity to anyone who has worked with dipmeter interpretation. The correlation logs from each well can be thought of as measurements from the pad of a gigantic dipmeter tool. The results of computer processing of microresistivity traces are then mimicked by automated correlation on a much larger scale. An example of gamma-ray logs processed in this manner from three wells in the Cold Turkey Creek field of North Dakota (Olea, 1993) is shown in Figure 6. The tadpole plot is presented for the lower Paleozoic section and shows systematic patterns that are readily reconciled with the stratigraphic history of the Williston basin. The trend in dips is primarily to the north and reflects the southerly location of the wells with respect to the basin center. The Tippecanoe formations collectively record a gradual shoaling of deposition. The Interlake Formation has an intraformational unconformity (Gerhard and Anderson, 1988) that appears to be picked up by a strong drop in dip magnitude. The regional unconformities at the base of both the Devonian and Mississippian (Gerhard and Anderson, 1988; Gerhard, Anderson, and Fischer, 1991) are also marked by distinctive changes in dip that were computed from the correlation.

The stratigraphic expert system operates on the edited correlations with a set of production rules that analyze various structural contingencies. These rules consider the rate of fluctuations in apparent dip, fluctuation amplitude, average correlation strength, and lithology (Olea, 1994). The expert system should then be able to locate systematic breaks or trends that are recorded by the logs used in automated correlation. The geometrical inference of a feature such as a break is only a first step to its geological *meaning*. Depending on the regional geology, a break could be assigned either to an erosional unconformity or a structural fault. Even this information could be coded in the expert system, although it represents "hearsay" rather than the geometrical reasoning of the other production rules. So, for example, the breaks in the correlation vector plot of Figure 6 appear to match regional unconformities very well. However, similar breaks could be modelled as faults where warranted by local geology. A fault model interpretation is shown in Figure 7 for SP–resistivity log correlations in a Tertiary clastic sequence between two wells. The correlated succession is from the Gulf Coast, where unconformities were ruled out in favor of faults as an explanation of structural breaks.

DYNAMIC SEQUENCE MATCHING OF LOGS

Considerable interest in recent years has been focused on the adaptation of a general procedure variously known as dynamic wave form matching, dynamic depth warping, string-to-string matching, and gene sequence matching to stratigraphic correlation. An optimal global match can be found by the dynamic programming algorithm applied to correlation by Smith and Waterman (1980), Howell (1983), Waterman and Raymond (1987), Wu and Nyland (1987), Lineman et al. (1987), Griffiths and Bakke (1990), and others. These authors have drawn on methods with proven success in sequence compari-

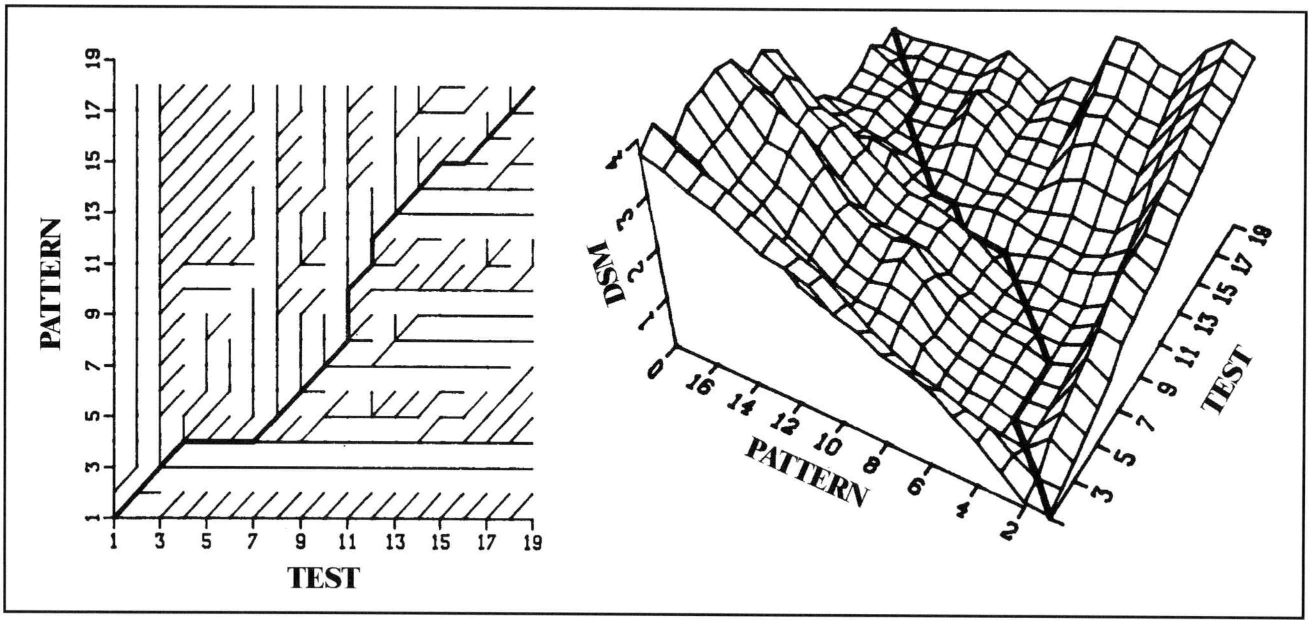

Figure 9. Tree of all local, optimal pathways resolved by dynamic programming algorithm (left) and final minimal cost route traced over landscape of cumulative cost (right). From Wu and Nyland (1987).

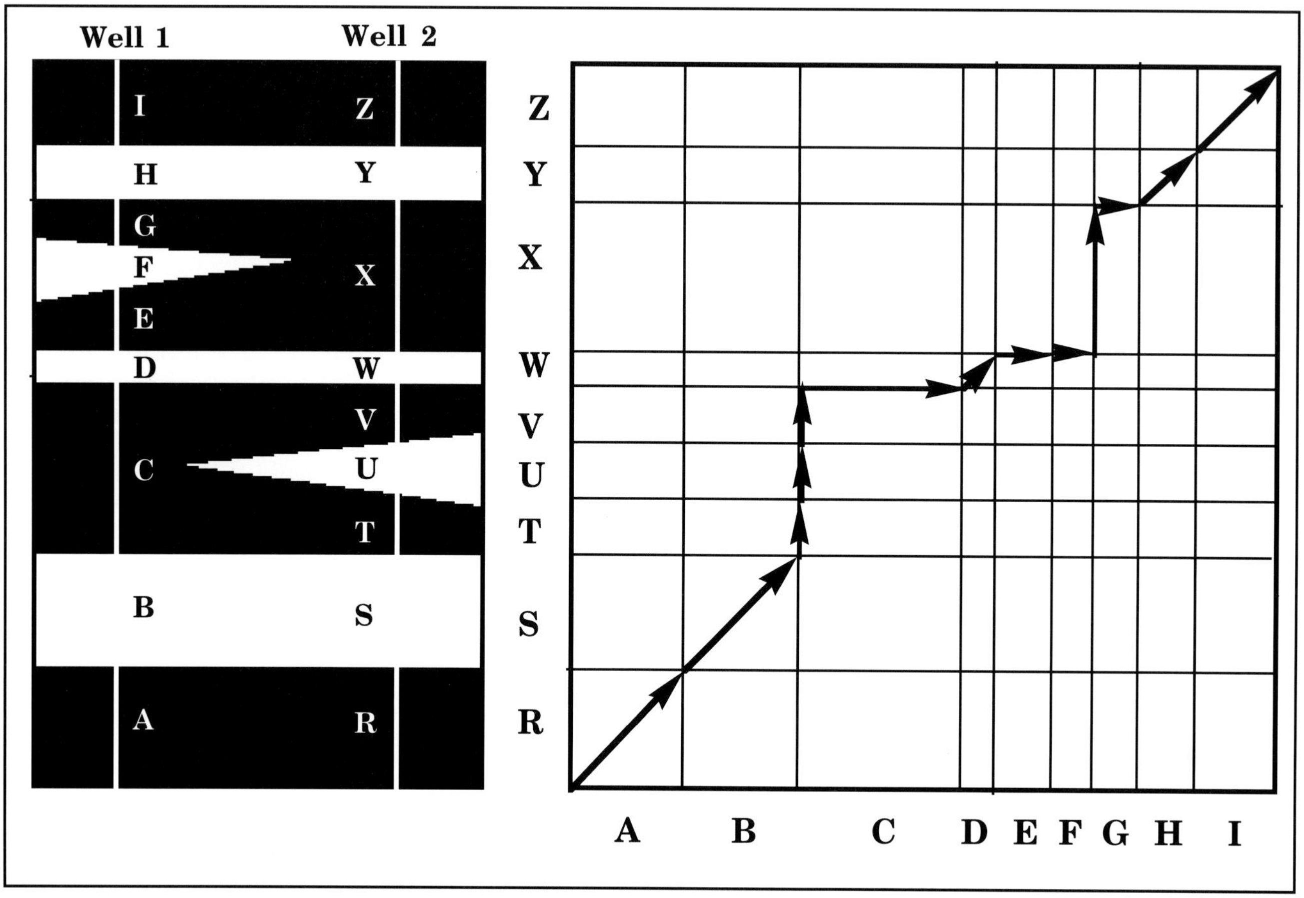

Figure 10. Path diagram showing correlation between two hypothetical sequences of sandstones (white) and shales (black). Diagonal vectors signify matches; horizontal shifts mark gaps in Well 1; vertical shifts show gaps in Well 2. Modified from Fang et al. (1992).

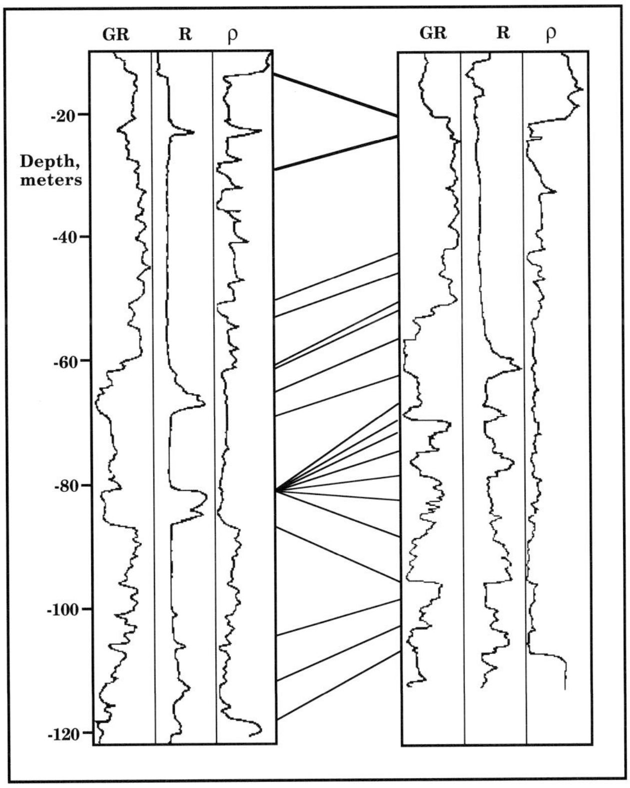

Figure 11. Results of automated matching of segmented sequences of gamma-ray, resistivity, and density logs of a Canadian Cretaceous clastic succession. Adapted from Wu and Nyland (1987).

son within linguistics, molecular chains, and birdsong characterization, which are summarized by Sankoff and Kruskal (1983). The approach is based on the original concept of Levenshtein (1966) and has several distinct advantages. It can handle multiple gaps, use any number of logs, accommodate stretching, and find a solution that is an optimum of all possible alternatives.

The procedure operates on stratigraphic successions that have been coded as sequences of units. Let us designate the sequence of n zones from Well A as $a_1...a_m$, and the sequence of n zones from Well B as $b_1...b_n$. The two sequences can be set as the axes of a graph to be used as a map to trace the best correlation between them (Figure 8). Collectively, the cells provide a framework to evaluate all possible contingencies of matching between the two sequences. The location of any particular cell sets the sub-strings in each sequence that can be considered. If we focus on one-to-one matching of units a_i and b_j, three possibilities can happen: a_i is gapped, b_j is gapped, or a_i and are matched (Figure 8).

The three alternatives can be matched with costs which determine which is the most likely. If a_i and b_j had identical log properties, then the cost of their match would be zero. In reality, stratigraphically equivalent units will have different properties, but these differences will generally be less than units that are not equivalent. Therefore, the cost of a match is some measure of the dissimilarity and will usually be smaller for true matches. The cost associated with a gap is more problematical. Most authors have chosen pragmatic values for gap penalties that are intermediate between the costs of acceptable match and unacceptable mismatch. Waterman and Raymond (1987) pointed out that the success of this algorithm is strongly dependent on good choices of cost functions and that these will be variable between applications. The three possible vectors associated with a gapped a_i gapped b_j, and matched a_i and b_j are shown on Figure 8. The vector that involves the least additional cost is the one selected as the optimal local pathway.

The entire procedure follows from the algorithm devised by Delcoigne and Hansen (1975). The dynamic programming framework uses the two sequences as axes, with the origin in the bottom left corner equated with a "null" state of ϕ. The algorithm seeks to find an optimum pathway through the matrix such that the cumulative cost is a minimum. The search is constrained to honor stratigraphic order by moving up and to the right and is applied recursively at each cell location. At each path step, the choice of either gap or match is made based on which leads to the lowest cumulative cost when the cost of the gap or match is summed with the cumulative cost at the previous position. By this means, the graph is progressively filled out by a branching system of possible pathways and a map of cumulative cost (Figure 9). When the procedure is completed, the optimum correlation can be found by a traceback of the minimal cost pathway.

The goal of the procedure can be seen by reference to the diagram shown in Figure 10 and adapted from Fang et al. (1992). There are two well sections in a hypothetical sequence of sandstones and shales, in which the shales and some of the sandstones are continuous between the wells. At the right, a path diagram shows the correct correlation linkage between the two well sequences which form the axes of the matrix. A horizontal shift marks a gap on the first well. A vertical shift shows a gap on the second well. A diagonal vector marks a match between the units referenced to the sequences of the horizontal and vertical axes. The diagram is highly simplified and alternative correlation routes would be possible, according to both the degree of dissimilarity between shales and sandstones in the sequences and the cost penalties assigned to gaps. This ambiguity is entirely appropriate, because several equally valid scenarios could be drawn on the relationship between the sequences in the two wells. However, the solution of the match that is geologically correct should be made if the matching costs are a good reflection of the actual geological similarity of the various units.

Dissimilarity of units can be estimated, based on one or any number of logs. In the case of a single log, the dissimilarity can be given by the absolute differ-

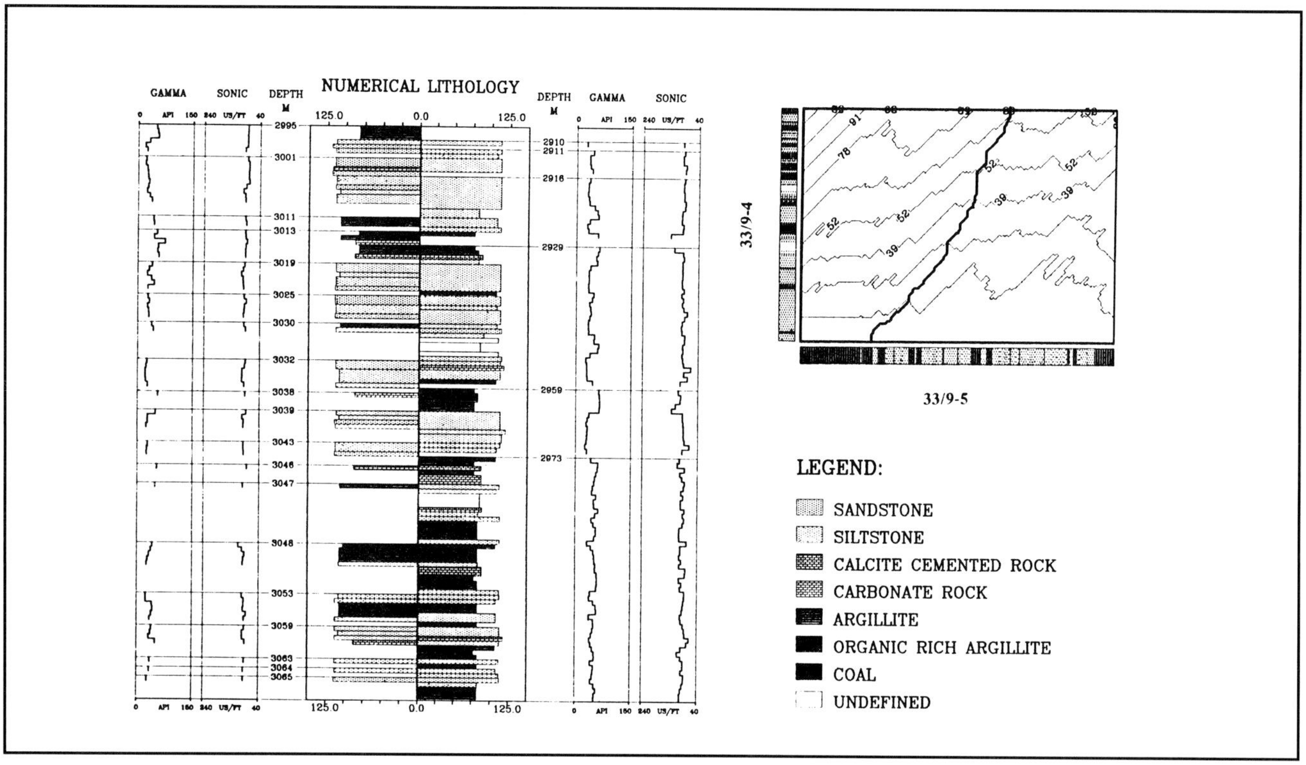

Figure 12. Modified alignment for matching of "numerical lithology" successions of the Statfjord Formation of the North Sea in wells 33/9-5 and 33/9-4 that is given by the traceback path across the cumulative cost matrix map (upper right). From Griffiths and Bakke (1990). Courtesy of the Geological Society of London.

ence between the log segments of the matched units as:

$$d(a_i, b_j) = \left| L(a_i) - L(b_j) \right|$$

where L is the log response in raw or standardized form for each unit. When several logs are used, Lineman et al. (1987) suggested the dissimilarity measure:

$$d(a_i, b_j) = \sqrt{\frac{\left\{ \sum \left[L_k(a_i) - L_k(b_j) \right]^2 \right\} \cdot w(k)}{k}}$$

where L_k is the response of the kth log and $w(k)$ is its associated weighting factor. Wu and Nyland (1987) matched segmented sequences of gamma-ray, density, and resistivity logs of Cretaceous sequences in western Canada (Figure 11) using this dissimilarity approach.

Sequence matching methods in stratigraphy have two powerful features: 1) dynamic programming ensures a global search of all possibilities; 2) the string-matching algorithm can accommodate gaps as well as matches. As mentioned earlier, the success of sequence matching in practical correlation is strongly controlled by the choice of match dissimilarity measure and gap penalty function. However, research continues in this area, with an emphasis on capturing useful geological measures within the cost statistics. So, for example, Griffiths and Bakke (1990) converted gamma-ray, neutron, density, and acoustic velocity log responses to "numerical lithologies" from their cell location in a *M-N*-gamma-ray crossplot space. The result is a coded alphabetical string for each sequence that is analogous to the lettered bases of the genetic code used in gene-typing applications. The discrete form of the code means that simple integers can be used to assign costs to matches and gaps.

An example of matching using the procedure of Griffiths and Bakke (1990) is shown in Figure 12 for a section within the Statfjord Formation in the North Sea. The blocked sequences in the center are the "numerical lithologies" for two wells that have been generated from their location within *M-N*-gamma-ray crossplot space. Assignations to conventional lithologies are given in the key to the right. The alignment of the two sections is shown at a local, rather than global, minimum cost position. This local minimum match conforms more closely with biostratigraphic markers in the two wells. In discussing the choice of this suboptimal alternative, the authors noted that there is no inherent philosophical conflict, because the global optimum represents the best lithostratigraphic matching, while the alternative marks a better stratigraphic correlation. In fact, the difference between the two solutions is minor, with a cumulative cost of 68 for the modified solution, as opposed to 64 for the global minimum. Inspection of the cost contours on the matrix map to the right of Figure 12 shows that several suboptimal paths are possible with costs that are close to the global minimum. This characteristic reflects the degree of ambiguity that commonly arises in realistic problems and the need for some external constraints on the final solution.

In a further refinement, Collins and Doveton (1993) suggested that vertical Markov transition probabilities estimated from the sequences themselves could be used as cost factors for lateral matches between units. By using the negative logarithm of the transition probabilities as matching costs, the pathway that would minimize the cumulative cost would simultaneously maximize the product of the transition probabilities of the linked units.

BOREHOLE MAGNETOSTRATIGRAPHY: TIME LINES FROM LOGS

Results from even the best techniques for log correlation will usually be at some variance with biostratigraphy based on characteristic fossil species or faunal assemblages. The discrepancy is caused by the fact that logs record physical properties of sedimentary rocks, thus log correlations are lithostratigraphic. Some rock types, such as ash deposits, can be considered to be geologically instantaneous and provide valuable common time markers if they can be traced laterally. However, most sedimentary rocks are markedly diachronous, because they were formed from deposits that migrated laterally during appreciable periods of geological time. The lithostratigraphic correlations that are deduced from logs therefore differ from the chronostratigraphic framework established by characteristic fossils. However, recent developments in borehole magnetostratigraphy (Bouisset and Augustin, 1993) suggest that absolute age dating and chronostratigraphic correlation can be made from logs of natural magnetism.

Bernhard Brunhes observed in 1906 that iron mineral particles in bricks aligned themselves with the Earth's magnetic field as the bricks cooled. He found that the same phenomenon was shown by iron minerals in lava flows and reasoned that measurements of ancient lavas could give useful information on the history of the magnetism of the Earth. Brunhes was surprised to discover that the magnetism in some ancient lava flows was opposite in polarity from that of modern times and he concluded that the Earth's magnetic field had reversed at various times in the past. This radical explanation was not fully accepted until the 1960's when it was finally put to use to establish a geomagnetic polarity time scale. Potassium-argon dating of lava samples was used to relate changes in polarity with absolute age dates (Cox et al., 1964). Equipment was developed to measure the relatively weaker magnetization of sediments in cores and establish a Pleistocene chronology linked with

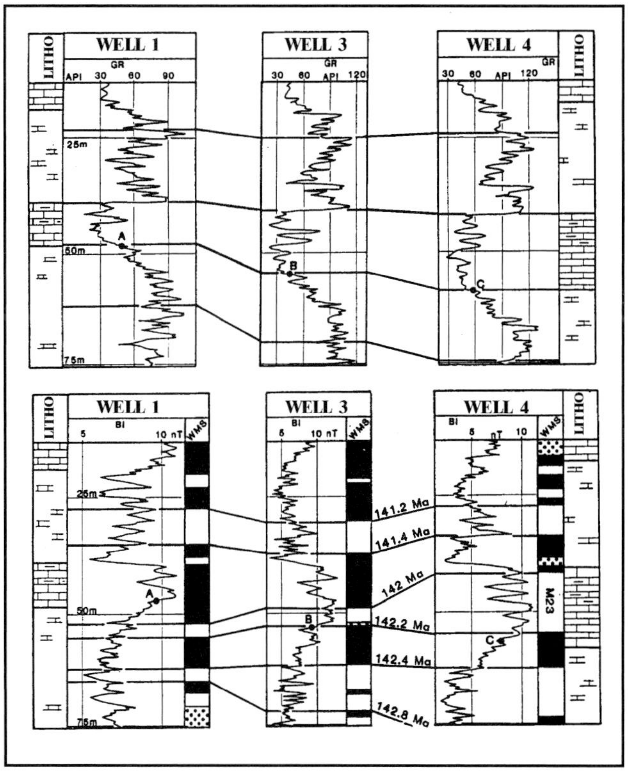

Figure 13. Gamma-ray logs (above) and magnetism logs (below) of a Kimmeridgian (Upper Jurassic) sequence in three wells from the Paris basin, France. Note the base of the middle limestone shown by depths A, B, and C on the lithostratigraphic correlation based on the gamma-ray logs, and marked on the magnetism logs. Induced magnetic intensity (nanoteslas) and the well magnetostratigraphic sequence (WMS) of polarities are shown together with interpreted absolute age time lines. From Bouisset and Augustin (1993).

microfossil successions. A number of geomagnetic polarity time scales have been proposed for older time intervals (see e.g. Harland et al., 1982).

The measurement of rock magnetization in core samples under controlled laboratory conditions is now done routinely. There are additional logistical and environmental problems to be surmounted when attempting the same procedure as a logging operation. However, advantages would include the recording of the magnetic properties of long and continuous successions. This would be particularly helpful in the common situation where core recovery is only partial. Experimental logging on Leg 102 of the Ocean Drilling Program demonstrated that both the intensity and direction of paleomagnetic polarization could be measured reliably by downhole tools in strongly magnetized volcanic rocks (Leg 102 Scientific Party, 1985).

Bouisset and Augustin (1993) described the development and performance of the first practical logging tools that allowed continuous logging estimates of both the magnetic intensity and polarity of weakly magnetized sedimentary rocks. The estimates were shown to be relatively precise and accurate and were derived from a nuclear magnetic resonance tool (a high-precision magnetometer) and a magnetic susceptibility tool. Data processing was used to remove drilling-rig effects and to correct for the geomagnetic drift and other factors. Evaluation of magnetic tool performance was made in four holes drilled by Total through an Upper Jurassic sequence in the Paris basin of France. Gamma-ray, sonic, and other more conventional logging tools were run in addition to the magnetic measurements. In each borehole, repeat sections were logged for comparison with the main runs. The magnetic logs were found to have better repeatability than the gamma-ray logs. The result was not completely unexpected because of the stochastic nature of gamma-ray counts, but it also indicated good precision in the magnetic measurements in spite of the weakly magnetic character of the logged formations.

Gamma-ray and magnetic logs are shown for an Upper Jurassic Kimmeridgian sequence in three of the boreholes from the Paris basin (Figure 13). Lithostratigraphic correlation of the limestones and calcareous shales is fairly straightforward, based on the gamma-ray logs shown in the upper part of Figure 13. The base of the limestone that occurs at levels A, B, and C in the three wells marks a third-order sequence boundary. The boundary was picked from logs from the region as a horizon that separated a highstand system tract and the succeeding deposition of a shelf margin wedge. Induced magnetic intensity and polarity logs of the same section are shown in the lower part of the same figure. Notice that the induced magnetism log (scaled in nanoteslas) is an approximate inverse image of the gamma-ray log. The reason for this is the sensitivity of both tools to shale content. Radioactive isotopes and magnetic minerals are both more abundant in the shaly sections. The magnetic log is therefore a useful lithology log in its own right in sequences characterized by interbeddings of shales.

The black and white bars on the lower three logs in Figure 13 mark the well magnetostratigraphic sequence (WMS)—black signifies normal (modern) polarity and white shows reverse polarity. The polarity sequences in the wells were correlated and matched with absolute dates from a magnetostratigraphic time scale. A variety of scales have been proposed for the Upper Jurassic, but Bouisset and Augustin (1993) elected to use the scale published by Haq et al. (1987) as a recent and reliable compilation of ties between magnetic polarities and both absolute ages and biostratigraphic markers. Several preliminary steps were taken before assigning absolute dates to the magnetic polarity sequence in the wells. First, biostratigraphy was used to define the broad time interval from palynological analysis of drill cuttings. Sequence stratigraphy was then applied to relate the well sections to the reference sequence of the Paris basin established by Vail et al. (1987) which had been calibrated to the time scale of Haq et al. (1987). Secondly, allowance had to be made for the fact that a magnetostratigraphic time

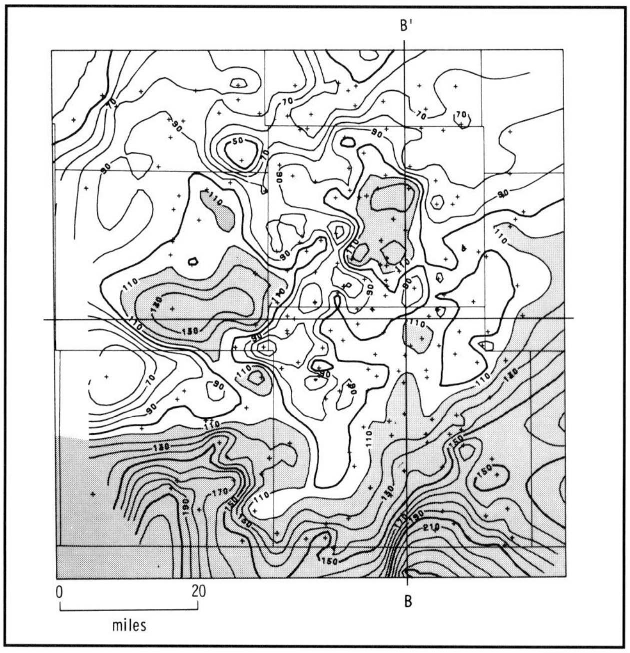

Figure 14. Isopach map of Middle Ordovician Simpson Group in south-central Kansas. Contours in feet and spaced at 10-ft intervals; thicknesses greater than 110 ft are shaded. An example of a map which makes limited use of logging information because it is based only on the elevations of the top and bottom of the unit. From Doveton et al. (1984).

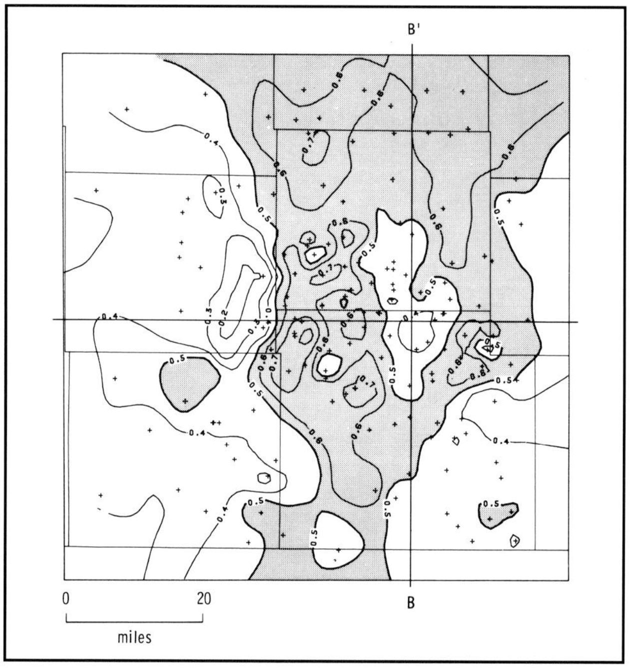

Figure 15. Gamma-ray shale ratio map of Middle Ordovician Simpson Group in south-central Kansas. Ratios greater than 0.5 are shaded. An example of a map that makes use a simple statistic of log trace variability within the unit. From Doveton et al. (1984).

scale is measured in time units but the well magnetostratigraphic sequence is recorded in units of depth. Consequently, allowances must be made for variable rates of sedimentation and compaction in the lithological sequence. Uncertainties concerning depth-to-time conversions are best resolved by defining narrow time slices of section where the boundaries have good biostratigraphic control.

The correlation lines that link the magnetic polarity events (Figure 13) are discordant with the lithostratigraphic correlation implied by the gamma-ray logs. The limestone in Well 4 appears to be older than its equivalent in Well 1 so that the facies appears to be measurably diachronous by this methodology. This conclusion is consistent with a depositional history of a major Jurassic transgression during which facies occurrence was determined by the interplay of sea-level changes with localized differences in subsidence rates and basin geometry.

Clearly, the technology of borehole magnetostratigraphy is new, but it has great potential for a wide variety of applications both at the scale of the reservoir and of the basin. For the first time, wireline logs can be used for chronostratigraphic purposes in addition to the matching of lithofacies between wells. In some ways the correlation of magnetic polarity sequences is likely to be more challenging than the correlation of conventional logs. This is because the record is only binary, with two states of normal or reverse polarity, while the depth axis is a variable function of time. Computer methods should be especially useful in the search for optimal correlations between magnetostratigraphic sequences. Among other reasons, automated correlation procedures can be designed to evaluate sequences from multiple wells and to incorporate estimates of sedimentation and compaction rates.

SPATIAL MAPPING OF LOG PARAMETERS BETWEEN WELLS

By interpolation between wells, the correlation of stratigraphic units described by wireline logs can be used to map spatial changes in subsurface formations. Most subsurface maps display either structure or thickness and make minimal use of log information. The mapped surfaces are estimated by linking structural elevations laterally, but take no account of the log trace between these stratigraphic picks. Consequently, both structure and thickness are geometrical descriptors of formation boundaries that, while extremely important, contain no direct measures of formation compositional changes (Figure 14).

Average formation values of the common logs may be mapped to show broad lateral changes in shale content or porosity. Estimates of the mean shale content can be computed easily from the gamma-ray log. Although they are indirect measures of shale, the gamma-ray values are often better than assessments from drill cuttings. Cuttings estimates are complicated by problems of infiltration of cavings from shallower levels and selective comminution by the drill-bit. An example of a map of average shale content in Figure 15 shows distinctive facies belts that are completely obscure on the isopach map in Figure 14. The mapped interval is the Middle Ordovician Simpson Group in a four-county area of south-central Kansas. The shale map shows three broad facies belts with a north-south orientation. Two sandstone features to the east and west are broken by a region of high shale content in the center.

A frustrating feature of average compositional maps is that while patterns may be revealed in aggregate lateral variation, absolutely no information is shown on fluctuations with depth. The limitation is caused by both the two-dimensionality of the map and the nature of the mapped variable. Various attempts have been made to display changes of lithology with depth across an area and the products are known collectively as "vertical variability maps" (Krumbein and Sloss, 1963). Most of these methods involve fairly arbitrary decisions in their design, application, and interpretation. However, the moment mapping introduced by Krumbein and Libby (1957) utilized basic parameters which were effective statistical measures and could be applied easily to wireline logs.

The calculation of moments is fairly straightforward; their meanings can be understood by reference to the gamma-ray log shown in Figure 16. The log is taken from a well in the area mapped in Figures 14 and 15. The mean value of the gamma-ray log in the Simpson Group section is given by:

$$\bar{g} = \frac{\sum g_i}{n}$$

where g_i is the ith gamma-ray log reading sampled from a total of n successive depth increments. The first moment can be calculated as:

$$V_1 = \frac{\sum g_i\, d_i}{\sum g_i}$$

where d_i is the depth of the ith reading. The depth of the top zone is usually set at a zero datum for computational convenience.

The first moment is a standard statistical parameter. It is also the mechanical center of gravity of the distribution, and so can be understood intuitively. If the gamma-ray trace was cut out as a solid silhouette and flipped sideways, then the balancing point of the cut-out would occur at the depth matched by the first moment (Figure 16). When this moment has a high depth value, then the shale content tends to be more concentrated in the lower part of the section. Vertical shifts in shale distribution will be tracked by corresponding movements in the center of gravity. Mapping of the first moment between well control provides a simple but effective measure of lateral change

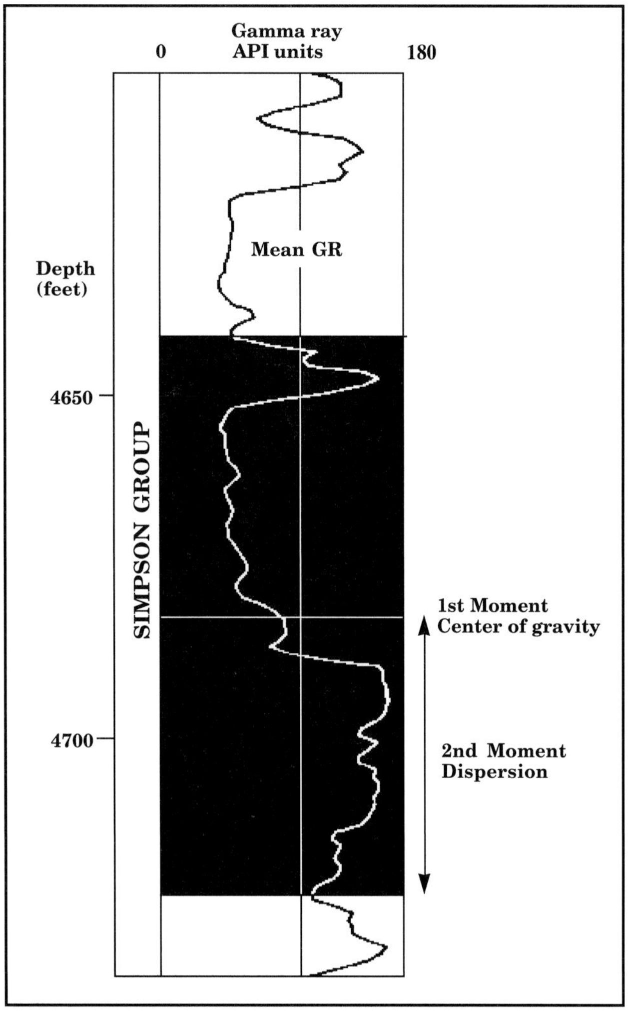

Figure 16. Gamma-ray log of Simpson Group from south-central Kansas, marked with mean gamma-ray value, depth of first moment, and dispersion depth range given by the second moment of the log. Modified from Doveton et al. (1984).

in vertical variability.

The second moment can be calculated from:

$$V_2 = \frac{\sum g_i \, d_i^2}{\sum g_i}$$

and has the mechanical meaning of a radius of gyration of the gamma-ray profile if it was spun on the axis of the center of gravity. Usually, the second-moment calculation is modified so that its origin is moved to the center of gravity:

$$M_2 = \sqrt{V_2 - V_1^2}$$

In this example, it represents the relative degree of dispersion of the shale about the depth marked by the first moment (Figure 16).

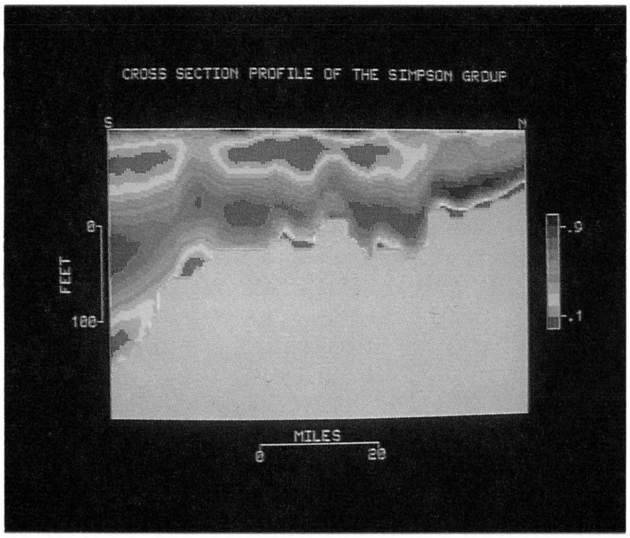

Figure 17. Computer graphic display of fourth-order trend slice of gamma-ray variability of the Middle Ordovician Simpson Group in south-central Kansas. The location of the section is shown as line B-B′ on the maps of Figures 14 and 15.

Higher order moments pick up progressively finer nuances of profile shape with depth. The third moment is a measure of skewness or asymmetry. The fourth moment reflects kurtosis, the relative "peakedness" or "squatness" of the distribution. The generic equation for an *m*th moment is given by:

$$V_m = \frac{\sum g_i \, d_i^m}{\sum g_i}$$

At very high orders, the moments become increasingly unstable due to roundoff errors caused by high powers of the depth variable. Moments of relatively low number are usually sufficient to capture trends in vertical variability that can be reasonably interpolated between wells.

Krumbein and Libby (1957) showed how the first two moments could be mapped and interpreted as aids in stratigraphic analysis. In order to extract the maximum information, several maps of thickness, mean value, and moments must be considered simultaneously. The fusion of these separate visual inputs into a unified picture can be physically and mentally challenging. Doveton et al. (1984) pointed out that the operation could be done by computer, because moments define curves of polynomial regression functions. So, for example, a fourth-order polynomial regression equation of a gamma-ray curve is given by the equation:

$$\hat{g} = a_0 + a_1 d + a_2 d^2 + a_3 d^3 + a_4 d^4$$

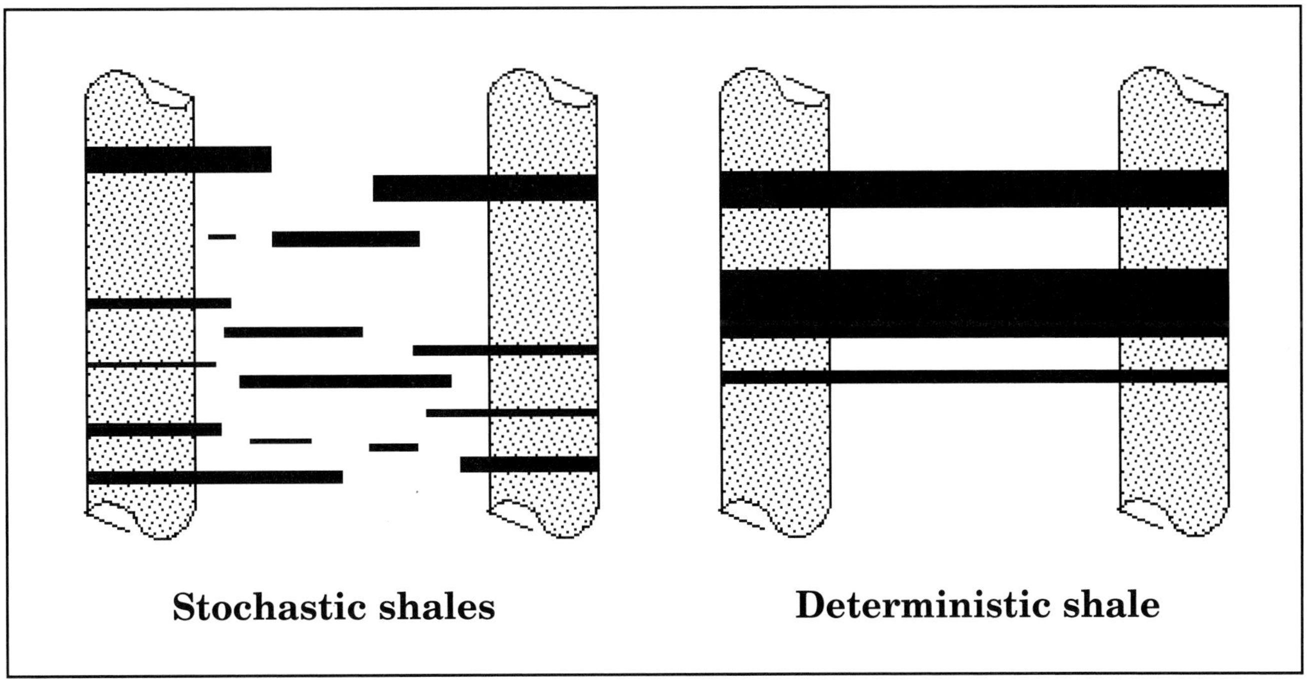

Figure 18. Diagrammatic concept of deterministic and stochastic shales. After Haldorsen and Lake (1984). Copyright SPE.

where $\hat{g}$ is the trend fit to the log at depth, d. The unknown regression coefficients, a_0 to a_4, are solved from the matrix algebra relationship:

$$\begin{bmatrix} n & \sum d & \sum d^2 & \sum d^3 & \sum d^4 \\ \sum d & \sum d^2 & \sum d^3 & \sum d^4 & \sum d^5 \\ \sum d^2 & \sum d^3 & \sum d^4 & \sum d^5 & \sum d^6 \\ \sum d^3 & \sum d^4 & \sum d^5 & \sum d^6 & \sum d^7 \\ \sum d^4 & \sum d^5 & \sum d^6 & \sum d^7 & \sum d^8 \end{bmatrix} \begin{bmatrix} a_0 \\ a_1 \\ a_2 \\ a_3 \\ a_4 \end{bmatrix} = \begin{bmatrix} \sum g \\ \sum gd \\ \sum gd^2 \\ \sum gd^3 \\ \sum gd^4 \end{bmatrix}$$

An interesting property of these matrices is that the unknown coefficients can be solved at any geographic location if a reasonable estimate of the section thickness and gamma-ray log moments can be made. The matrix on the left is completely determined by the thickness. If the vector at the right is divided by Σg, then the column will contain a unit value, followed by the first three moments.

The method is perfectly general, so that lower or higher order versions of this matrix construct will solve the complete range of polynomial functions. At the lowest end, a first-order polynomial specifies a straight line and is defined by the thickness, the mean gamma-ray value, and the first moment. An mth-order polynomial equation requires m moments. So, a polynomial curve can be created at an undrilled location provided that there is sufficient local well control to interpolate reasonable estimates of thickness and gamma-ray moments. If the matrix system is symbolized as:

$$DA = G$$

then the solution of the curve parameters is given by:

$$A = D^{-1}G$$

The mapping of lower order moments is intrinsically reasonable because they selectively capture the lower frequency components of logs that are more likely to persist laterally Higher order moments account for progressively higher frequency elements with less likelihood of lateral continuity. When necessary, the spatial autocorrelation of moments between well locations can be analyzed rigorously by kriging methods.

Because the curves can be generated at any location, it follows that a trend prediction of the log can also be made at any depth. This property makes it possible to assign trend values to all cells in a three-dimensional framework bound by axes of depth and geographic space. The validity of these estimates is controlled by both the section variability and the density of neighboring well control in just the same manner that conventional mapping is determined by these same factors.

An example of a fourth-order trend slice of the Simpson Group based on gamma-ray logs is shown in Figure 17. The slice is oriented north-south on the transect line marked on Figure 15. The basal sand is probably the northernmost extension of the Oil Creek Formation, which is picked up locally in drill cuttings from wells in the south of the area. The thick shale of the McLish Formation is capped by sandstones of the Bromide Formation. Small, localized oil fields occur in the area, generally at the top of the Bromide. Only

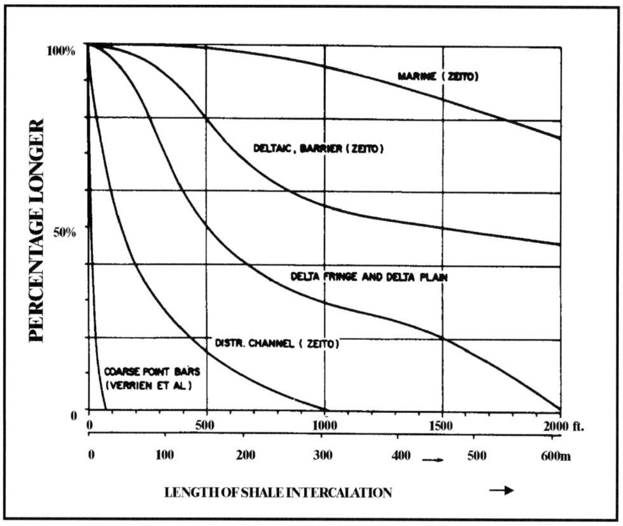

Figure 19. Lateral shale continuity widths as a function of depositional environment. From Weber (1982). Copyright SPE.

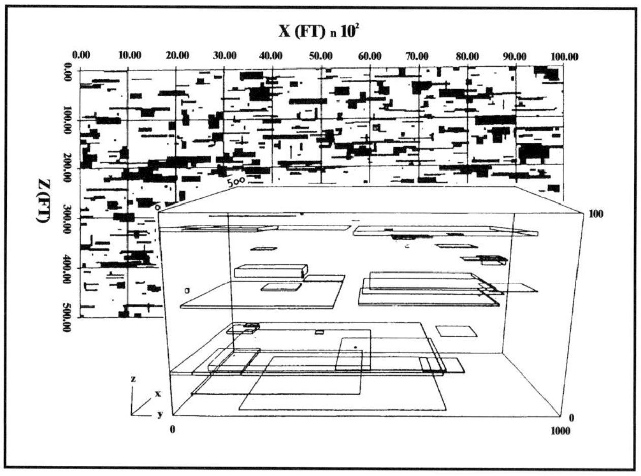

Figure 20. Two- and three-dimensional stochastic simulation models of shale distributions in reservoirs. From Haldorsen and Lake (1984). Copyright SPE.

the trend of variability given by the moments is shown by a slice map. How much variability is lost in such a representation? An estimate can be given from the typical degree of fit to gamma-ray logs in the area by fourth-order polynomial regression. As would be expected, the degree of fit is variable, but is generally on the order of 50%. Consequently, the trend image shows about half of the total variability, with the remainder contained in more localized features and thin bed events.

This graphic technique can be applied to cross-sections or maps angled at any orientation; examples are shown by Doveton et al. (1990). True three-dimensional models are easy to compute, although more difficult to display on the two dimensions of a computer monitor. In this treatment, we have discussed the use of moment mapping in conjunction with gamma-ray logs and a specific example of spatial variability within the Simpson Group. However, the method is general in purpose and can be applied to any type of log. The methodology might be particularly useful as a graphic trend descriptor of porosity developments and flow units within relatively thick and complex reservoir units.

STOCHASTIC RESERVOIR MODELING FROM LOGS

Interpolation and processing of moments results in an image that shows broad trends of three-dimensional variability. The moments capture the broader statistics of log variability that can most reasonably be expected to be traced between wells. The log variability that is not absorbed by the moments are finer scaled features of the thinner units. The moment trends will also tend to smooth sharp bed contacts and average thinly interbedded sequences. The three-dimensional images generated from moments may be a good medium to display the overall geometry of regional lithofacies patterns or the flow stratigraphy of a reservoir unit. However, the bland characterization by moments is inadequate as a means of producing a reasonable geological model for complex reservoir simulation and management. Here, the need is to make a construct that mimics the distribution of thin units. Many of these units are laterally discontinuous, but are a critical component of any realistic reservoir model. Fluids will migrate through tortuous routes of linked permeable units and will be deflected around impermeable layers or be trapped where the impermeable layers form an effective seal.

In clastic reservoirs, the representation problem is often a matter of the relative spatial distribution of sandstones and shales. For reservoir modeling purposes, shales can be equated with fine-grained clastics that have no effective porosity or permeability. They form aquitards to fluid flow within a reservoir complex. Shales are easily recognized on logs, but while some can be traced between well control, others lens out laterally at short distances from the well bore. It also follows that many shales that occur between well control are never penetrated by drilling, but are still an integral part of the reservoir. Within a reservoir model, how can we locate shale units that are not only impersistent but are often invisible?

Haldorsen and Lake (1984) suggested a new approach to the problem of shale management in reservoir simulation by treating it as a stochastic problem. We may not be able to specify the location and dimension of all the shales but we can create a hypothetical model in which the number and extent of the shales are the same as in the real reservoir. The model and reality will match in the aggregate sense. Each individual shale will have no meaning, but taken together they will emulate the texture of shale distribution within the actual reservoir. If the model

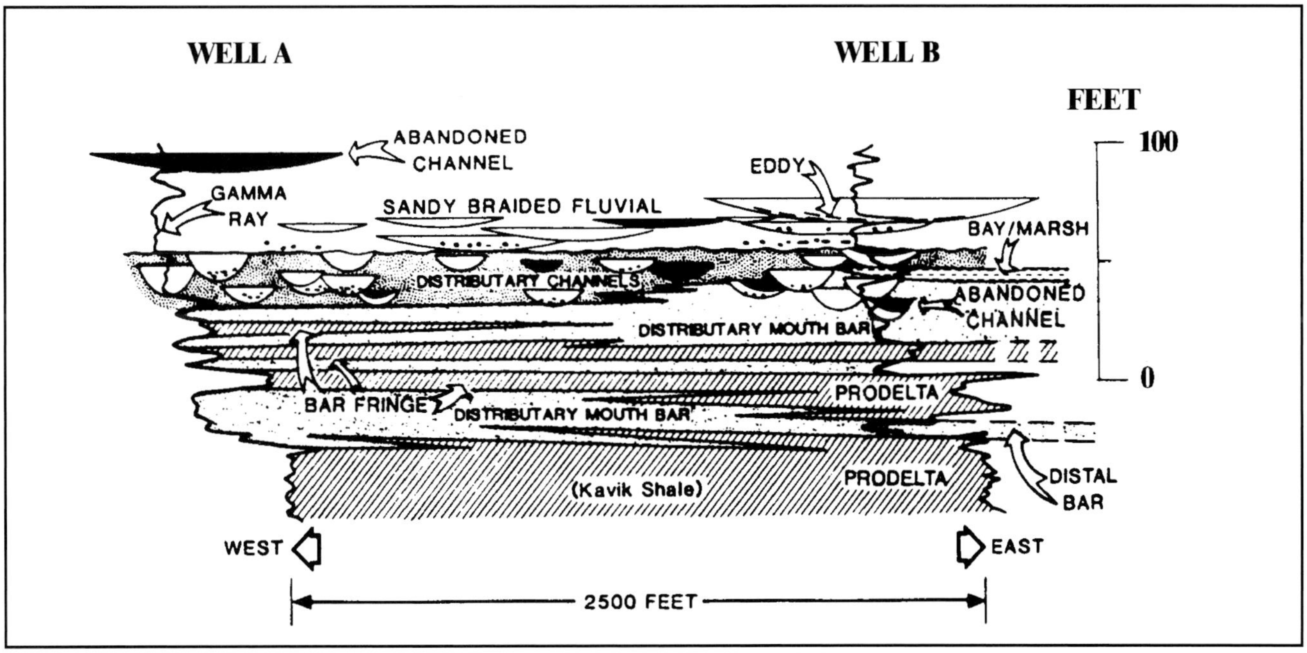

Figure 21. Interpreted depositional environments and shale types in a section of the Ivishak Formation in the Prudhoe Bay field. Well A was not cored; Well B was cored and described. The shale type assignations in Well A were made from discriminant analysis classification of wireline log measurements. From Geehan et al. (1986).

is designed correctly, then we can expect a reasonable match between model predictions and field measurements, because the overall flow will be determined by the totality of the shale distribution.

A distinction was made by Haldorsen and Lake (1984) between "deterministic shales" and "stochastic shales." Deterministic shales are those that extend laterally over large distances and so can be correlated easily between wells. Stochastic shales have small lateral dimensions and cannot be traced between wells. The basic concept is shown diagrammatically in Figure 18. Obviously, the distinction between the two types is determined to some degree by the spacing between wells. However, the major control on shale "width" is the depositional environment of the original muddy sediment. Marine shales commonly have a large lateral extent. They are contrasted with shales with much more localized distributions characteristic of fluvial and deltaic facies. Some broad statistics on relative shale widths as related to genesis are shown in Figure 19. As an additional complicating factor, the thickness of shales generally shows no relationship to lateral width. Thin shales can be ephemeral or persist for hundreds of miles. Thick shales may be localized or form regional stratigraphic units.

Most reservoirs contain a hybrid mixture of deterministic and stochastic shales. The centers of individual shales can often be represented adequately by random coordinates. The number of shales per unit volume and the thickness and width distributions of shales can be predicted from well sequences and outcrop measurements. The outcrop successions must obviously represent the same genetic type as the reservoir to be modelled. Both two- and three-dimensional reservoir models can then be generated by the relatively simple but practical procedure described by Haldorsen and Lake (1984).

In a two-dimensional model the shales known to be deterministic are first located on a scaled cross-section of the reservoir. The remaining section is then infilled with stochastic shales up to the point at which the proportion of shale in the cross-section conforms with shale proportions in nearby control wells. The center of each shale is found by two values from a random number generator. Its thickness is allocated by a random selection from the thickness distributions of shales drawn from the reservoir well sections. The width of the shale is determined by a random selection from the distribution measured from outcrop. The generation scheme allows shales to overlap, so account must be taken of this when assessing the stopping point of shale generation. The algorithm is easily extended to the creation of a three-dimensional model. Examples of both two- and three-dimensional stochastic models are shown in Figure 20.

Current computer modeling of fluid flow in reservoirs operates on gridblock representations of the total variability. Consequently, the averaging of shale content within individual gridblocks can be expected to produce results reasonably close to actual averages, even though each simulated shale has no match with a real shale. Shale averaging can be used to generate predicted values of porosity, as well as horizontal and vertical permeabilities. This is a more difficult operation because it involves scaling up hydraulic properties that are measured at much finer scales.

However, this scaling problem is common to all models of numerical reservoir simulation (Haldorsen, 1986). Finally, any discrepancies between the stochastic model and reality can be tested by standard validation procedures, either by history matching or hydraulic tests in individual wells. Modifications and fine-tuning of the model should improve performance to an acceptable level for most applications.

Haldorsen and Lake (1984) pointed out that many enhancements could be made to the basic approach. A primary necessity is the acquisition of more and improved outcrop measurements of shale dimensions. Also, more research should be directed to the conversion of shale distributions in gridblocks into effective measures of gross hydraulic properties. Refinements to the characterization of stochastic shales were outlined by Haldorsen and Chang (1986). In particular, they described the use of geostatistical methods to verify that modelled sections were acceptable emulations of shale distributions observed in outcrop.

Outcrops of analogous facies provide the necessary information on shale widths, but well logs are the primary source for data on shale thicknesses and frequencies in the depth dimension. Haldorsen and Chang (1986) used a simple measure of shale density:

$$s(x,y) = \left(\frac{N_{sh}}{H}\right)$$

where x,y are the geographic coordinates of the well, N_{sh} is the number of shales in the section, and H is the length of the section in feet or meters. The shales were usually picked from a gamma-ray log, following an empirical partition between sandstone and shale subdivisions. The shale-density values at the field well control provide the first criterion to be honored by the model between the wells.

This stochastic shale model is only conditional to the degree that the genetic type of reservoir geology determines the appropriate facies class of outcrop shales to be used as a source of shale width distribution. However, beyond this point no distinction is made between different types of stochastic shale. Geehan et al. (1986) described a case study where log data were used to estimate shale continuity both for improved mapping of shale distributions and predictions of reservoir performance. Their terminology used "continuous shale" and "discontinuous shale" to differentiate shales that were narrower than interwell spacing and those that were not. The slight change in semantics reflected a tighter focus on the probability of lateral extent of *individual* shales, rather than on discontinuous shales of the section taken as a group.

Six different shale facies that could be determined from core observations in the Ivishak Formation of the Prudhoe Bay field were recognized by Geehan et al. (1986). The shales were interpreted to have formed in prodelta, bar-fringe, bay/marsh, abandoned channel, flood-plain, and drape environments (Figure 21). If the shales can be distinguished on the basis of litho-

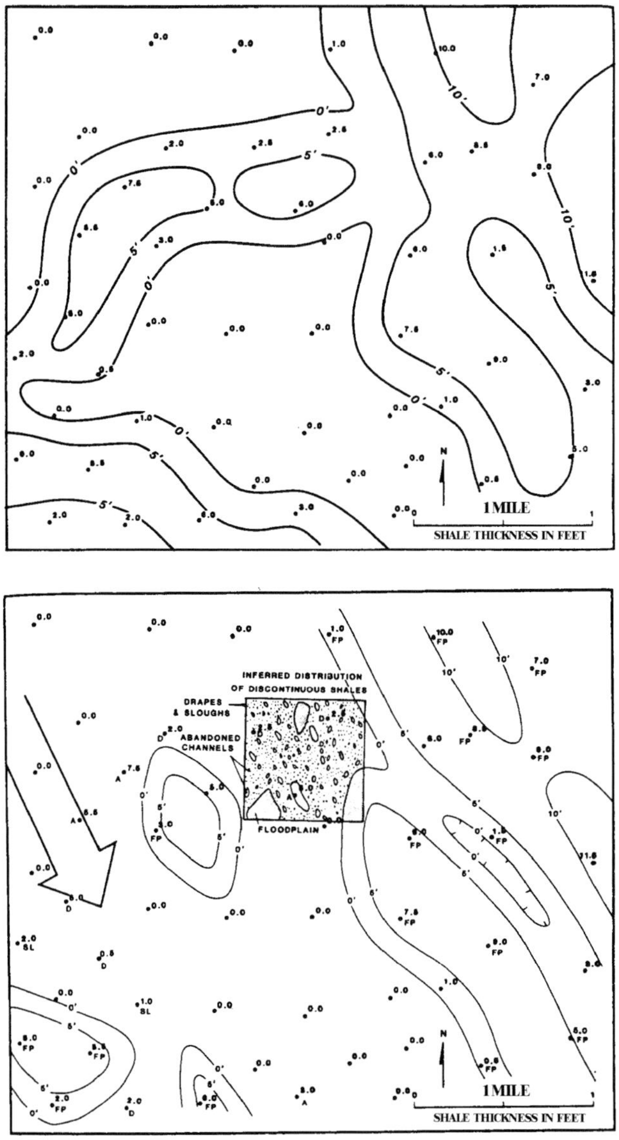

Figure 22. Two alternative maps of contoured shale thickness in a thin correlative interval within the Ivishak Formation in the Prudhoe Bay field. The top map shows the thickness as contoured by automated interpolation with no geological bias. The lower map shows the same data contoured to conform to continuity/geometry interpretations of shale types predicted from discriminant analysis of wireline log measures. The inset of the lower map shows a hypothetical areal distribution of discontinuous shales. The large arrow shows the generalized channel flow direction inferred from the orientation of shale boundaries. From Geehan et al. (1986).

logical characteristics, it seems reasonable that the distinctions may be reflected in their log characteristics. If so, then shale classification could be extended to complete logged sections, rather than restricted to sporadic core intervals.

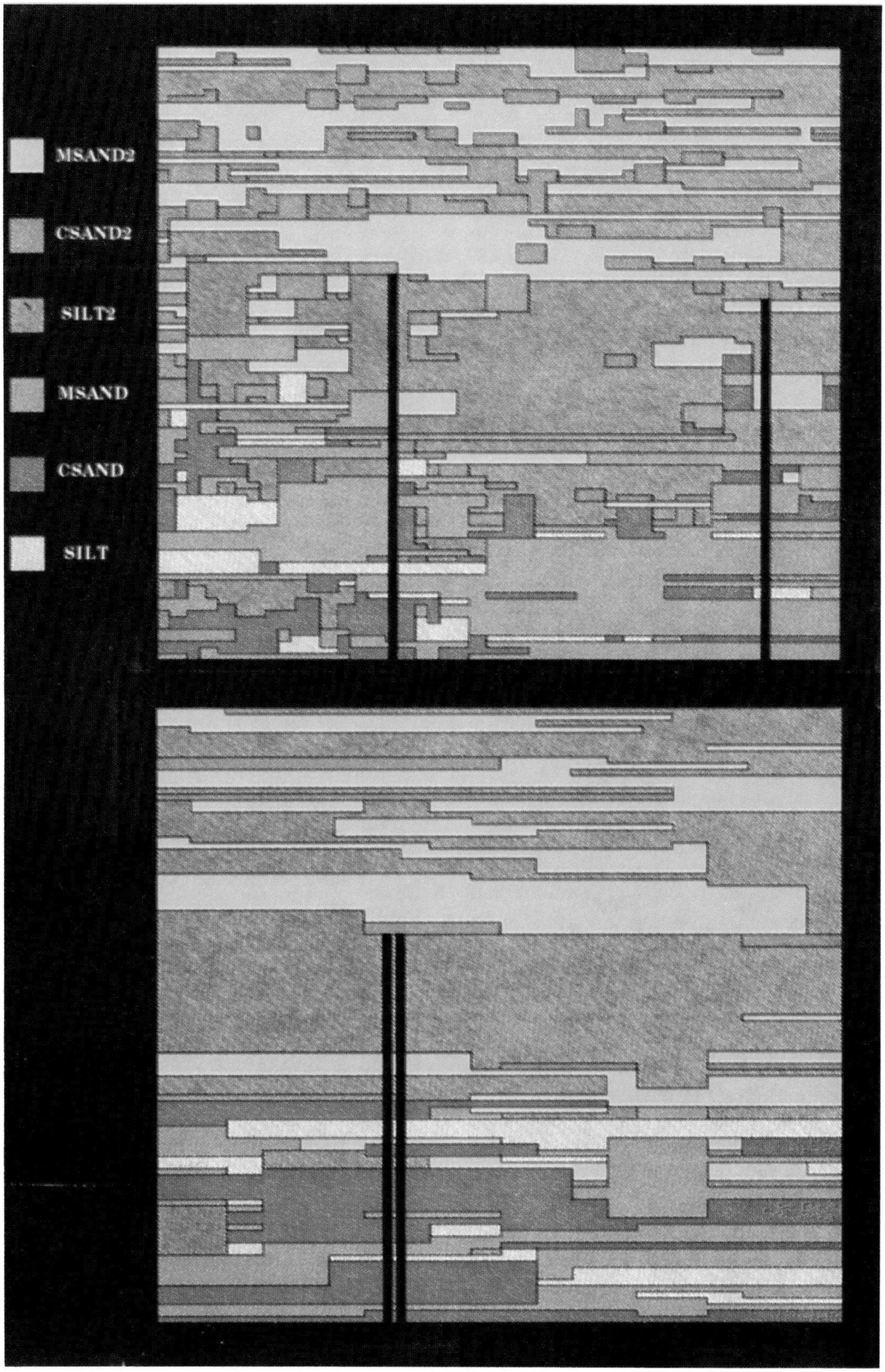

Figure 23. Sections of stochastic reservoir model of the North Sea Statfjord Formation, parallel to depositional strike (above) and orthogonal to strike (below). The vertical black bars represent logged sections of wells that coincide with the line of section. The model was generated by repeated sampling from Markov transition probabilities of vertical successions of lithotype electrofacies observed on logs, coupled with sampling of areal dimensions suggested by the sedimentological model. Adapted from Moss (1990). Courtesy of the Geological Society of London.

Geehan et al. (1986) used core control to supervise discriminant function analysis of the shale classes based on log characteristics. The problem is then another example of electrofacies discrimination using a methodology similar to those described in Chapter 4. By close collaboration with the project geologist they learned that the nature of the associated sandstones was used routinely as an aid in shale classification. Consequently, log information was encoded from both the shales and their contiguous sandstones. Core and log data from nine wells were used as calibration sets in the development of discriminant functions; data from a tenth well were used for classification validation. The two most important log variables in discrimination proved to be the normalized gamma-ray response multiplied by the shale thickness and the difference in the volume of shale between the shale and the underlying sandstone. Prodelta shales were easily distinguished from other shales by their high products of normalized gamma-ray response and thickness. Differences in shale volume differentiated abandoned channel fills that tended to overlie abruptly relatively shale-free sandstones, in contrast with other shales that had gradational lower contacts.

The shales in the nine calibration wells were correctly classified in 90% of the cases using discriminant analysis applied to their log measures. One out of 11 shales was misclassified in the validation test well. An advantage of a discrimination between various types of discontinuous shales is that differences in areal shape can be considered. In simple stochastic shale models, stochastic shales are both compounded in a single group and assumed to be equilateral in width. In part, this assumption also addresses a physical limitation of outcrops in that they provide shale lengths in the plane of the exposure, but generally no information on the lengths normal to the outcrop. In the Ivishak Formation case study, the various types of shales classified by discriminant analysis could be matched with different expectations on relative degree of continuity and shape. Both prodelta and bar-fringe shales have high continuity, as do marsh/bay shales, although marsh/bay shales are more variable. Shales in abandoned channel fills have much lower lateral continuity and are also more elongate parallel to the channel axis. Geehan et al. (1986) used this information primarily to improve on shale mapping between well control. The resulting maps integrated geometry and continuity interpretations as a refinement of conventional mapping that presumes that shale can be interpolated between well control as a continuous phase (Figure 22).

Moss (1990) extended the stochastic modeling approach in a consideration of multiple stochastic lithotypes and the use of Markov chain analysis to regulate lateral relationships from vertical transitions observed on logs. Clastic lithotypes were defined on the basis of log measurements from sections of the Statfjord Formation in the Statfjord field of the Norwegian North Sea. The formation was interpreted by Roe and Steel (1985) to have formed in a coastal fan and fan delta complex on a low-energy coastline during Late Triassic and Early Jurassic times. Lithotypes were developed from electrofacies based on compensated neutron, calculated effective porosity, corrected gamma-ray, and true resistivity logs. Four fundamental types were recognized and interpreted: lagoon bay shales, distal fan and flood-basin bay siltstones, medium-grained fan-plain sandstones, and coarse-grained channel sandstones. After experimenting with an initial model based on these four types, Moss (1990) expanded the set to two types each of siltstone, medium sandstone, and coarse sandstone to obtain a more realistic simulation.

The simulation model was developed by sampling from a Markov chain based on transition probabilities between the lithotypes observed on the logs. The geographic axes of the model were aligned with the depositional dip and strike of the Statfjord Formation. The orientation allowed lithotype width selections to be keyed easily with dip or strike elongations of some lithotypes. So, for example, siltstones were modelled with equal dimensions in both directions, but medium-grained fan-plain sandstones were elongated parallel to the strike, and coarse channel sandstones were lengthened parallel to the dip. In assigning wedges of lithotypes to coordinates in the model, overlaps were handled by a protocol of erosional rules. Coarse-grained sandstones eroded all other lithotypes; medium-grained sandstones eroded siltstones and shales. A completed model run was a realization of a stochastic reservoir architecture of the Statfjord Formation such as the strike and dip sections shown in Figure 23.

In his paper, Moss (1990) stressed that the modeling procedure was in a provisional state of development. However, the basic strategy shows a way forward in stochastic modeling in its use both of multiple stochastic facies and vertical transition probabilities to direct the building of the model. Further developments can capitalize on electrofacies characterization methods described in Chapter 4. It should be noticed that the stochastic reservoir models described up to this point have all been drawn from clastic successions. There is no reason why the same procedures should not be extended successfully to reservoirs characterized by thick, heterogeneous carbonate, although there are many additional challenging problems to be tackled. Many clastic reservoirs can perhaps be represented adequately by an essentially binary split between shaly aquitards and sandy flow units with simple granular pore structure. Heterogeneous carbonate reservoirs are generally much more complex, with porosities created and modified by both depositional and diagenetic events. Porosity types can be highly variable, both in shape and size, while the occurrence of several types in a single zone is not uncommon. In addition, recent studies suggest that porosity layers and patches exist over a wide range of scales. Important heterogeneities can be averaged by conventional logging tools, although borehole electrical images have proved to be valuable in discriminating porosity variation at fine scales linked with fabric changes (Nurmi et al., 1990).

Recently there has been increasing interest in the potential of fractal geometry as a means for modeling variability of reservoir properties between well control. The concept of fractal geometry was originally popularized by Mandlebrot (1977) and has proved to be a powerful descriptor of irregular patterns that are common in nature. Hewett (1986) concluded that petrophysical properties showed fractal character that could be analyzed either from core or well log data. He described ways to evaluate the effects of fractal property distributions on overall reservoir performance. Mathews et al. (1989) claimed that reservoir modeling based on fractal geostatistics resulted in improvements in predictions of the results of miscible flooding.

Research into the fractal properties of wireline logs is in the formative stage. However, the results have useful implications for improved modeling procedures because reservoir simulations are heavily dependent on the effective use of petrophysical logs. A particularly interesting application is the modeling of fractured reservoirs, because evidence from logs has been used to suggest that fractures are described by fractal distributions (Leary, 1991).

REFERENCES CITED

Bouisset, P. M., and A. M. Augustin, 1993, Borehole magnetostratigraphy, absolute age dating and correlation of sedimentary rocks, with examples from the Paris Basin (France): AAPG Bulletin, v. 77, no. 4, p. 569-587.

Collins, D. R., and J. H. Doveton, 1993, Automated correlation based on Markov analysis of vertical successions and Walther's Law, *in* Davis, J. C., and U. C. Herzfeld, eds., Computers in geology—25 years of progress: Oxford University Press, Studies in Mathematical Geology 5, p. 121-132.

Cox, A., R. R. Doell, and G. R. Dalrymphe, 1964, Reversals of the Earth's magnetic field: Science, v. 163, p. 1537-1543.

Delcoigne, A., and P. Hansen, 1975, Sequence comparison by dynamic programming: Biometrika, v. 62, no. 3, p. 661-664.

Doveton, J. H., Ke-an Zhu, and J. C. Davis, 1984, Three-dimensional trend mapping using gamma-ray well logs, Simpson Group, south-central Kansas: AAPG Bulletin, v. 68, no. 6, p. 690-703.

Doveton, J. H., R. R. Charpentier, and E. P. Metzger, 1990, Lithofacies analysis of the Simpson Group in south-central Kansas: Kansas Geological Survey, Petrophysical Series 5, 34 p.

Fang, J. H., H. C. Chen, A. W. Shultz, and W. Mahmoud, 1992, Computer-aided well log correlation: AAPG Bulletin, v. 76, no. 3, p. 307-317.

Geehan, G. W., T. F. Lawton, S. Sakurai, H. Klob, T. R. Clifton, K. F. Irman, and K. E. Nirtzberg, 1986, Geologic prediction of shale continuity, Prudhoe Bay Field, *in* L. W. Lake, and H. B. Carroll, Jr., eds., Reservoir characterization: San Diego, Academic Press, p. 63-82.

Gerhard, L. C., and S. B. Anderson, 1988, Geology of the Williston Basin (United States portion), *in* L. L. Sloss, ed., Sedimentary cover—North American Craton, U. S., Vol. D-2: Geological Society of America, p. 221-241.

Gerhard, L. C., S. B. Anderson, and D. W. Fischer, 1991, Petroleum Geology of the Williston Basin, *in* M. Leighton, D. Kolata, D. Oltz, and J. Eidel, eds., Petroleum geology of interior cratonic basins: AAPG Memoir 31, p. 507-559.

Gill, D., 1970, Application of a statistical zonation method to reservoir evaluation and digitized-log analysis: AAPG Bulletin, v. 54, no. 5, p. 719-729.

Gill, D., A. Shomrony, and H. Fligelman, 1993, Numerical zonation of log suites by adjacency-constrained multivariate clustering: AAPG Bulletin, in press.

Griffiths, C. M., and S. Bakke, 1990, Interwell matching using a combination of petrophysically derived numerical lithologies and gene-typing techniques, *in* A. Hurst, M. A. Lovell, and A. C. Morton, eds., Geological applications of wireline logs: Geological Society of London, Special Publication 48, p. 133-151.

Haldorsen, H. H., 1986, Simulator parameter assignment and the problem of scale in reservoir engineering, *in* L.W. Lake, and H. B. Carroll, Jr., eds., Reservoir characterization: San Diego, Academic Press, p. 293-340.

Haldorsen, H. H., and D. M. Chang, 1986, Notes on stochastic shales: From outcrop to simulation model, *in* L. W. Lake, and H. B. Carroll, Jr., eds., Reservoir characterization: San Diego, Academic Press, p. 445-485.

Haldorsen, H. H., and L. W. Lake, 1984, A new approach to shale management in field-scale models: SPE Journal, v. 29, no. 4, p. 447-457.

Haq, B. U., J. Hardenbol, and P. R. Vail, 1987, The chronology of fluctuating sea level since the Triassic: Science, v. 235, no. 4793, p. 1156-1167.

Harland, W. B., A. V. Cox, P. G. Llewellyn, C. A. G. Pickton, D. G. Smith, and R. Walters, 1982, A geological time scale: Cambridge University Press, Cambridge, 131 p.

Hawkins, D. M., and D. F. Merriam, 1974, Zonation of multivariate sequences of digitized geologic data: Mathematical Geology, v. 6, no. 3, p. 262-269.

Hawkins, D. M., and D. F. Merriam, 1975, Segmentation of discrete sequences of geologic data: Geological Society of America, Memoir 142, p. 311-315.

Hewett, T. A., 1986, Fractal distributions of reservoir heterogeneity and their influence on fluid transport: Paper SPE 15386 presented at the 1986 SPE Annual Technical Conference and Exhibition, New Orleans, October 5-8.

Howell, J. A., 1983, A FORTRAN 77 program for automatic stratigraphic correlation: Computers & Geosciences, v. 9, no. 3, p. 311-327.

Jordan, D. W., R. M. Slatt, R. H. Gillespie, A. E. D'Agostino, and C. G. Stone, 1993, Gamma-ray logging of outcrops by a truck-mounted sonde: AAPG Bulletin, v. 77, no. 1, p. 118-123.

Kerzner, M. G., 1986, Image processing in well log analysis: Boston, IHRDC Press, 140 p.

Krumbein, W. C., and W. G. Libby, 1957, Application of

moments to vertical variability maps of stratigraphic units: AAPG Bulletin, v. 41, no. 2, p. 197-211.

Krumbein, W. C., and L. L. Sloss, 1963, Stratigraphy and sedimentation (2nd Edition): San Francisco, W. H. Freeman, 660 p.

Kwon, B. D., and A. J. Rudman, 1979, Correlation of geologic logs with spectral methods: Mathematical Geology, v. 11, no. 4, p. 373-390.

Lang, W. H., 1984, Conductivity and interval transit time as correlation tools: The Log Analyst, v. 25, no. 3, p. 21-33.

Lang, W. H., 1986, Correlation with multiple logs: The Log Analyst, v. 27, no. 1, p. 43-52.

Leary, P., 1991, Deep borehole log evidence for fractal distribution of fractures in crystalline rock: Geophysical Journal International, v. 107, no. 3, p. 615-627.

Leg 102 Scientific Party, 1985, Old hole yields new information: Geotimes, v. 30, no. 12, p. 13-15.

Levenshtein, V. I., 1966, Binary codes capable of correcting deletions, insertions, and reversals: Cybernetics and Control Theory, v. 10, no. 8, p. 707-710.

Lineman, D. J., J. D. Mendelson, and M. N. Toksoz, 1987, Well to well log correlation using knowledge-based systems and dynamic depth warping: Transactions of the SPWLA 28th Annual Logging Symposium, Paper UU, 25 p.

Mandlebrot, B. B., 1977, Fractals: Form, Chance, and Dimension: W. H. Freeman and Company, San Francisco, 286 p.

Mann, C. J., 1979, Obstacles to quantitative lithostratigraphic correlation, *in* D. Gill, and D. F. Merriam, eds., Geomathematical and petrophysical studies in sedimentology: New York, Pergamon Press, p. 149-165.

Mann, C. J., and T. P. L. Dowell, Jr., 1978, Quantitative lithostratigraphic correlation of subsurface sequences: Computers & Geosciences, v. 4, no. 3, p. 295-306.

Mathews, J. L., A. S. Emanuel, and K. A. Edwards, 1989, Fractal methods improve Mitsue miscible predictions: Journal of Petroleum Technology, v. 41, no. 11, p. 1136-42.

Matuszak, D. R., 1972, Stratigraphic correlation of subsurface geologic data by computer: Mathematical Geology, v. 4, no. 4, p. 331-343.

Moran, J. H., M. A. Coufleau, G. K. Miller, and J. P. Timmons, 1962, Automatic computation of dipmeter logs digitally recorded on magnetic tapes: Journal of Petroleum Technology, v. XIV, no. 7, p. 771-782.

Moss, B. P., 1990, Stochastic reservoir description: A methodology, *in* A. Hurst, M. A. Lovell, and A. C. Morton, eds., Geological applications of wireline logs: Geological Society of London, Special Publication 48, p. 57-75.

Nurmi, R., M. Charara, M. Waterhouse, and R. Park, 1990, Heterogeneities in carbonate reservoirs: Detection and analysis using borehole electrical imagery, *in* A. Hurst, M. A. Lovell, and A. C. Morton, eds., Geological applications of wireline logs: Geological Society of London, Special Publication 48, p. 95-111.

Olea, R. A., 1988, CORRELATOR—an interactive computer system for lithostratigraphic correlation of wireline logs: Kansas Geological Survey, Petrophysical Series 4, 85 p.

Olea, R. A., 1993, Interpretation of wireline logs from Cold Turkey Creek, North Dakota, using the program CORRELATOR: Kansas Geological Survey, Open File Report 93-6, 11 p., 14 plates.

Olea, R. A., 1994, Expert systems for automatic correlation and interpretation of wireline logs: Mathematical Geology, in press.

Poelchau, H. S., 1987, Coherence mapping—an automated approach to display goodness-of-correlation between wells in a field: Mathematical Geology, v. 19, no. 8, p. 833-850.

Robinson, J. E., 1975, Transforms to enhance correlation of mechanical well logs: Mathematical Geology, v. 7, no. 4, p. 323-334.

Roe, S. L., and R. Steel, 1985, Sedimentation, sea-level rise and tectonics at the Triassic-Jurassic boundary (Statfjord Formation), Tampen Spur, Northern North Sea: Journal of Petroleum Geology, v. 8, no. 2, p. 163-186.

Rudman, A. J., and R. F. Blakely, 1976, Fortran program for correlation of stratigraphic time series: Indiana Geological Survey, Occasional Paper 14, 31 p.

Sankoff, D., and J. B. Kruskal, eds., 1983, Time warps, string edits, and macromolecules: The theory and practice of sequence comparisons: London, Addison-Wesley, 328 p.

Shaw, B. R., 1977, Parametric interpolation of digitized log segments: Computers & Geosciences, v. 4, no. 3, p. 277-283.

Shaw, B. R., and J. M. Cubitt, 1979, Stratigraphic correlation of well logs—an automated approach, *in* D. Gill, and D. F. Merriam, eds., Geomathematical and petrophysical studies in sedimentology: New York, Pergamon Press, p. 127-148.

Smith, T. F., and M. S. Waterman, 1980, New stratigraphic correlation techniques: Journal of Geology, v. 88, no. 4, p. 451-457.

Testerman, J. D., 1962, A statistical reservoir-zonation technique: Journal of Petroleum Technology, v. 14, no. 8, p. 889-893.

Vail, P. R., J. P. Colin, R. J. du Chene, J. Kuchly, F. Mediavilla, and V. Trifilieff, 1987, La stratigraphie sequentielle et son application aux corrélations chronostratigraphiques dans le Jurassique du basin de Paris: Bulletin de la Société Géologique de France, 8, Tome III, no. 7, p. 1301-1321.

Waterman, M. S., and R. Raymond, Jr., 1987, The match game: New stratigraphic correlation algorithms: Mathematical Geology, v. 19, no. 2, p. 109-127.

Weber, K. J., 1982, Influence of common sedimentary structures on fluid flow in reservoir models: Journal of Petroleum Technology, v. 34, no. 3, 665-672.

Webster, R., and L. F. T. Wong, 1969, A numerical procedure for testing soil boundaries interpreted from air photographs: Photogrammetria, v. 24, no. 1, p. 59-72.

Wu, X., and E. Nyland, 1987, Automated stratigraphic interpretation of well log data: Geophysics, v. 52, no. 12, p. 1665-1676.

Chapter 7

Applications of Artificial Intelligence in Log Analysis

What, exactly, is "artificial intelligence"? On the face of it, the answer is very simple. The two words suggest that AI is a mechanical system that emulates the systematic thought processes of the human mind. Any expansion on this definition is very difficult because of the controversy concerning the nature of human intelligence and how it works. The raging debates in this area are not simply fueled by the intellectual challenges of an exceedingly complex subject, but the heavy emotional freight that is carried by something that is so close to the core of our being.

In this chapter we will review the two current major paradigms of artificial intelligence and some of the rudimentary applications that have recently been developed within log analysis. Why should log analysts even be concerned with AI? Because unlike the observations of many geological subdisciplines, wireline logs swamp the analyst with numerical data. If the petrophysical models to resolve the data are relatively simple, then computer processing is the obvious means to ease the heavy workload of manual analysis of long sections. Alternatively, if the analytical models are very complex, then analysts need all the help that they can get. Either way, computer methods can be designed to aid analysis and are particularly appropriate for geological applications involving subtle petrophysical interrelationships. Ultimately, wireline log measurements are remotely sensed physical properties of rocks that are often the product of an extended history of depositional and diagenetic processes. Artificial intelligence offers the potential of computer procedures incorporating rudimentary reasoning and pattern recognition features that, while simple individually, can be compounded into powerful systems.

Many interesting books and articles have appeared on the political and scientific history of artificial intelligence developments. We will restrict ourselves to a fairly cursory examination that retains the explanation of ideas that may be useful. Relevant concepts are those that help us understand the way the AI systems work and disclose their strengths and limitations. Whether or not the system is a good representation of the human mind is immaterial to our purpose. The airplane is a poor imitation of a bird but is a successful means of flight. However, research on brain functions cannot be dismissed as irrelevant, because it continues to be the key to important developments in artificial intelligence. In the same way, studies of bird flight revealed basic aerodynamic principles that have been incorporated in the design of aircraft.

There are two paradigms that currently compete as models of the human mind. The first is a symbolic system that views thinking as the management of information to arrive at logical conclusions. Philosophically, the symbolic system approach takes thinking to be an abstract chain of reasoning, and its roots are traced to Cartesian notions of mind/body separation. The second paradigm is a connectionist model grounded in the findings of biological studies. Thinking and learning are considered to result from electrochemical changes in the massively parallel neural architecture of the brain. If we use the same terminology developed earlier in this book, the symbolic model can be looked upon as a deductive system in which external logic is brought to bear on the solution of a problem. The connectionist model is inductive in operation because its learning process is not set by predetermined rules, but evolves from complex interactions with the problem data.

The symbolic paradigm was the first to receive major support as a vehicle for artificial intelligence applications in the late sixties. The reasons are both historical and political. One of the earlier connectionist models, the "perceptron," was found to have severe limitations. This weakness, coupled with the notion that AI research funding could only support one paradigm, tipped the balance to a commitment to the symbolic model. Meanwhile, the connectionist approach stalled until major conceptual breakthroughs were made in the mid-eighties. By this time, the unrealistic expectations and rather feeble practical results of symbolic models had cooled the ardor of its commercial practitioners. The stage was set for an enthusiastic promotion of the new wave of connectionist models.

The overheated description in the last paragraph gives some of the flavor of the volatile and charged aspects of AI development. Serious academic practitioners of artificial intelligence are generally cautious both in their assessments of model capabilities and projections for the future. Bitter experience has taught them that there is a major step to be overcome in scaling up prototypes from the "toy domain" to working systems that can be used routinely for real-world applications. By contrast, overzealous promotions by venture capitalists and product claims by commercial vendors have sometimes obscured the modest but real gains of AI applications.

In this chapter we will examine the operation of the two products that have resulted from the symbolic and connectionist approaches: the expert system and the neural network. Both have their strengths and weaknesses. Both are applicable to log analysis problems. As we shall see, the choice between them is dictated both by the form of the petrophysical problem and the goals of the user. As might be expected, hybrid systems of the two approaches also have a great potential as aids to log analysis in the real world.

SYMBOLIC ARTIFICIAL INTELLIGENCE: THE EXPERT SYSTEM

The idea that systematic thought is equivalent to the manipulation of symbols can be traced back to the development of symbolic logic by George Boole and other Victorian mathematicians. Statements concerning real-life situations can be coded as strings of symbols that have an appearance similar to algebraic equations. Logical conclusions can then be drawn from these statements through the operation of simple rules of inference. In 1937, Claude Shannon showed that the operation of electrical relays and switching circuits were equivalent to the basic logical relations, TRUE, FALSE, NOT, AND, and OR. Although the earliest computers were developed primarily as giant calculating machines, their ability to manipulate symbols was recognized and led to the first experiments in artificial intelligence.

The first successful AI program was able to prove theorems in symbolic logic. In order to make it work, some heuristics, or rules of thumb, had to be made part of its operation. Otherwise, in its search for the correct answer, the program would simply evaluate all possibilities regardless of their likelihood. Such an exhaustive search is clearly not intelligent behavior, particularly because the number of possibilities in a system of any complexity increase exponentially. However, the introduction of heuristic reasoning showed the way forward for more ambitious AI products. Encouraged by the success in symbolic logic proofs, work was started on a general problem solver. The results were disappointing. Heuristic reasoning was generally found to be insufficient because the solution of many real-world tasks was keyed to local rules of thumb, known as domain-specific knowledge.

In order to solve real-world problems, powers of reasoning needed to be linked with knowledge. This realization marked a significant shift in approach from what had been known as "cognitive science" to "knowledge engineering." The words also show a change in emphasis from attempts to model thought patterns to attempts to construct systems capable of solving real problems. In some way, everything known by an authority on the topic at hand had to be captured and coded as a knowledge base to be linked with inferential procedures. This concept became known as the "expert system" or "knowledge-based system." The specialized knowledge of the expert could be represented as a set of rules of the IF–THEN variety that relate observations to consequences. Many of the conclusions would be phrased in terms of likelihood along the lines of that well-known inferential rule: "IF it looks like a duck AND it quacks like a duck THEN it probably is a duck."

The basic outline of a simple expert system is shown in Figure 1. The inference engine is the active core of the process that infers new conclusions concerning the problem presented to it. The knowledge base contains the facts and rules that are drawn from the problem or task domain. Production rules within the knowledge base codify relationships distilled from the experience of the experts. These production rules consist of an antecedent that represents some observational pattern and a consequent that dictates an action that should be taken. The antecedent is the IF rule component that is often subdivided into clauses linked by logical connectives of AND and OR. The consequent is the THEN component that can specify one or more actions or consequences. The rule base contains the symbolic logic operations that are the grammar of inference. Basically, the inference engine draws on rules and facts from the knowledge domain to work with observations on the current problem. The inference engine applies a control strategy for inference that determines its use of symbolic logic operations.

The engine may forge a chain of inferences that leads to a conclusion. When run in this direction, the procedure is one of forward chaining. Alternatively, inference may proceed in the reverse direction as backward chaining, when a conclusion is stated and the procedure locates the matching set of conditions in the knowledge base. Forward chaining is most commonly used to construct a product based on certain attributes. Backward chaining is a diagnostic strategy that seeks to find the cause of an observation set. Backward chaining is usually easier to constrain to simple possibilities because it moves counter to the flow of exponentially increasing possibilities that can be generated by uncontrolled forward chaining. Backward chaining is a more common strategy in expert systems and is sometimes described as "goal driven." By contrast, forward chaining is "data driven." Hybrid strategies may use forward chaining to generate initial alternative hypotheses and backward chaining to evaluate the likelihood of each alternative.

Simple expert systems work with hard facts and rigid rules. Other expert systems allow numerical certainty factors to be applied to facts and rules. Each certainty factor is an estimate of the reliability of the

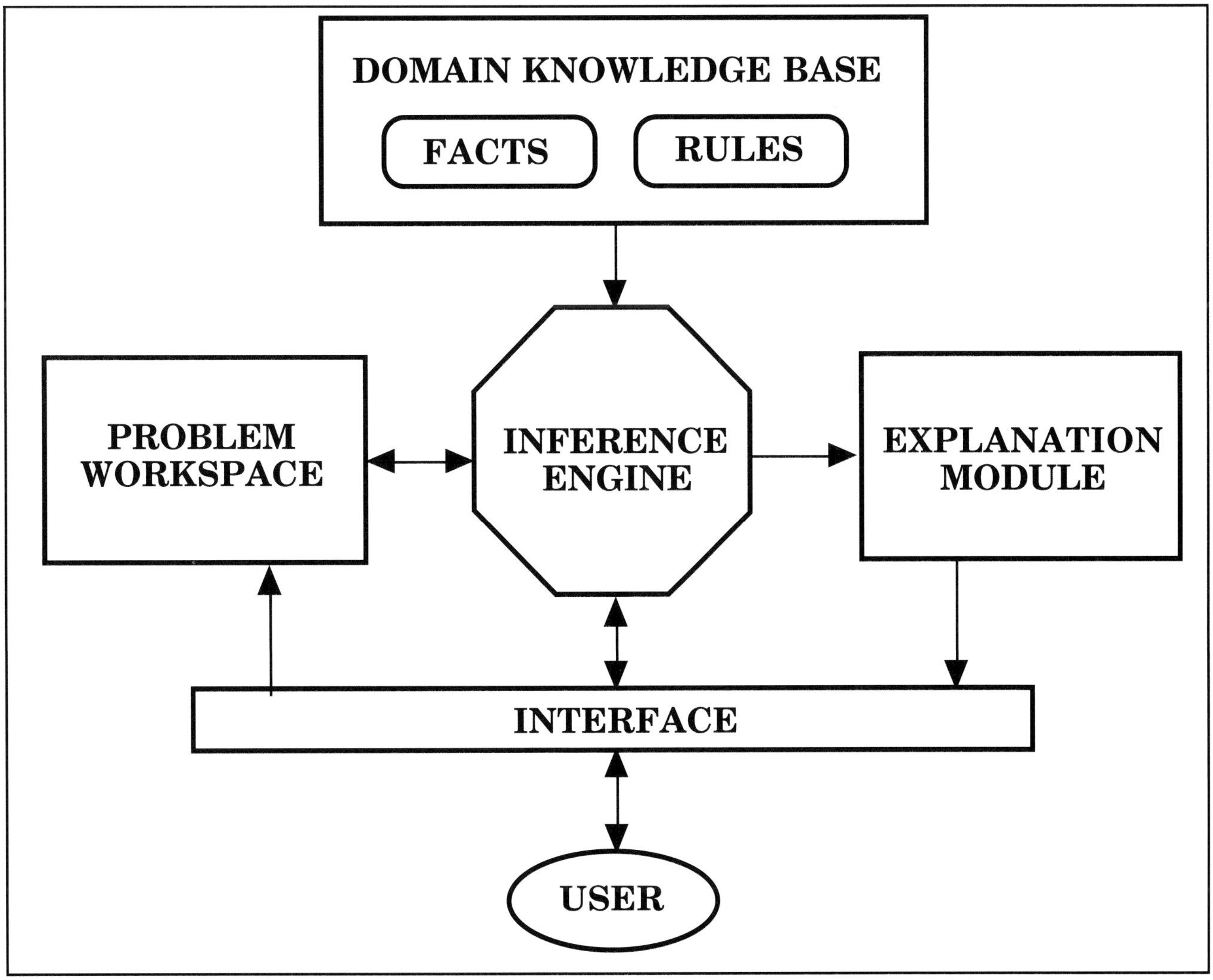

Figure 1. Schematic of the basic structure of an expert system.

conclusion to be associated with the rule. Reasoning by the expert system then becomes a process of tracing the most plausible chain in the face of uncertainty or incomplete information.

A particularly valuable feature of expert systems is their ability to recall the sequence of inferences that was the basis for the conclusion. The sequence is presented to the user at the interface by an explanation module. This audit "paper trail" allows the user to be the final judge of any result. It also provides the means to fine tune the expert system in its developmental stages. Faulty conclusions concerning known cases can be corrected through appropriate modifications of the facts and rules in the inference sequence.

Because the operation of an expert system is one of symbol manipulation, most of the advanced expert systems are written in the specialized symbolic computer languages of either LISP or Prolog. This makes the design of expert systems a more efficient process, but is not essential for the construction of a modest yet useful expert system in a PC environment. Part of the reason for the original development of symbolic languages was the limited ability of conventional languages such as FORTRAN to extend beyond simple numerical processing to symbol manipulations and recursive procedures. Improvements in these languages have made it possible to code expert system "shells" that can be run on standard microcomputers.

Watney et al. (1990) described the development of Porosity Advisor, built from a commercially available expert system package. The shell consists of all the system elements shown on Figure 1, with the exception of the domain knowledge base. The purpose of Porosity Advisor is to aid in the interpretation of porosity origins in carbonate rocks. Watney et al. (1990) supplied facts and production rules for the knowledge base. The facts were the pore types and cements. The production rules were drawn from geological expertise concerning the origins and historical sequences of pore and cement formation in Pennsylvanian carbonate reservoirs of the central Midcontinent. Porosity Advisor can be used for making decisions in the field because it is installed on a microcomputer. The user enters information on the occur-

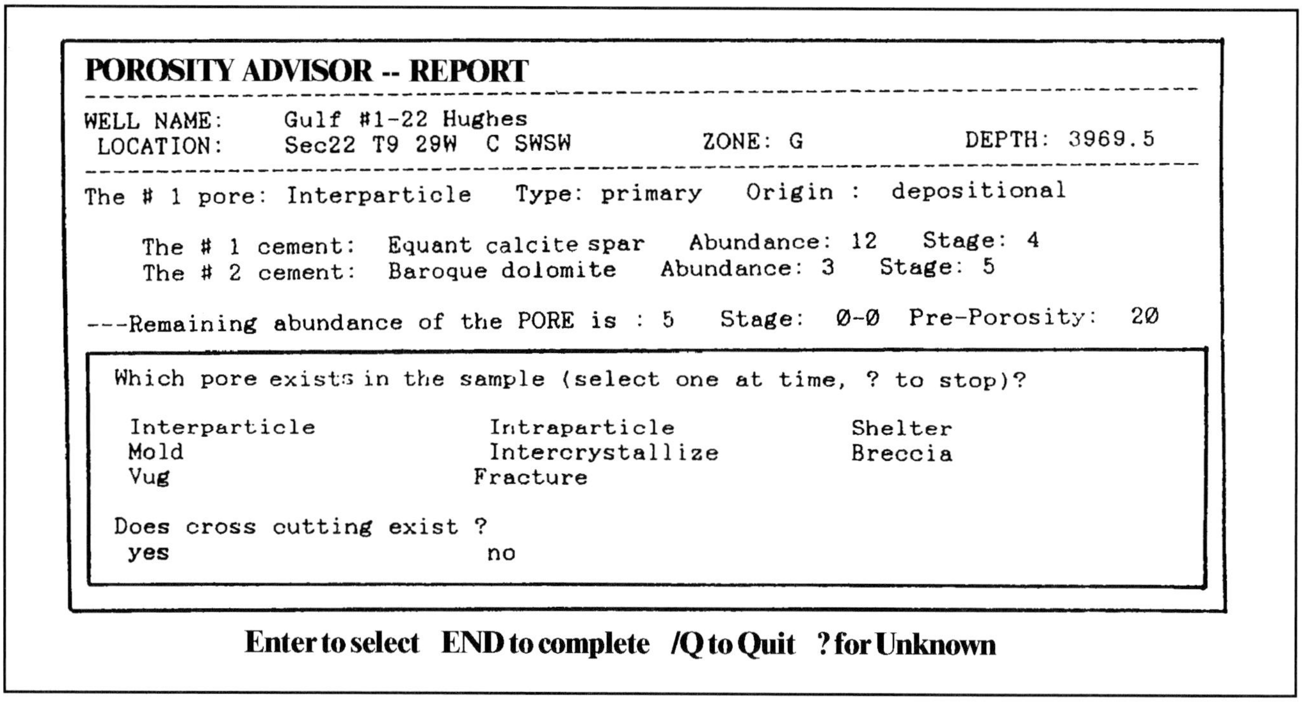

Figure 2. Example of a query for user interaction at the interface of the Porosity Advisor expert system. Reprinted from Watney et al. (1990), Copyright 1990, p. 286, with permission from Pergamon Press Ltd., Headington Hall, Oxford OX3 0BW, UK.

rence and abundance of cement and pore types from the problem reservoir rock. Interaction with the system is in the form of a consultation in which both the user and the "advisor" engage in a dialogue of questions and answers (Figure 2). The final output consists of a database that summarizes and tabulates results and interpretations.

Many of the criteria for the appropriate application of an expert system are a matter of common sense. The problem area should be one in which human experts exist and are in substantial agreement. The area should have known boundary conditions and a structure that lends itself to cut and dried rules. Inferences can be made only as a result of logical steps, so that intuitive leaps and pragmatic decisions are excluded. By and large, the task should not be overly complex. The pattern that emerges from all this is that expert systems are best suited for routine tasks that call on the simple experience of the astute practitioner and would be much more difficult, if not impossible, for the novice to perform. The key element is knowledge.

Probably the earliest and most famous geological expert system is Prospector, developed to advise whether a site was likely to have ore-grade deposits. The decision process is similar to that of the diagnosis of a disease based on observed symptoms, but using geological observations in its inference chain (Hart et al., 1978; Duda, 1980). Prospector identified a deposit of ore-grade porphyry molybdenum at Mount Tolman in the state of Washington. The system was originally designed around five alternative mineral deposit models. In its current successor, Prospector II, the number of models has been expanded to 89 (McCammon, 1993).

Oil exploration drilling operations are an obvious potential area for useful expert system applications. Drill-sites are often in remote locations which can make it difficult to manage effectively the time commitment of a small cadre of trained specialists, especially in the face of unexpected delays and other events. Lost time at the drill site is expensive and gives the major motivation to develop an expert system provided that it can capture the core expertise of the specialist. Drilling Advisor was developed as a prototype system for Elf-Aquitaine as a means to assist rig supervisors in the resolution and avoidance of problem situations. Generally speaking the supervisor is able to deal with the more common problems, so the function of Drilling Advisor is to give the extra advice that would normally be provided by an expert flown out to the rig. The prototype was developed to handle the specific problem of down-hole sticking, when rotary and vertical movement of the drill is badly affected. A total of 250 rules was used to construct the knowledge base. Experiments with the prototype were encouraging, particularly when the system coped successfully with some problem scenarios that were not included in the training set.

The expert system Mud emerged from an interesting collaboration between fluids experts at the drilling mud company, Baroid, and AI researchers at Carnegie-Mellon University (Kahn and McDermott, 1986). The expert system, marketed commercially as Mudman, emulates the decision procedures of a trained mud engineer. Its knowledge base draws on the accumulated experience of the company in addition to the expertise of individual engineers and comprises a set of over a thousand decision rules. The aim of the system is to reduce risks in drilling triggered by problems such as stuck pipe and lost circulation.

Similar economic incentives suggest that expert systems would be helpful in routine log analysis interpretations that can be formulated as a set of facts and rules. The training of competent log analysts takes time and in most companies their numbers have been cut back so that the availability of sufficient trained manpower is a real consideration. The expert system has a potential role at the well site as a way to provide the resources of a large knowledge base and a means to expedite decisions concerning testing, completion, or abandonment. Expert systems should also be useful for more complex tasks such as the analysis of log suites from multiple wells within fields as part of a reservoir management package. An example of such an application is provided by Playmaker, a knowledge-based system built around a comprehensive basin analysis strategy for prospect analysis (Cheong et al., 1993). Playmaker draws on a complex hierarchy of geological information ranging in scale from depositional settings down to grain size, porosity, and permeability characterizations.

EXPERT SYSTEMS IN LOG ANALYSIS

The area of log analysis that so far has received the largest commitment of effort to the production of a commercial expert system is probably that of dipmeter interpretation. Dipmeter Advisor, an early AI program developed by Schlumberger, is widely known in the artificial intelligence community (Davis et al., 1981). The knowledge and rules of this expert system were compiled mostly from the wisdom of a single, highly experienced dipmeter interpreter. Behind the effort was the realization that if successful, the rules of thumb from a lifetime of experience could be captured and used to advise and train novices. Dipmeter interpretation appears to be well suited for an expert system because the task involves a complex blend of geology, pattern recognition, and geometry that mixes natural constraints with user heuristics. Production rules in Dipmeter Advisor were keyed to both structural and stratigraphic information and concepts.

On a typical run, the system first identifies "green" (similar dip and azimuth) patterns of dip vectors to identify segments of constant structural dip. Systematic dip vector patterns with similar azimuth bearing are color-coded using the conventions of standard dipmeter interpretation. In addition to green patterns, red (increasing dip with depth) and blue (decreasing dip with depth) are differentiated. In several passes of structural analysis, features such as faults are located and associated characteristic vector patterns are analyzed to determine the geometry of structural features. The system then makes a provisional lithologic analysis based on other open-hole logs and draws conclusions concerning the stratigraphic meaning of green, red, and blue patterns when coordinated with the user's input concerning likely depositional environment. The stages in the analysis make variable use of machine–human interaction. At some points, rules can be applied directly to the unaided processing of patterns; at other times, the system requires human intervention to make choices among broad intangibles such as depositional settings. However, this is not a limitation of the system since, as is the case with most expert systems, it is designed to be an assistant rather than a replacement.

Following several years of development, Dipmeter Advisor was tested in the field (Smith and Baker, 1983: Baker, 1984; Frank, 1986; Josso, 1986; Shanor, 1987). An interesting and candid summary of some of the lessons that were learned is provided by Crain (1985). Dipmeter pattern analysis is a highly interpretative skill that mixes perceptions concerning the intangibles of depositional models with the geometric constraints of structural features. Since Dipmeter Advisor was constructed from the input of a single expert, it seems inevitable that some conflicts would emerge when the system was matched against the interpretations of other experts. In addition, the knowledge engineers noted that the original expert sometimes appeared to abandon his own rules when faced with a markedly different example and reasoned from geological and geometric models. None of these conclusions are surprising if the basic conditions of a successful expert system are recalled. Dipmeter analysis stretches the envelope of tractable problems when experts disagree among themselves; some facts are highly interpretative assertions and correct behavior by the system is difficult to evaluate. The switch of the original expert to a different mode of reasoning when confronted with a novel problem is understandable in AI terms. Many practitioners in the field of artificial intelligence differentiate between expert systems built from "superficial knowledge" and those based on "deep knowledge." The first generation of expert systems has used a superficial knowledge base drawn from generalized rules of experience. A second generation is projected that will operate from deep knowledge: first principles of proven models and the laws of nature.

Automated correlation is a good area of application for expert systems for a variety of reasons. The repetitive character of many successions often causes a variety of possible correlation solutions to be generated. There will be occasions when the mathematically optimum choice does not match the geologically correct answer. Consequently, an expert system can be useful as an assistant to guide automated correlation results towards geologically meaningful models. An expert system forms an integral component of automated correlation systems developed by Wu and Nyland (1986), Lineman et al. (1987), Startzman and Kuo (1987), and Olea and Davis (1989). A crucial aspect of the correlation problem domain is that real correlations cannot defy the geometry of the natural

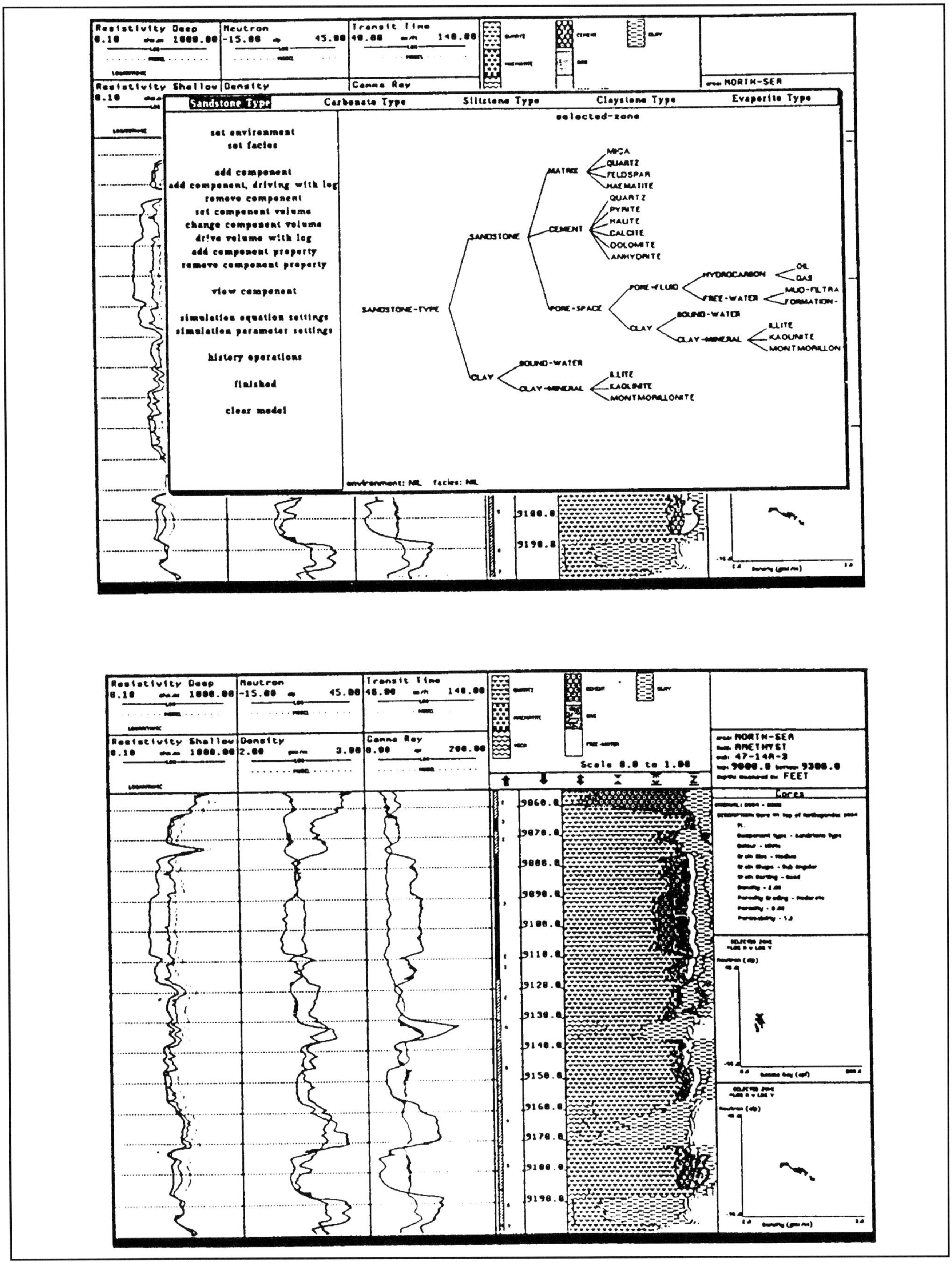

Figure 3. User interface of the HESPER expert system showing model manipulation display (above) and model results of log display, crossplots, and compositional analysis. From Peveraro and Lee (1988). Copyright SPE.

world. The rule base of an expert system is particularly well suited for keeping track of the geometrical consequences of any correlation. The explanation module can then explain to the user in terms that he or she can understand why certain solutions are either implausible or impossible. Human frailty being what it is, manual correlation must typically be restricted to the consideration of interwell correlation taken two (possibly three) wells at a time. This constraint does not apply to expert systems which can trace correlative ties between sets of wells and check them for geometrical validity. Used in this way, the expert system is already "second generation" in part and is successful because it draws on "deep knowledge" of the natural constraints of the physical world.

Mainstream log analysis is concerned with the evaluation of effective pore volumes and hydrocarbon saturations in potential reservoir zones. The skills of the experienced log analyst are based on simple physical models and algorithms acquired during initial training, tempered by the lessons of real-life evaluation. For a conventional example, the results of a routine log analysis should not differ substantially between experts. Differences will be caused by variations in the choice of parameter values, but there is usually broad similarity in the procedures used. If this is not the case, then the problem domain is poorly defined and conflict between experts makes pointless any work on a practical expert system. The transfer of ideas to real systems also requires effective cooperation between log analysis practitioners and computer software and hardware specialists. Zivy (1984) provided an interesting review of experiences gained in the development of expert system software for log interpretation. The advantages of an expert system include the ability to train novices according to the accepted norms of company petrophysicists without squandering manpower. In addition, use of the system should improve consistency in log interpretation.

The ELAS expert system was built by Amoco as a front end for their conventional log analysis package (Crain, 1985). It is an example of a so-called surface level expert system, because it is linked as an advice and control module to an existing software system. ELAS is made up of a set of production rules that keep track of actions taken and the numerical results of the actions; in addition they suggest analysis steps and alert the user to inconsistencies. Interaction takes place through the display of a master panel which shows the current status of the analysis, the parameters selected, and the zones analyzed. If needed, ELAS can be prompted to recommend actions or make interpretations. Alternatively, the user can make the entire analysis unaided, in which case ELAS continuously updates the status and automatically makes any changes to the logs or parameters that have been triggered by the actions of the user.

Einstein and Edwards (1990) made a comparison of log analysis results from an expert system compared to those produced by human experts. In their evaluation they audited results and methodologies in case studies from China, Colombia, the United States, and Morocco. They considered that the overall performance of the expert system was better than initially expected. The expert system sometimes showed surprising signs of "intelligent" behavior and highlighted blunders made by the human analysts. Failures most commonly occurred when the system did not have access to additional information available to the analysts, such as boundary values drawn from log traces higher in the section or values of some petrophysical parameters. Realistically, the expert system could be considered to make a good emulation of a first-pass analysis of a routine formation evaluation problem by a moderately proficient log analyst. This, in itself, holds promise for economic benefits because it provides a way to train and monitor less experienced personnel for routine analyses. At the same time, it frees up the experienced log analysts, leaving time to tackle the more challenging aspects of anomalous reservoir analysis.

Peveraro and Lee (1988) described a more ambitious expert system that integrates information from a variety of sources in a formation evaluation procedure that is tied closely to a selected geological model. The prototype of HESPER was tested on the analysis of logs run through the upper Rotliegend sandstones in the Southern Permian basin of the North Sea. The geological model was therefore equated with lithologies and minerals associated with sandstones and shales deposited in wadi, eolian, desert, sabkha, and lacustrine environments. Operation of the system focusses on the task of creating synthetic logs that are the closest possible match with the real logs recorded in a problem well. The interpretation process that makes this possible is carried out at a graphic interface (Figure 3) and involves finding the rock sequence and modelling the petrophysical properties of each formation. These properties are varied by following rules and suggestions contained within the geological model of the system. Advice can be solicited at any point from a module built on a goal-directed rule interpreter. The advice is offered together with an outline of supporting evidence.

The Rotliegend model obviously limits the HESPER prototype to the analysis of a subset of potential logging problems. However, as pointed out by Peveraro and Lee (1988), although the model is limited, it is realistic. Furthermore, an expert system that could process logs from every conceivable geological setting may be beyond the capabilities of current hardware and software. However, the approach of the system is closer to the philosophy of this book than that of expert systems which are extensions of conventional log analysis programs. Although classic log interpretation calculations are an integral part of the system, they are subordinated to the framework of the geological model. This is an advance from simple systems constructed from heuristics collected from experts. Petrophysical reasoning within a broad geological framework would seem to be a better and more flexible means to evaluate novel and complex reservoir zones than the aphorisms of a venerable log analyst.

CONNECTIONIST SYSTEMS: THE NEURAL NETWORK

As discussed at the beginning of this chapter, connectionist systems originate from a different model for thought than that of the knowledge-based expert systems. Much of the theory has been inspired by functional studies of animal brains by neuroscientists. The brain is now known to consist of a massively interconnected system of neurons that do sensory processing, control motor functions, and engage in patterns of thought. Thought originates in the cerebral cortex and is still an enigma. However, enough is now known concerning the operation of neurons to design useful artificial emulators that have the potential to learn patterns from data presented to them.

Most neurons consist of a central nucleus that is surrounded by short fibrous dendrites and an axon which is a single long extension with branches at its end. Typically, each neuron receives inputs from other neurons though its dendrites and transmits information along the axon. When a neuron is fired by the impulses from its dendrites, an electrical pulse is sent down the axon. The pulse causes a chemical change in the synapse, the gap that separates the end of the axon from the dendrites of another neuron. The neurotransmitter chemicals change the charge of receptors in the dendrites of the neighboring neuron. The synapses are the communication medium between neurons and are made either weaker or stronger by these electrochemical processes. It is theorized that we learn, store memories, and modify our behavior because of changes in the strengths of synapses within our brains. This property gives the core concept of artificial neural networks.

Obviously we cannot hope to create a model that is even a feeble approximation of the human brain because of its immense structural complexity that links about a hundred billion neurons. However, the central idea has been put to work for artificial intelligence applications. Neural networks differ from expert systems in several fundamental ways. Neural networks are "trained" by being exposed to a large number of input patterns that cause them to learn from experience. Therefore, they operate from the bottom up, in contrast to the top-down style of symbolic expert systems which are directed by an external knowledge base. A neural network may be slower in its ability to work with data, but it has the potential for useful change. An expert system has the advantage of built-in inferential rules and domain facts, but these may trap it in an outmoded procedure if there is a fundamental shift in the expert paradigm.

The structure of a neural network is usually drawn as a hierarchy of layers in which nodes (representing neurons) are connected by arcs (Figure 4). The arithmetic value of any node is equal to the sum of the values of the preceding nodes each multiplied by the weight of the connecting arc:

$$y_i = \sum w_{ij} x_j$$

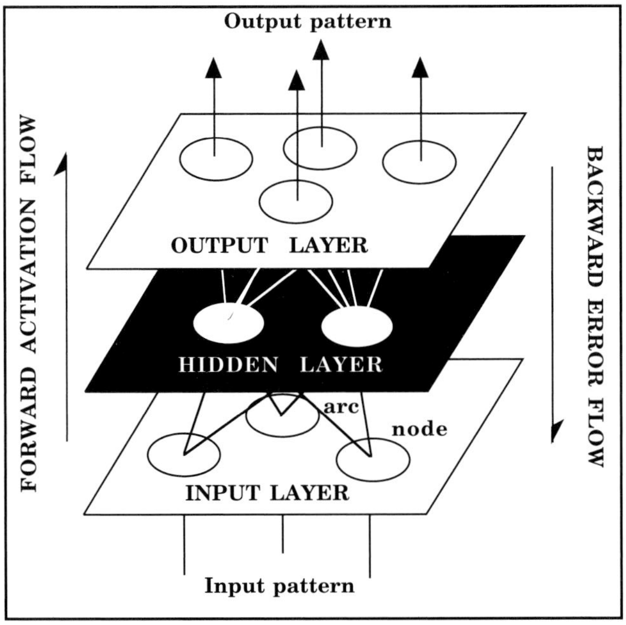

Figure 4. Schematic structure of a simple neural network with a single hidden layer, showing network nodes linked by connecting arcs.

where y_i is the value of the ith node, x_j is the value of the jth node in the preceding layer, and w_{ij} is the weight associated with the arc that connects the two nodes.

The output of a node is governed by an activation function of the summed input and a threshold that determines the initiation of output. In simpler networks, a node fires and passes output when the node value, y_i, exceeds a given threshold, U (Figure 5). The firing state of a node is either unity or zero, determined by whether the activation, a, is positive or negative, where:

$$a = y_i - U$$

More recent neural networks commonly use a sigmoidal function to model the transfer from input to output signals (Figure 5). The general equation of the sigmoidal function takes the form:

$$P = \frac{1}{1 + e^{\frac{-a}{t}}}$$

where P is the probability of the node firing, t is a constant that determines the steepness of the function (see Figure 5), and a is the activation of the node. This feature attempts to imitate the behavior of real neurons which often tend to be either active or inactive. The steepness of the activation function will determine whether most of the input is transferred onwards through the nodes, or whether output is only initiated by stronger inputs.

A basic model uses three layers of nodes. The first is an input layer that receives input data. The middle or hidden layer draws stimulation from the input nodes and transmits onwards to the final output layer which is the result of the system. In training the network, a set of patterns is presented repeatedly and the weights of the arcs are modified so that the output makes a better match with a desired result. Training is usually accomplished by the back-propagation of errors through the network that distributes the difference between the desired result and the actual output as small incremental adjustments in the interconnection weights. The process is gradual and iterative; weights gradually converge to an equilibrium setting and the network is trained.

The speed of the training operation is controlled largely by a learning rate set by the user. The learning rate determines the size of incremental correction at each iteration. If too high, then the network learns very quickly, but the weights may oscillate wildly with an unstable solution. Very low learning rates ensure a smoother passage to stability but may take up excessive computer time. At the completion of training, an unknown pattern can be entered for purposes of classification or prediction.

Verification of performance by a validation set is particularly important when dealing with neural networks. It provides a more realistic test of effective prediction power than the statistics of learning generated by the training set. There are times when a network can be made sufficiently complex that it reproduces outputs from the training sets almost perfectly. However, this same network will perform worse than conventional statistical methods in its predictions about new observations. This apparent paradox dissolves after a little reflection. The purpose of the training is to acquire the ability to generalize from observations rather than to regurgitate them by rote. Generalization will absorb the systematic trends that link observations and screen out the random error component. Rote learning will result in exact reproduction of training input. This is not simply a philosophic difference. In absorbing the errors associated with the training input the weight configuration is distorted and may show erratic behavior in the intervals between the observation data. Neural network specialists are well aware of this problem and commonly suggest the insertion of a little error noise into the training process in order to encourage the network to generalize rather than memorize (Caudhill, 1991).

The design of an effective neural network is usually a matter for experimentation with the problem data set at hand, although some basic guidelines can be followed. Most practitioners recommend that the structure be made reasonably simple. Caudhill (1991) suggested that one hidden layer is often adequate, but two hidden layers may be necessary; he considered that multiple hidden layers beyond two could adversely affect error correction in back-propagation models. The number of nodes within each layer must

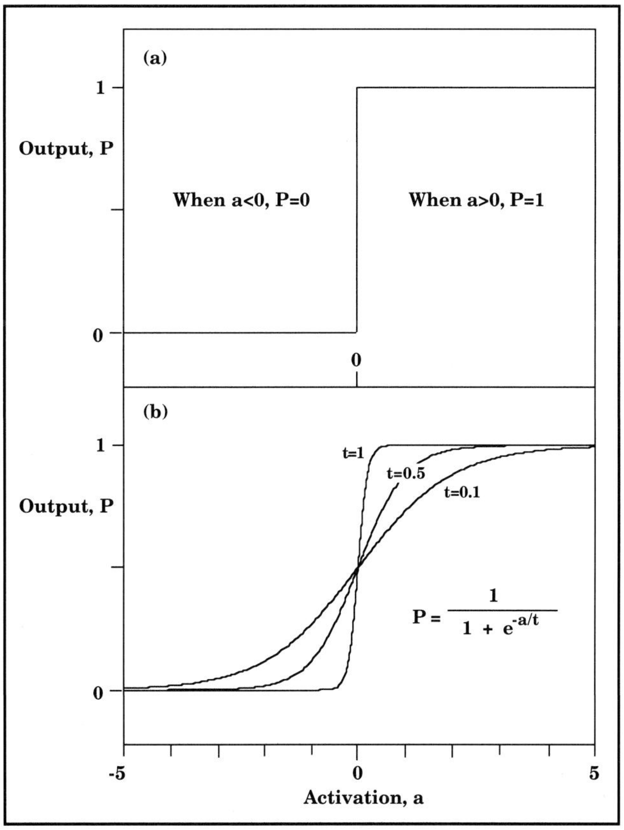

Figure 5. Control of the firing of a neural network node by an activation function either by (a) a threshold value determination of positive activation or by (b) a sigmoidal function, where the value of *t* influences the relative ease of node firing.

also be monitored carefully. Too large a number will cause the network to lock into the memorization mode of learning. A modest number of nodes will tend to encourage generalization in the allocation of connecting weights.

A common problem in the use of neural networks is the matter of deciding whether the solution that is generated is the best possible or whether it is simply a runner-up among a whole variety of solutions. This ambiguity does not occur in conventional parametric methods (the statistical techniques discussed in earlier chapters). When a problem is condensed into a form described by parameters of means, variances, and covariances there is a unique solution. In nonparametric methods, a number of alternative solutions are usually possible, because the total variability of the data is under scrutiny. Obviously some solutions will be better than others, while one should be the best, as determined by some criterion chosen by the user. How can we be sure that the solution found by the neural network is the best?

When working with a simple neural network, the problem of multiple solutions must be approached in a pragmatic manner. The arrival at any given solution is determined by iterative correction of the connection weights that cause the weights to drift from their initial values to a better and stable configuration by a steepest descent route. The likelihood that this is the global solution can be checked by repeating the training process, but starting with a different set of connection weights for each trial.

More systematic ways to find the optimum solution are used as the basis for Boltzmann machines. These exotic-sounding neural networks are named after the 19th-century Austrian physicist who discovered that the energy of random motion of molecules was a direct function of temperature. Mathematically, the machine imitates the process of annealing in metallurgy. Annealing involves raising the temperature of a metal close to the melting point, then lowering the temperature gradually to relieve localized stresses in the metal. Random configurations at high temperatures gradually adjust to an ordered low-energy configuration if the rate of cooling is moderate. The analogy is carried through by changing the ease with which neural nodes are activated at the start of the process to increasingly more inhibited firings. This can be done through the progressive modification of the steepness of a sigmoidal activation function that governs the transfer of input to output at each node (see Figure 5). The form of the sigmoidal function is taken directly from Boltzmann physics.

Finally, most neural networks operate as supervised procedures, that is, they are trained on known material before being applied to problem data. Unsupervised learning is obviously a much more difficult process to emulate in a realistic and convincing manner. However, some useful progress has been made on self-organized networks, largely due to the efforts of Teuvo Kohonen of the Helsinki Technology University. Both supervised and unsupervised neural networks have been applied to log analysis problems with encouraging results.

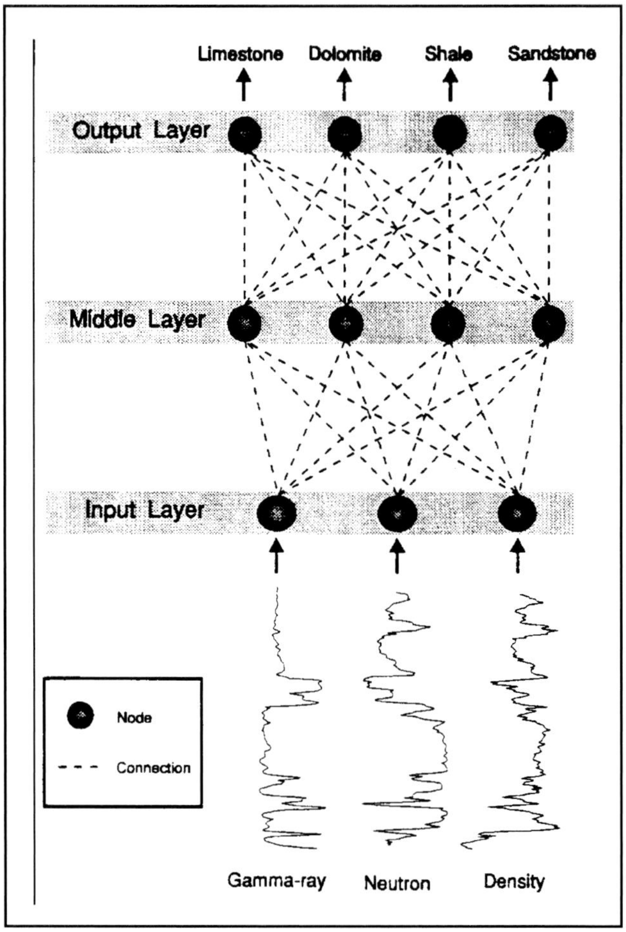

Figure 6. Schematic diagram of back-propagation neural network used to determine lithologies of zones based on their log responses. The logs are taken from a Middle Ordovician sequence from northern Kansas and the solution is shown in Figure 7. From Rogers et al. (1992).

APPLICATIONS OF NEURAL NETWORKS TO LOG ANALYSIS

Wiener et al. (1991) provided a useful example of a neural network applied to a log-analysis problem in the estimation of permeabilities in carbonates based on wireline log responses. This is a problem that we tackled in Chapter 1 when we explored the use of regression analysis for this purpose. What potential advantages does a neural network offer that are an advance over the classical estimation procedures of statistical regression? The parallel-processing structure of the network with its rich interconnections allows a complex nonlinear solution to be located. Furthermore, both the weightings and their distributions are determined by the input data themselves. By contrast, regression analysis is generally linear and is computed from the parameters of the data which are modelled as a multivariate normal distribution.

In their application, Wiener et al. (1991) used a training set of responses from shallow and deep laterologs, neutron porosity, acoustic velocity, bulk density, computed porosity, water saturation, and bulk volume water to predict permeabilities measured in cores of the Mississippian Canyon formation in North Dakota. They experimented with a variety of network designs, log suites, input and output scalings, learning rates, error transfer functions, and training iterations in order to find the best match between output predictions and actual permeabilities. Their final model configuration had two hidden layers and took 6000 iterations to converge on a stable result. The prediction results had a correlation coefficient of 0.96 with measured permeabilities, which was a substantial improvement on the maximum correlation coefficient of 0.76 that was generated using multiple linear regression. More importantly, they achieved a correlation coefficient of 0.90 when they applied the neural network to predictions in a validation set outside the calibration set used for training.

Another interesting application of neural networks to wireline logs is in the recognition of geological facies. The facies can take the form of conventional lithofacies or associations that are diagnostic of features linked with depositional environments or diagenetic episodes. How does the connectionist approach differ from the methods described earlier in this book for the same purpose? Neural networks are data driven and induce relationships and associations. The design philosophy is therefore the same as the bottom-up methods described in Chapter 4. However, as was the case with the neural network permeability predictor, two principal differences are that the approach is nonparametric and it is nonlinear. The neural network considers the individual values of all the input observations, and the network weights are not constrained to a linear representation. This contrasts with a classical statistical classification method, such as discriminant function analysis, which is parametric, because it works with means, variances, and covariances, and is linear in structure.

Rogers et al. (1992) provide a useful demonstration of neural networks applied to lithology recognition from logs. In one example, they used data tabulated by Doveton (1986) to train a network to classify limestones, dolomites, shales, and sandstones from gamma-ray, neutron, and density logs. They used a conventional design of one input and one output layer separated by a single hidden layer. The number of nodes in the input and output layers is straightforward and set by the logs and lithology types. However, the number of nodes in the hidden layer usually must be determined empirically through comparative learning trials guided by common sense. Rogers et al. (1992) noted correctly that too many nodes in the hidden layer would cause the network to simply memorize the input patterns. They found that four nodes appeared to be the best number—enough to build an effective system, but sufficiently restrictive to force the network to generalize.

The network learned to discriminate between the lithologies through repeated presentation of the training set of log data as input patterns that were linked with the desired output classification patterns (Figure 6). At each step, outputs computed from the current network weights were compared with the desired classification binary output pattern of lithology presence or absence. Corrections were made to the weights through standard back-propagation and the process repeated until the weights converged on a stable and useful set of values. Once the network was trained, it was used to transform gamma-ray, neutron, and density logs from a Middle Ordovician section into a sequence of lithological types. At each depth level the network generated a four-node output of values between zero and one that signified the lithological classification (Figure 7). The output node with the highest value determined the output lithology. In a number of instances the result might appear to be ambiguously divided between two lithologies, but these occasions generally mark the occurrence of mixed rock types, such as dolomitic limestones.

Derek et al. (1990) made a comparative study of the performance of a neural network in contrast with standard statistical methods for the discrimination of sandstone lithofacies. The reservoir chosen for their case study was composed of sheet-like, lenticular,

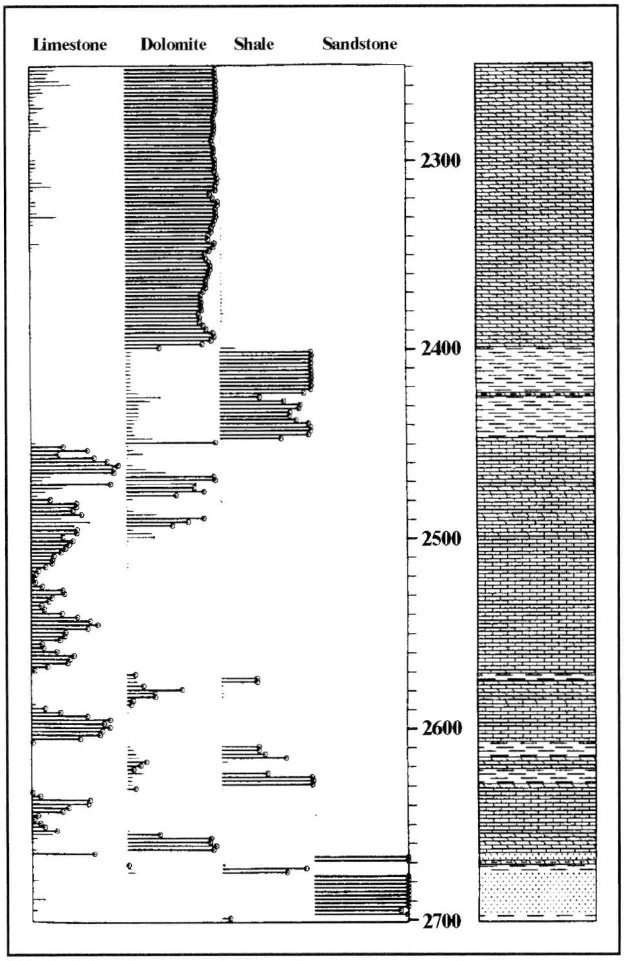

Figure 7. Neural network determination of lithologies from gamma-ray, neutron, and density logs in a Middle Ordovician succession. From Rogers et al. (1992).

and shingled sandstone units. Two reservoir-quality sandstone facies could be recognized and appeared to have been formed as episodic storm deposits. Four non-reservoir sandstone types were characterized by high clay content and small-scale sedimentary structures. Derek et al. (1990) used gamma-ray, density, and neutron logs as data inputs to differentiate the various sandstone facies.

It should be noted that although neural networks are more exotic than traditional approaches, the aim is exactly the same as that of the discrimination methods discussed at length in Chapter 4. Comparisons in performance can be made both in terms of relative success rates in classification and the structure of the decision function. When trained, the discriminant characteristics of the neural network can be graphed as decision boundaries that cut the crossplot space of the input log variables. If a neural network has no hidden layer, then it can only generate a linear plane as a partition between classes (Figure 8). It was this limitation that caused progress in neural networks to be slowed for a decade. The addition of a hidden layer allows the neural network to create convex par-

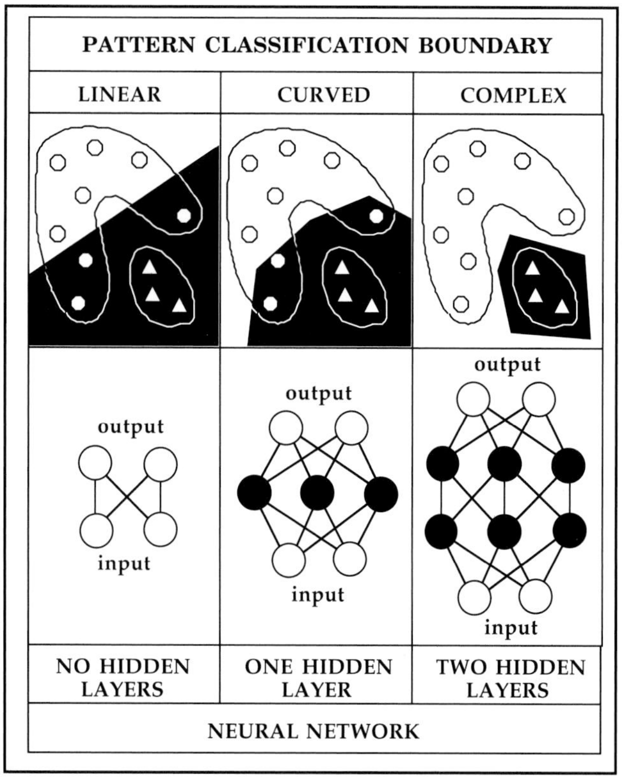

Figure 8. Basic relationship between number of neural network layers and the complexity of partition boundaries that can be generated in discrimination space. Note that the most complex solution may not necessarily be the best if the result tends to memorize the training patterns. This can cause poor generalization and weak predictive powers when applied to a test data set.

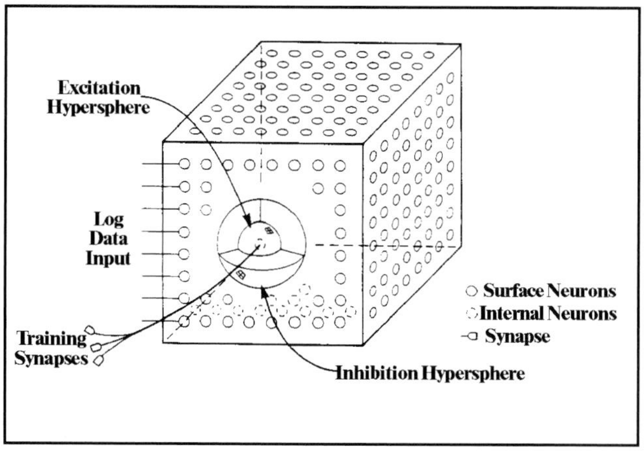

Figure 9. Self-organizing neural network hypercube (only three dimensions can be shown). An activated neuron is shown surrounded by its excitation and inhibition hyperspheres that determine which of the neighboring neurons are affected. From Baldwin et al. (1989). Copyright SPE.

tition surfaces. Finally, two hidden layers free the network to model any function and thus produce complex partition shells (Figure 8).

Derek et al. (1990) experimented with a basic neural network design of one hidden layer, but varied the number of nodes in a search for improved performance. The network was evaluated with test data after learning with a training data set. For this particular test set, there was little difference in predictive power between the neural network and three other methods. Use of a Bayes' classifier, k-nearest neighbor classifier, linear discriminant function, and the neural network all made correct predictions in over 90% of the test cases. In fact, linear discriminant function analysis performed best, with a score of 94.6%. The authors considered this to be remarkable, when taking into account the fact that the method can use only linear decision boundaries to identify the various lithofacies. Their conclusion is correct and points to some useful morals to be drawn as guidelines for a well-tempered analysis.

The success of the linear discriminant function showed that the data were linearly separable to a high degree. Any improvement by a neural network would come from its ability to warp the decision surface to a better accommodation of the remaining 5% of misclassified zones. However, an overly complex partition would lead to memorization of the input patterns and damage useful generalizing properties for predictions in the test data set. In cases where there is a systematic overlap in the data-point clusters, there also may be no way to trace a complex partition surface in the training set that will have useful predictive powers in test sets. Allocation of an observation vector in this mixed zone to one or another class is then a matter of probability. Notions of Bayesian probability discussed in Chapter 4 are applicable in this situation.

The choice of the appropriate method(s) for analysis depends on the nature of the problem and the properties of the method: choose the right tool for the job. For simple problems, simple methods are preferable because they are generally robust, easy to implement, and are effective in such cases. The linear discriminant function is a good example of such an approach. Because it is computed from data parameters, the function is generalized and thus has useful predictive powers when moving from training to test data sets. Neural networks, by contrast, are nonparametric and can make significant improvements when dealing with more complex problems where linear parametric techniques perform poorly. The user can make some judgments about the complexity of the classification problem by an examination of log crossplots. The shapes and relative separations of the zone clusters give insights that enable the analyst to determine whether simpler techniques are likely to be effective, or if their assumptions will limit the analysis severely.

Up to this point, all neural networks that have been described are examples of supervised systems. In each case, the network has been trained with a known data set before being used for purposes of prediction or classification. By contrast, self-organiz-

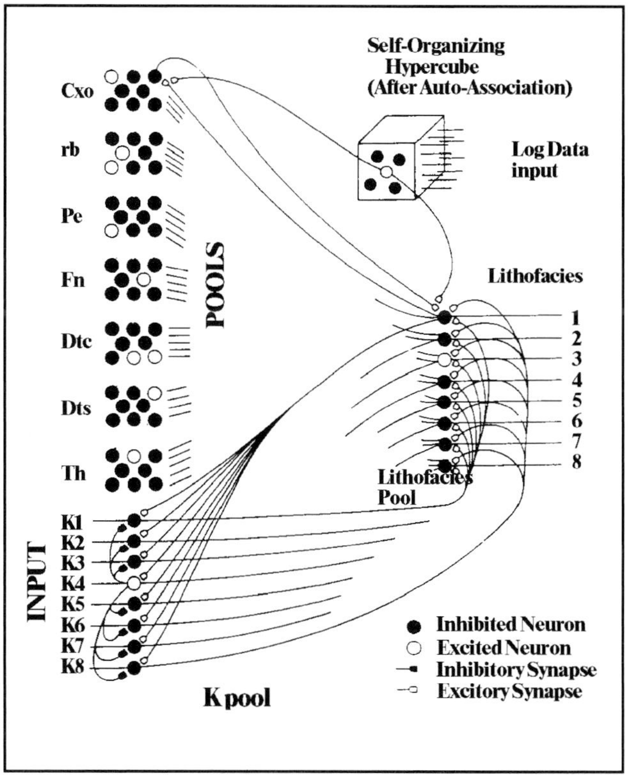

Figure 10. Schematic diagram of lithofacies pattern identification using synaptic connections trained by a self-organizing neural network hypercube. From Baldwin et al. (1989). Copyright SPE.

ing neural networks are unsupervised systems that are designed to be sensitive to "natural" divisions within the data. Most attempts at self-organizing networks have built on the seminal concepts of Kohonen (1984). Unlike conventional neural networks, a simple Kohonen network consists of a single layer. However, the nodes within this layer are highly interconnected and transmit information to one another as well as taking input from data patterns.

Every node of a Kohonen network has a weight vector associated with it and the initial set of weights are randomized. Each time an input pattern is presented to the layer, the nodes compete for learning through a comparison of the weight vectors with the input. The winning node is that whose vector is most similar to the input, as determined mathematically by the fact that the dot product of the input vector and the node weight will be a maximum when they are most alike. This is the only node allowed to generate an output signal. However, the weights of both the winning node and its immediate neighbors are adjusted by an incremental amount to be closer to the input vector. The neighborhood is initially large, but is steadily reduced during training. Each node that is allowed to output its activation transmits it as an inhibitory input to all the other nodes. This moderation of competition is known as lateral inhibition. In the learning process, competing outputs resonate around the layer and move towards a situation where localized pieces of the network are activated and generate output, while other parts are quiescent.

Baldwin et al. (1989) used a self-organization model in an eight-dimensional construct whose axes were set by eight logs of shallow guard conductivity, bulk density, neutron porosity, photoelectric factor, compressional and shear acoustic travel times, and thorium and potassium content. The system was set to recognize eight lithofacies from a logged sequence of mixed lithologies. The number chosen was arbitrary, but other than this constraint, the network was free to locate facies associations from the input logs. The learning procedure was made within the eight-dimensional hypercube as input log vectors were used to activate nodes in a hypersphere centered on their locations (Figure 9). At the end of the self-organization process, the winning neurons located eight-dimensional hyperclusters of log patterns that represented eight different facies.

The facies neuron patterns were then used to train synaptic connections in a competitive-activation pattern classification environment (Figure 10). The trained connections were then used to classify inputs from logs recorded in test sequences. Pattern classification was made by allowing the pools of neurons to compete through excitation and inhibition until the network became stable. The stable state of the lithofacies neurons gives the expectation of the lithofacies classification as determined by the system. The confidence of the system in its classification is reflected in the relative activity of each lithofacies neuron. In some cases, moderate activations in several neurons would reflect either that one or more lithofacies could be possible outcomes of the same log data vector, or that the zone was a mixture of several lithofacies.

The general approach of the self-organizing neural network model has some exciting possibilities, although both philosophical and technical problems must be considered carefully. If the system is allowed to identify facies by itself, how many facies should be chosen? Commonly, the number will not be known, so what criterion should be used to determine the point at which the system no longer subdivides the data set into smaller clouds? In most cases, subdivision will follow a gradational hierarchy in which a stopping point is arbitrary. In the absence of supervised training, special attention will have to be paid to the physical nature of the input logs, since they will determine the character of any facies associations that are isolated by the network.

Clearly, this kind of work is at its earlier and more formative stages. However, the major reasons to move forward are pragmatic rather than esoteric and were discussed by Baldwin et al. (1989). The number of experienced petrophysicists is unlikely to keep pace with the demand for log analysis. As new tools are introduced, the variety of logs continually expands. The processing of logs is now aimed towards considerably more sophisticated characterization of rock properties than was the norm at the well site in former times. Therefore, automated methods with some rudimentary features of intelligence could be of considerable help as aids in a first pass through the huge volumes of data that are generated during a modern logging run.

If expert and neural network system methods can be devised and shown to work credibly in field trials, then they are likely to move from grudging acceptance to uncritical dependence in a relatively short time. Schank (1991) noted that "artificial intelligence" is often a label applied to a task that no machine has done before. Once the task is achieved and becomes a routine function, it ceases to be artificial intelligence. A good example for this point of view is given by optical character recognition. There was a time when the consistent recognition of an alphabetic letter by a machine was considered to be at the cutting edge of artificial intelligence. Now, OCR software is commonly installed on microcomputers and is used routinely to scan lengthy documents. The initial novelty wears off quickly and the user finds that scanning text is not much different or any more exciting than operating a toaster.

Hopefully a similar progression from elaboration to common acceptance and everyday use will happen with artificial intelligence applications to log analysis. We have seen how AI has come to be thought of in terms either of symbolic programming (expert systems) or connectionism (neural networks). However, at a broader level the concept can be expanded to the notion of control systems that coordinate a variety of computer techniques of graphic display and analysis. Many of these methods have been described in earlier chapters of this book. Part of the incentive for the development of such systems is the increasing demands on limited manpower for more solutions to complex problems in exploration and production. But the future is by no means as bleak as this might suggest. The increase in measurement categories stemming from recent developments in tool technology has strained our abilities to analyze the mass of new data using extensions of the old manual methods. These additional measurements have the potential to give unique new insights into subsurface geology on a wide range of scales. Computer methods are the obvious key to their analysis.

REFERENCES CITED

Baker, J. D., 1984, Dipmeter Advisor—an expert log analysis system at Schlumberger, in P. H. Winston, and K. A. Prendergast, eds., The AI business: The commercial uses of artificial intelligence: Cambridge, MA, MIT Press, p. 51-65.

Baldwin, J. L., D. N. Otte, and R. M. Bateman, 1989, Computer emulation of human mental processes; application of neural network simulators to problems in well log interpretation: SPE Paper 19619, 64th Annual Fall Meeting], in SPE annual technical conference and exhibition, proceedings, volume omega, Formation evaluation and reservoir geology: Society of Petroleum Engineers, p. 481-493.

Caudhill, M., 1991, Neural network training tips and techniques: AI Expert, v. 6, no. 1, p. 56-61.

Cheong, D.-K., J. Strobel, G. Biswas, G. Lee, C. G. St. C. Kendall, R. Cannon, and J. Bezdek, 1993, Playmaker, a knowledge-based expert system: Geobyte, v. 7, no. 6, p. 28-41.

Crain, E. R., 1985, A primer on artificial intelligence and expert systems in the petroleum industry: Canadian Well Logging Society Journal, v. 14, p. 17.

Davis, R., H. Austin, I. Carlbom, B. Frawley, P. Pruchnik, R. Sneiderman, and J. A. Gilreath, 1981, The Dipmeter Advisor--Interpretation of geological signals, in Proceedings 7th International Joint Conference on Artificial Intelligence, Vancouver, British Columbia, Canada, p. 846-849.

Derek, H., R. Johns, and E. Pasternack, 1990, Comparative study of back-propagation neural network and statistical pattern recognition techniques in identifying sandstone lithofacies, in Proceedings 1990 Conference on Artificial Intelligence in Petroleum Exploration and Production: College Station, TX, Texas A & M University, p. 41-49.

Doveton, J. H., 1986, Log analysis of subsurface geology—concepts and computer methods: New York, John Wiley & Sons, 273 p.

Duda, R. O., 1980, The Prospector System for mineral exploration: Menlo Park, CA, Stanford Research Institute, Final Report SRI Project 8172, 120 p.

Einstein, E. E., and K. W. Edwards, 1990, Comparison of an expert system to human experts in well log analysis and interpretation: SPE Formation Evaluation, v. 5, no. 1, p. 39-45.

Frank, W., 1986, The Dipmeter Advisor is field tested, in T. Bernold, ed., Expert systems and knowledge engineering (essential elements of advanced information technology): Amsterdam, Elsevier, p. 111-120.

Hart, P. E., R. O. Duda, and M. T. Einaudi, 1978, Prospector—a computer-based consultation system for mineral exploration: Mathematical Geology, v. 10, no. 5, p. 589-610.

Josso, M., 1986, The Dipmeter Advisor—an expert system comes to South-East Asia, in Proceedings 6th SPE Offshore South East Asia Conference, Singapore: South East Asia Petroleum Exploration Society, p. 461-466.

Kahn, G., and J. McDermott, 1986, The Mud System: IEEE Expert, v. 1, no. 1, p. 23-32.

Kohonen, T., 1984, Self-organization and associative memory: New York, Springer-Verlag, 255 p.

Lineman, D. J., J. D. Mendelson, and M. N. Toksoz, 1987, Well to well log correlation using knowledge-based systems and dynamic depth warping: Transactions of the SPWLA 28th Annual Logging Symposium, Paper UU, 25 p.

McCammon, R. B., 1993, Recent experiences with Prospector II, in J. C. Davis, and U. C. Herzfeld, Computers in geology—25 years of progress: New York, Oxford Press, Studies in Mathematical Geology 5, p. 45-54.

Olea, R. A., and J. C. Davis, 1989, An expert system for the correlation of geophysical well logs, in M.

Simaan, and F. Aminzadeh, eds., Artificial intelligence and expert systems in petroleum exploration: Greenwich, CT, JAI Press, Inc., Advances in Geophysical Data Processing, v. 3, p. 279-307.

Peveraro, R. C. A., and J. A. Lee, 1988, HESPER—an expert system for petrophysical formation evaluation, Paper R, in Transactions 11th European Formation Evaluation Symposium: SPWLA, Norwegian Chapter, 22 p. [also published in 1988 in Proceedings, SPE European Petroleum Conference, London: Society of Petroleum Engineers, SPE 18375, p. 361-370].

Rogers, S. J., J. H. Fang, C. L. Karr, and D. A. Stanley, 1992, Determination of lithology from well logs using a neural network: AAPG Bulletin, v. 76, no. 5, p. 731-739.

Schank, R. C., 1991, Where's the AI?: AI Magazine, v. 12, no. 4, p. 38-49.

Shanor, G. G., 1987, The Dipmeter Advisor: A dipmeter interpretation workstation: Bulletin of the Geological Society of Malaysia, v. 21 no.1, p. 37-54.

Smith, R. G., and J. D. Baker, 1983, The Dipmeter Advisor system, a case study in commerical expert system development, in Proceedings 8th International Joint Conference on Artificial Intelligence, p. 122-129.

Startzman, R. A., and T. B. Kuo, 1987, An artificial intelligence approach to well log correlation, The Log Analyst, v. 28, no. 2, p. 175-183.

Watney, W. L., J. E. Anderson, and J.-C. Wong, 1990, Porosity Advisor—an expert system used as an aid in interpreting the origin of porosity in carbonate rocks, in J. T. Hanley, and D. F. Merriam, eds., Microcomputer applications in geology II: Oxford, Pergamon, p. 275-288.

Wiener, J. M., J. A. Rogers, J. R. Rogers, and R. F. Moll, 1991, Predicting carbonate permeabilities from wireline logs using a back-propagation neural network, CM1.1: Society of Exploration Geophysicists, Expanded Abstracts with Biographies, 1991 Technical Program, v. 1, p. 285-289.

Wu, X., and E. Nyland, 1986, Well log data interpretation using artificial intelligence technique: Transactions of the SPWLA 27th Annual Logging Symposium, Paper M, 16 p.

Zivy, G. M., 1984, The role of expert systems in producing log interpretation software: Expert Systems, v. 1, no. 1, p. 57-62.

Index